Calf Nutrition and Management

Calf Nutrition and Management

Guest Editor

Farhad Ahmadi

MDPI

Basel • Beijing • Wuhan • Barcelona • Belgrade • Novi Sad • Cluj • Manchester

Guest Editor
Farhad Ahmadi
School of Agriculture,
Food and Ecosystem
Sciences (SAFES)
Faculty of Science
The University of Melbourne
Melbourne
Australia

Editorial Office
MDPI AG
Grosspeteranlage 5
4052 Basel, Switzerland

This is a reprint of the Special Issue, published open access by the journal *Animals* (ISSN 2076-2615), freely accessible at: https://www.mdpi.com/journal/animals/special_issues/AA6NHPLD8C.

For citation purposes, cite each article independently as indicated on the article page online and as indicated below:

Lastname, A.A.; Lastname, B.B. Article Title. *Journal Name* **Year**, *Volume Number*, Page Range.

ISBN 978-3-7258-4181-3 (Hbk)
ISBN 978-3-7258-4182-0 (PDF)
https://doi.org/10.3390/books978-3-7258-4182-0

Contents

About the Editor

Farhad Ahmadi

Farhad completed his Ph.D. in Animal Science from Konkuk University and worked as Postdoctoral Scientist at the Forage and Agronomy Department at Seoul National University, South Korea. Farhad primarily works on the chemical and biological treatment of underutilized agricultural wastes, including palm pruning waste, fruit and vegetable discards, spent mushroom substrate, and liquid whey to develop value-added products (notably animal feed applications) with successful commercial outcomes. He has led multiple R&D initiatives that have advanced technological developments and commercialization within the livestock industry. Farhad actively engages in multiple research projects that focus on improving the productivity, health, and sustainability of farm animals via innovative nutritional and management strategies. His current research pertains to the extraction, encapsulation, and characterization of phytochemicals, targeting their effective use in livestock feeding, particularly under environmentally challenging conditions.

animals

MDPI

Article

Stochasticity Highlights the Development of Both the Gastrointestinal and Upper-Respiratory-Tract Microbiomes of Neonatal Dairy Calves in Early Life

A. Nathan Frazier [1,*], Logan Ferree [2], Aeriel D. Belk [3], Khalid Al-Lakhen [2], M. Caitlin Cramer [2] and Jessica L. Metcalf [2,4]

1 United States Department of Agriculture—Agricultural Research Service (USDA-ARS), Bushland, TX 79012, USA
2 Department of Animal Science, Colorado State University, Fort Collins, CO 80523, USA
3 Department of Animal Science, Auburn University, Auburn, AL 36849, USA
4 Canadian Institute for Advanced Research (CIFAR) Azrieli Global Scholars Program, CIFAR, Toronto, ON M5G 1M1, Canada
* Correspondence: anthony.frazier@usda.gov

Simple Summary: Many factors influence the development and establishment of the neonatal gut and upper-respiratory-tract microbiomes. Importantly, these microbiomes play a significant role in host maintenance and health physiologies in host–microbe systems. In dairy calf systems, the ecological factors impacting microbial assembly are not fully elucidated. Thus, the aim of this pilot study was to characterize the establishment of early-life fecal and nasal microbiomes in dairy calves and evaluate the impact of disease states on microbial development. In addition, we investigated the governing forces of microbial composition to better understand the factors that shape microbial community populations. Herein, we observed dynamic changes in diversity and composition in the neonatal calf microbiomes and demonstrated that disease state impacts the gastrointestinal microbiome. Our results further suggest that early-life microbiomes are stochastically driven, governed by neutral theory-based dynamics. These results together provide researchers with a roadmap to further investigate how early-life microbiomes develop and how they are affected by disease. Together, this study and future studies could help improve the quality of life in dairy calf management.

Abstract: The microbiome of dairy calves undergoes extensive change due to various forces during the first weeks of life. Importantly, diseases such as bovine respiratory disease (BRD) and calf diarrhea can have profound impacts on the early-life microbiome. Therefore, a longitudinal, repeated-measures pilot study was designed to characterize the establishment of nasal and fecal microbiomes of dairy calves, assess the governing forces of microbial assembly, and evaluate how disease states impact these microbial ecologies. Dairy calves (n = 19) were clinically evaluated for gastrointestinal and respiratory disease across three weeks beginning at age $\leq$ seven days old. Fecal (n = 57) and nasal (n = 57) microbial samples were taken for paired-end 16S rRNA gene amplicon sequencing. Taxonomy and diversity analyses were used to characterize early-life nasal and fecal microbiomes. Stochasticity and determinism were measured using normalized stochasticity testing (NST) and Dirichlet multinomial model (DMM). All analyses were tested for statistical significance. Clinical diarrhea was observed in 11 of the 19 calves. Clinical BRD was not independently observed among the cohort; however, two calves presented clinical signs of both BRD and diarrhea. Taxonomic analysis revealed that fecal samples were highlighted by *Bacteroidaceae* (40%; relative abundance), *Ruminococcaceae* (13%), and *Lachnospiraceae* (10%), with changes in diversity (Kruskal–Wallis; $p < 0.05$) and composition

 https://doi.org/10.3390/ani15030361

(PERMANOVA; $p < 0.05$). Clinical diarrhea reduced diversity in the fecal microbiome but did not impact composition. Nasal samples featured *Moraxellaceae* (49%), *Mycoplasmataceae* (16%), and *Pasteurellaceae* (3%). While no diversity changes were seen in nasal samples, compositional changes were observed ($p < 0.05$). NST metrics (Kruskal–Wallis; $p > 0.01$) and DMM (PERMANOVA; $p < 0.01$) revealed that stochastic, neutral theory-based assembly dynamics govern early-life microbial composition and that distinct microbial populations drive community composition in healthy and diarrheic calves.

Keywords: microbial ecology; dairy calf; disease; stochasticity; determinism

1. Introduction

Mammals and microorganisms have evolved a fundamental, symbiotic relationship crucial to mammalian life that is known as the microbiome. The microbiome is the collection of all microbes and the genomes that naturally inhabit their host. Host–microbe interactions influence both the host and its microbial partners. Currently, a large effort to characterize animal-associated microbial ecologies has resulted in an improved understanding of the importance of the microbiome [1]. In humans, studies have indicated the importance of the microbiome to health [2–4], digestion [5], growth and development [6], and behavior [7,8]. In parallel, recent research has increased the understanding of the microbiomes of domesticated livestock such as dairy cattle. Research efforts in dairy cattle have revealed the relationship between the microbiome and diet [9], methane emissions [10], stress and physiology [11], and overall health [12,13]. The vast importance of the dairy cow microbiome has given way to a significant effort to understand how their microbial partners establish and assemble. In humans, and likely in all mammalian life, microbial establishment is thought to begin at the time of birth via vertical transmission governed by various stochastic forces [14]. Similarly, early-life gut microbiomes in dairy calves assemble in the same stochastic manner with variations in composition due to vaginal birth or Cesarean section birth [15]. While the establishment of the early dairy calf microbiome is beginning to be understood in greater detail [15,16], the colonization of early-life microbiomes is a complex process involving both the host and its surrounding environment. Thus, further investigation is crucial to fully understanding the establishment of the dairy cow microbiome.

Most investigative studies in cattle characterize the microbiome in adults, while the importance of the early-life microbiome has only recently seen an increase in research attention [17,18]. The early-life microbiome could have profound implications for its symbiotic host, as evident in other ruminant livestock functionalities [19]. Many factors influence the establishment of the neonatal microbiome and can be defined as either stochastic or deterministic. Importantly, recent research has indicated that early-life microbial structures could be primarily stochastically driven [15]. In real ecological systems, stochastic forces are often constrained by deterministic processes such as age and diet, leading to a unification of both niche and neutral theory [20,21]. Thus, while neutral theory is inherently stochastic, unified neutral theory can reasonably be applied to real ecological environments. In young ruminants such as dairy calves, there is little evidence supporting the existence of neutrality. Furman et al. [15] demonstrated that the deterministic factors diet and age, and stochastic factors, act in concert to shape the assembly of the rumen microbiome. Additionally, Pan et al. [22] revealed neutral theory-based assembly in neonatal calves with or without clinical diarrhea onset. However, to date, no other studies have directly investigated the stochastic neutrality of microbial assembly, nor have the factors that can affect microbial assembly been fully elucidated.

Many factors can disrupt the establishment of the microbiome, creating altered microbial ecologies (Figure 1). These factors include husbandry, calf management, environmental and/or farm conditions, and diet. Importantly, disease is one of many factors that can perturb microbial assembly and functionality. These altered microbiomes could impact the effort to diagnosis and treat important diseases for the dairy industry. Specifically, preweaned dairy calf morbidity and mortality have remained mostly unchanged over the last 18 years due to diseases such as bovine respiratory disease (BRD) and calf diarrhea [23–26]. Thus, there is a need to develop novel ways to identify animals at a greater risk for disease to address the economic impact of disease on the dairy industry [23]. The microbiome could prove beneficial in addressing these issues by assisting in understanding the role microbes play during perturbation events such as disease. Surprisingly, recent research efforts have reported conflicting findings when assessing the microbiome and disease. In fecal microbiomes, there is evidence of diversity and taxonomic differences between healthy and diarrheic calf microbiomes [27]. In contrast, studies have also reported no changes between healthy and diarrheic calves [28]. Similar patterns can be seen in nasal microbiomes when assessing BRD events in dairy calves [29,30]. Additionally, differences in methodologies between studies creates variation, making comparison between studies difficult. Studies using a defined methodological approach could prove valuable to reducing cross-study variation [31]. Further investigation is warranted and necessary to untangle the cross-study variation in results. Overall, it remains unclear which mechanisms and factors regulate divergent microbial communities during perturbation events such as disease.

While the establishment of the neonatal calf microbiome has received an increase in research interest, there remains a need to investigate the governing processes of microbial community development as well as the impact of disease state in early-life microbiomes. Additionally, many studies are constrained by age in longer-term studies, resulting in diversity changes being attributed to animal age rather than disease state. Thus, investigating temporal patterns on a narrower timescale could elucidate key microbial factors that either are impacted by perturbation events such as disease or have an impact on later development stages. Therefore, we hypothesized that early-life (<3 weeks of life) microbiome establishment is governed by stochasticity and that these forces could be impacted by disease onset. To test this, a pilot study was designed to provide proof-of-concept analysis of the development and assembly of gastrointestinal (GI) and upper-respiratory-tract (URT) microbiomes in dairy calves (the outcome) during the first three weeks of life and to understand the correlations between disease states and their associated microbiomes in early life (the exposure). Our study sought to answer three primary questions: (i.) how microbiome establishment occurs in neonatal calves in early life, (ii.) which forces (i.e., stochastic or deterministic) are driving community assembly during the first three weeks of life, and (iii.) whether disease impacts microbial development during the first three weeks of life. We utilized inexpensive and rapid 16S rRNA gene sequencing to assess GI and URT microbial ecologies and combined this with health assessments to identify calves with respiratory disease and diarrhea.

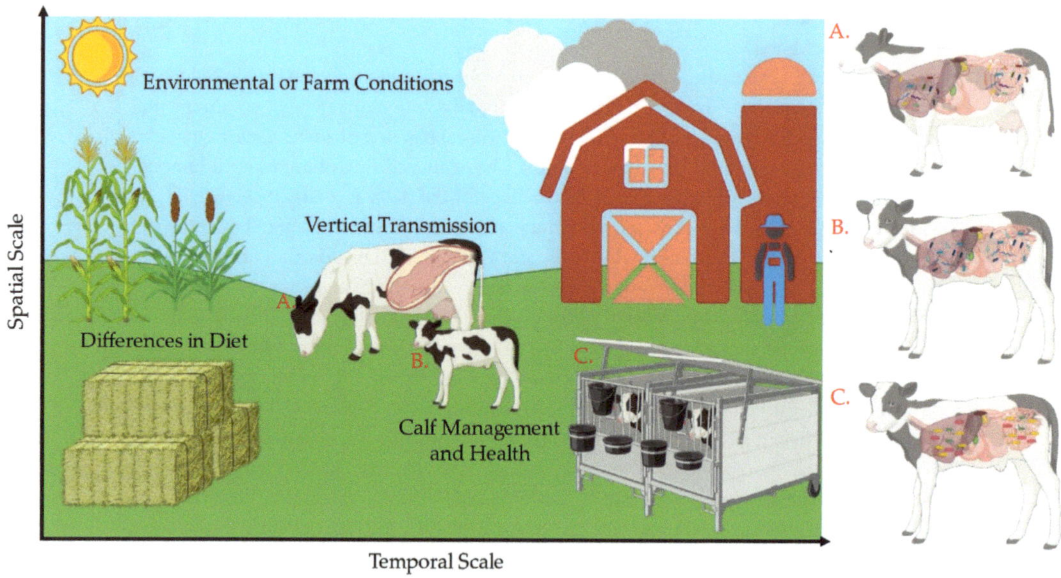

Figure 1. Factors contributing to microbial assembly differences. Stochasticity and determinism are important forces in microbial assembly patterns. Stochastic, random forces such as vertical transmission are important in early-life microbial compositions. Deterministic forces such as diet contribute to composition shifts that often constrain the stochastic forces across a temporal scale. As the calf ages, it will inevitably have a different microbiome (**A**) than it did in early life (**B**). Additionally, factors like calf management, health, and environmental filtering can have marked effects on microbial assembly. Calves reared in hutches (**C**) can have different microbial ecologies than those allowed access to their dam (**B**). These forces are important to the microbial assembly process that is seen in later life.

2. Materials and Methods

2.1. Animals, Facilities, and Experimental Design

The present study was part of a larger longitudinal, repeated-measures observational cohort recruited from a larger study population investigating respiratory disease at a commercial dairy farm located in Colorado spanning from March to early April 2021 [32]. For this pilot study, the dairy was chosen based on proximity to Colorado State University and previous successful engagement in research. Calf enrollment was not randomized and was based on days of age. Calves were enrolled on a weekly basis as they were born, with the enrollment period beginning 4 March and ending on the 25th of the month. Calves aged $\leq$ seven days were enrolled in the study at sampling time point one and evaluated for two additional sampling time points. Six calves were enrolled on 4 March, three calves were enrolled on 11 March, six calves were enrolled on 18 March, and seven calves were enrolled on 25 March. Final evaluations occurred on 8 April. Calves were enrolled regardless of health status. This method was used to capture microbiome establishment under various states of health. In total, 19 calves (n = 19) were utilized for microbiome analyses. Of the 19 calves used in the study, eight were Jersey, eight were Holstein, and three were cross-bred; additionally, there were ten males and nine females. Calves were immediately removed from the dam following birth and individually housed throughout the pre-weaning period. The calves were not completely isolated, as they were able to see their neighbors. All calves were only fed pooled whole-line milk via a bucket and did not have access to forages during the first three weeks of life. All animal husbandry duties were performed by dairy personnel; however, they were not directly involved in the collection of samples, health assessments, or administration of treatment.

2.2. Microbial Sampling and Clinical Health Observations

Microbial sampling was performed by trained research personnel (A.N.F.) at each sampling time point. Each calf was sampled one time per week, which coincided with health exams. The experimental unit was the calf, and all data were collected at the calf level. Sampling consisted of both a nasal and a fecal swab using labelled, double-headed sterile cotton tip swabs (BD 220135; Franklin Lakes, NJ, USA). The nasal swab was inserted into the calf's left nasal cavity, approximately five to seven cm inside, and nasal contents were gathered from the nasal vault (n = 57); fecal swabs were collected by placing a sterile swab approximately five to seven cm into the rectum of the calf and removing fecal content (n = 57). Following sampling, the swabs were placed into freezer bags by sample type, immediately placed on ice, transported to a −20 °C freezer located at the Metcalf microbiome laboratory (Colorado State University, Fort Collins, CO, USA), and stored until DNA extraction following all sampling time points. No research personnel involved in sample collection or data analyses were blinded to health status because microbiome data results are not considered subjective.

The calves were observed for clinical health parameters once per week by trained research personnel (L.F.) during the three-week sampling time frame using the Wisconsin Calf Health Scoring System [33]. The Wisconsin Calf Health Scoring System uses the following parameters to assess calf health: nasal discharge, eye discharge, ear position, cough (based on number of coughs per minute, e.g., one cough per minute equated to a score of 1 or mild), rectal temperature, observation of fecal consistency for calf diarrhea, and navel and joint palpation. All parameters were scored from zero to three, with zero being healthy, one being mild, two being moderate, and three being severe. A lung ultrasound was performed, and lung lesions were identified for determination of respiratory disease as previously described [34]. Lung ultrasounds were conducted by one trained research team member to reduce bias and the subjectivity of score reporting (L.F.). Respiratory disease was defined using both clinical observations and lesions found during ultrasound. Calves considered to have respiratory infections were treated with tulathromycin (Draxxin®, Zoetis Services, LLC, Parsippany, NJ, USA; 1.1 mg/kg subcutaneously once) during a health exam if one or both of the following were present: (1) two or more abnormal clinical signs and/or (2) the number of lung lesions found was ≥two [32]. Following health scoring at each time point, calves were assigned to one of four health categories: (i). healthy—calves presenting no signs of clinical disease, (ii). BRD—calves presenting clinical signs consistent with respiratory illness, (iii). diarrhea—calves presenting clinical signs consistent with GI disorder, or (iv). BRD plus diarrhea—calves presenting clinical signs of both respiratory and GI disorder. Researchers worked with farm staff to maintain treatment records; however, farm records were unreliable, and therefore, only health information collected by researchers was used for analysis.

2.3. DNA Extraction and Sequencing

At the conclusion of the sampling time frame, microbial communities were characterized via paired-end 16S rRNA gene sequencing similar to previously published methods [35]. All DNA extraction was carried out by trained research lab personnel within the Metcalf lab. DNA was extracted from all fecal samples (n = 57) and nasal samples (n = 57) using the Qiagen PowerSoil Kit (Qiagen 27000-4-KF; Hilden, Germany) per the manufacturer's instructions, with one modification during the binding step (2 band steps were used to utilize all the supernatant in step 11). DNA was extracted from all fecal and nasal swabs using 96-well plates, with nine negative extraction control samples and three positive mock microbial community controls (ZymoBIOMICS D6300; Irving, CA, USA) per plate. Controls were used to confirm no contamination in the sequencing run (negative controls) and provide verification for sequencing accuracy (positive controls). Once DNA was extracted, the DNA was amplified and sequenced using 515f/806r primers targeting the V4 region of the 16S rRNA subunit following the Earth Microbiome Project protocols (www.earthmicrobiome.org; accessed on 4 March 2021) [36]. Sequences were multiplexed using error-correcting Golay barcodes within the PCR primers and quantified using Picogreen Quant-iT (Invitrogen P7589, Life Technologies; Grand Island, NY, USA). Quantified PCR products were pooled at equimolar concentrations for sequencing. Pools were then sequenced using a 500-cycle kit on the Illumina MiSeq sequencing platform (Illumina; San Diego, CA, USA) on Colorado State University's campus. Samples were placed in the sequencing wells at random to avoid being confounded due to technical artifacts. The 16S rRNA gene sequencing data are publicly available in the EBI repository under study accession PRJEB76862 and on the QIITA study platform under study ID 13813 [37].

2.4. Analysis of Fecal and Nasal Microbiomes

Following sequencing, data processing and analyses were performed using Quantitative Insights into Microbial Ecology (QIIME2) version 2020.8 [38]. Sequence data were demultiplexed, quality filtered, and denoised using the QIIME2 dada2 plugin version 2021.8.0, creating amplicon sequence variants (ASVs) [39,40]. Taxonomic classification was achieved using the SILVA 138 99% database via the QIIME2 feature-classifier plugin, and ASVs assigned to mitochondria and chloroplasts were subsequently filtered [41,42]. Sequences were rarefied to a sampling depth of 9060 ASV/sample to filter low-quality reads based on initial analysis of the sequencing run. Filtering steps were also conducted to remove sequences that appeared in less than 10% of all samples. A phylogenetic insertion tree was constructed via the QIIME2 fragment-insertion plugin, which utilizes the SEPP program, with the SILVA 128 tree used as a reference tree backbone [40,43]. The phylogenetic tree was used in downstream analysis.

Prior to downstream analyses, samples were split into two groups using the QIIME2 filter-samples plugin: fecal and nasal. This was based on the sample type and was carried out to analyze the fecal and nasal microbiomes individually as well as analyze their associations with disease state. Breed and sex were assessed for differences in community composition using unweighted UniFrac distance, with statistical comparisons carried out using permutational multivariant analysis of variance testing (PERMANOVA) with multiple testing correction using the "pairwise" parameter in QIIME2 (q-value), which assesses significant differences between the sample groups and controls false discovery rates (FDR) [44,45]. Linear regression models were fit to establish the correlation between alpha diversity using breed and sex as predictors.

Differential abundance for taxonomic analysis was assessed for both fecal and nasal samples using analysis of composition of microbiomes with bias corrections (ANCOM-BC) across sampling time points [46]. ANCOM-BC analysis measures the differential abundance of taxa between groups while accounting for compositionality and controlling both Type I and Type II error, producing an FDR-adjusted p-value (q-value). The top three enriched or depleted differentially abundant bacteria were recorded at alpha level = 0.001. Phylogenetic diversity analysis was run using the core metrics pipeline within QIIME2 [38,42,43]. Alpha diversity was measured for fecal and nasal samples using Shannon's Index, Faith's Phylogenetic Diversity (PD), and observed richness metrics [47–49]. Briefly, Shannon's Index measures diversity by accounting for richness (number of species in a sample) and evenness (distribution of species abundances), Faith's PD considers species richness and their phylogenetic relationships, and observed richness simply counts the total number of unique species. Alpha diversity metrics were then tested statistically using a Kruskal–Wallis test with a Benjamini–Hochberg multiple-testing correction, providing both a *p*-value and an FDR-adjusted *p*-value (fdr_bh) [50]. Linear regression models were fit for both nasal and fecal samples to establish the correlation between alpha diversity, with sampling time point as the predictor, using the "stats" package in R version 4.4.2 and using RStudio version 2023.12.1 [51]. Linear regression models producing significant results were retested using multiple regression models, with sampling time point, calf age, and individual calf as predictors, to increase model applicability. Beta diversity was analyzed for each sample group using unweighted UniFrac distance with PERMANOVA multiple-testing corrections. Fecal samples were further analyzed based on health status. Calves were grouped into either healthy calves or diarrheic calves. Health status was derived from fecal severity scores, where healthy calves corresponded to fecal severity scores of "healthy" or "mild", and diarrheic calves corresponded to fecal severity scores of "moderate" and "severe". Taxonomic, diversity, and composition testing was analyzed using the methods described above for health status groups. Linear regression models were also fit for health status using sampling time point as the predictor for any correlations to alpha diversity. All diversity and compositional statistical testing were considered significant at alpha level = 0.05.

Next, normalized stochasticity testing (NST) was performed using the Bray–Curtis distance matrix, providing the modified Raup–Crick distance (β_{RC}) [52], the standardized effect size index (SES) [53,54], the normalized stochasticity index (NSTi), and the modified stochasticity ratio (MST) [55]. Since stochastic and neutral metrics rely on abundance-based comparisons, Bray–Curtis was chosen, as it captures abundance changes better than unweighted UniFrac distance. These metrics were calculated in R using cNST within the "NST" package to determine the stochastic or deterministic nature of assembly in both fecal and nasal microbiomes following similar previously published methods [22,55]. For β_{RC}, values approaching zero indicate stochastic, random assembly driven by neutrality. The magnitude of the SES index determines the strength of deterministic forces on community assembly such that SES values > 2 and <−2, indicating deterministic processes, are favored. Like β_{RC}, values close to zero indicate neutrality. For NSTi, values greater than 0.5 indicate stochasticity and values approaching 1 indicate that stochastic processes are the primary drivers of community assembly. Similarly, MST provides a percentage where >50% favors stochasticity. Nonmetric multidimensional (NMDS) scaling plots were used to visualize the ordination of community compositions by β_{RC} values. NSTi, MST, SES, and β_{RC} fecal values were tested for significance across sampling time points and health status using Kruskal–Wallis with Benjamini–Hochberg multiple-testing corrections [50]. Failing to reject the null model ($p > 0.01$) indicates that the values are not changing across time. A significance of alpha level 0.01 was used to reduce the risk of false-positive rates, as microbiome data can contain high variability. This adjustment ensured that only robust and biologically meaningful differences were considered significant. Only the sampling time point was used for nasal samples.

We then evaluated whether different taxa were driving microbial community structure for healthy calf and diarrheic calf groups. Fecal samples were filtered into either the healthy or the diarrheic group using the QIIME2 filter-samples plugin and analyzed. The Dirichlet multinomial model (DMM) was used in R ("DirichletMultinomial" package) to identify key changes in microbiome development at the genus level. The DMM method utilized the lowest Laplace approximation score to cluster samples based on microbial structure similarity. The top three genera with the highest approximation scores were considered main drivers of microbial community structure for each group (i.e., healthy and diarrheic) [22,56,57]. The clusters were tested for significance against the Bray–Curtis distance matrix with PERMANOVA using the adonis2 function within the "vegan" package in R [58,59]. PERMANOVA adonis2 testing used 999 permutations to mitigate Type I error by generating a null distribution. Significant results ($p < 0.01$) demonstrate that the clusters derived from the DMM are not random and reflect true differences in community structure. Thus, significant results indicate that the top three genera indeed have a role as primary drivers. Similar to NST testing, alpha level was set at 0.01 to reduce false-positive rates, ensuring only biologically meaningful differences were captured. All visualizations were performed with R using the "ggplot2" package in RStudio [60].

3. Results

3.1. Calf Characteristics and Sequencing Results

Nineteen calves were swabbed for fecal and nasal microbiome characterization as well as observed for clinical health assessment (Table 1) over a three-week sampling time frame. A total of 22 calves were initially screened and sampled for analysis; however, three calves were removed from the study due to lameness, farm management errors by farm staff, and repeated antibiotic treatment. For the analyzed population ($n = 19$), calf breed, sex, and health characteristics were recorded. For breed, there were eight Jersey, eight Holstein, and three cross-bred calves enrolled in this study; additionally, there were ten males, and nine females enrolled. Overall, there were 17 incidences of observed clinical disease. A total of 11 calves were observed with clinical diarrhea, with four calves having clinical signs at two or more sampling time points. At time point one, clinical signs of GI illness (i.e., diarrhea) were observed for three calves only. All other calves were reported healthy. At time point two, a distinct increase could be noticed for observed clinical disease. Clinical signs of diarrhea were observed in eight calves while 10 remained healthy. BRD/diarrhea was reported in one calf. At time point three, clinical signs of diarrhea were observed in four calves, while once more, one calf was reported to show clinical signs of BRD/diarrhea. The remaining 14 calves were reported to be healthy.

Table 1. Calf breed and sex characteristics group by health category [1].

Health Category *			Time Point 1	Time Point 2	Time Point 3
Healthy			16	10	14
	Breed				
		Jersey	7	3	7
		Holstein	6	5	5
		Cross	3	2	2
	Sex				
		Male	9	4	8
		Female	7	6	6
Diarrhea			3	8	4
	Breed				
		Jersey	2	5	2
		Holstein	1	2	2
		Cross	0	1	0
	Sex				
		Male	2	6	2
		Female	1	2	2
BRD [¥]			0	0	0
BRD/Diarrhea			0	1 [¶]	1 [!]

[1] [61], * Health categories were determined by scores from the Wisconsin Calf Scoring System [33]. [¥] No observations of clinical BRD were recorded. [¶] The only calf observed with both clinical diseases was a Jersey male. [!] The only calf observed with both clinical diseases was a cross-bred male.

In total, there were 114 combined samples taken for microbial analysis. Paired-end sequencing yielded 7,063,380 sequencing reads. Denoising, quality filtering, removing chimeras, and filtering steps yielded a total of 3425 unique ASVs, with a total frequency of 4,016,483 with a mean feature frequency per sample of 35,232.30 (range: of 209 to 59,874 ASVs/sample). Quality-control checks of negative controls indicated low sequencing numbers (0–285 feature count), signaling contamination of samples during DNA extraction was not likely. Based on the positive control taxonomic mock community, the sequencing run was determined to be valid (Figure S1). Quality-control checks of the positive and negative sample controls indicated it was reasonable to remove control samples and utilize the data to answer the three proposed questions.

3.2. Characterization of the Early-Life Microbiome

Fecal and nasal samples had distinct microbial compositions (Figure 2A; unweighted UniFrac PERMANOVA, $p = 0.001$, q = 0.001), with nasal samples having higher diversities (p-value < 0.05). Therefore, for further analysis, the fecal and nasal samples were separated. Analysis of sex and breed using PERMANOVA revealed a significant difference for breed in fecal samples between cross-bred calves and Holstein calves ($p = 0.045$, q-value = 0.0675 corrected for multiple comparisons) and between cross-bred calves and Jersey calves ($p = 0.019$, q-value = 0.0570). For fecal data only, PERMANOVA analysis revealed no significant differences in microbial communities for sex. For nasal data only, PERMANOVA analysis revealed no significant difference in microbial communities for sex; however, a significant difference was observed for breed between cross-bred calves and Holstein calves ($p = 0.004$ q-value = 0.006 corrected for multiple comparisons) and cross-bred calves and Jersey calves ($p = 0.003$, q-value = 0.006). Unfortunately, the small sample size of cross-bred calves prevented the consideration of breed as a factor within this study. Additionally, linear regression models indicated that breed (Figures S2A and S3) and sex as predictors (Figure S4) were not correlated with diversity metrics or changes.

The taxonomic profiles, diversity levels, and compositions for fecal and nasal samples were analyzed over the first three weeks of life. *Bacteroidaceae* (40%), *Ruminococcaceae* (13%), *Lachnospiraceae* (10%), *Enterobacteriaceae* (7%), and *Fusobacteriaceae* (5%) were the most relatively abundant bacteria in fecal samples at the family level throughout the sampling time frame (Figure 3A). Additionally, *Bifidobacteriaceae* (6%) was higher in relative abundance at time point one. ANCOM-BC analysis at the ASV level revealed that *Bacteroidaceae* and *Bifidobacteriaceae* were depleted at sampling time point two (Figure 3B; q-value < 0.001). At sampling time point three, *Marinifilaceae*, *Lachnospiraceae*, and *Rikenellaceae* were the top three enriched families, and *Bifidobacteriaceae*, *Peptostreptococcaceae*, and *Lachnoclostridium* (*genus*) were the top three depleted bacterial groups (Figure 3C; q-value < 0.001). Fecal diversity changes were detected during the first three weeks of life (Figure 2B). The fecal sample microbiota was different in terms of Faith's PD (p-value = 0.001, fdr_bh = 0.004) and richness (p-value = 0.003, fdr_bh = 0.008) between sampling time points one and three. Between sampling time points two and three, Faith's PD ($p = 0.003$; fdr_bh $p = 0.005$) and richness ($p = 0.007$; fdr_bh $p = 0.001$) were significantly different, indicating a change in alpha diversity. Correlation analysis using linear regression models using sampling time point as a predictor indicated that sampling time point two was significantly correlated with fecal diversity changes in Faith's PD only (Figure S5A; p-value = 0.002, $R^2 = 0.20$). Given the low R^2 value of the previous model, a second multiple regression model was generated, with the predictors being sampling time point, individual calf, and age in days, to determine the influence of multiple predictors on Faith's PD. The new model indicated that these predictors were not correlated with Faith's PD (Figure S6A; p-value = 0.3033, $R^2 = 0.57$).

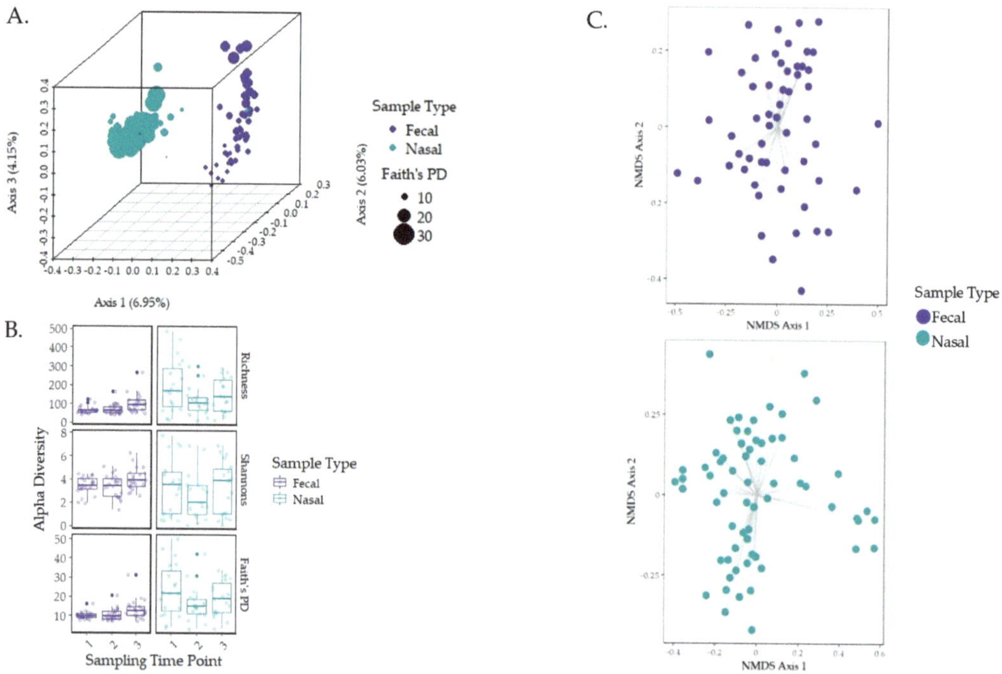

Figure 2. Alpha and beta diversity metrics for nasal and fecal samples over the sampling time frame. Community assembly was also assessed. (**A**) Beta diversity of each sample type distance was measured using unweighted UniFrac distance and revealed that separation was driven between sample type. (**B**) Alpha diversity was measured using Faith's PD, Shannon's Index, and richness, with statistical significance analyzed using Kruskal–Wallis pairwise testing across the three sampling time points [61]. (**C**) The ordination of community compositions based on β_{RC} dissimilarity distance is represented by NMDS plots.

The most relatively abundant bacteria in nasal samples were the families *Moraxellaceae* (49.42%), *Mycoplasmataceae* (16.24%), *Pasteurellaceae* (3.4%), *Streptococcaceae* (2.83%), and *Weeksellaceae* (2.06%; Figure 3A). By week three, *Mycoplasmataceae* and *Bacteroidaceae* had increased in relative abundance (24.85% and 4.1%, respectively). ANCOM-BC analysis at the ASV level revealed that *Ruminocccaceae, Sutterellaceae*, and *Lachnospiraceae* were the top three enriched families, while *Neisseriaceae* and *Moraxellaceae* were depleted at sampling time point three (Figure 3D; q-value < 0.001). Alpha diversity and composition analysis produced non-significant results. However, linear regression analysis indicated that sampling time point as a predictor was correlated with nasal diversity changes in Faith's PD (Figure S5B; *p*-value = 0.0003, R^2 = 0.22) and richness (*p*-value = 0.0005, R^2 = 0.2) at sampling time point two. As with fecal samples, new multiple regression models were generated due to the low R^2 values of the previous models, with the predictors being sampling time point, individual calf, and age in days. The new regression models revealed that the predictors were not correlated with Faith's PD (Figure S6B; *p*-value = 0.2369, R^2 = 0.59); however, the predictors were correlated with richness (Figure S6B; *p*-value = 0.016, R^2 = 0.72).

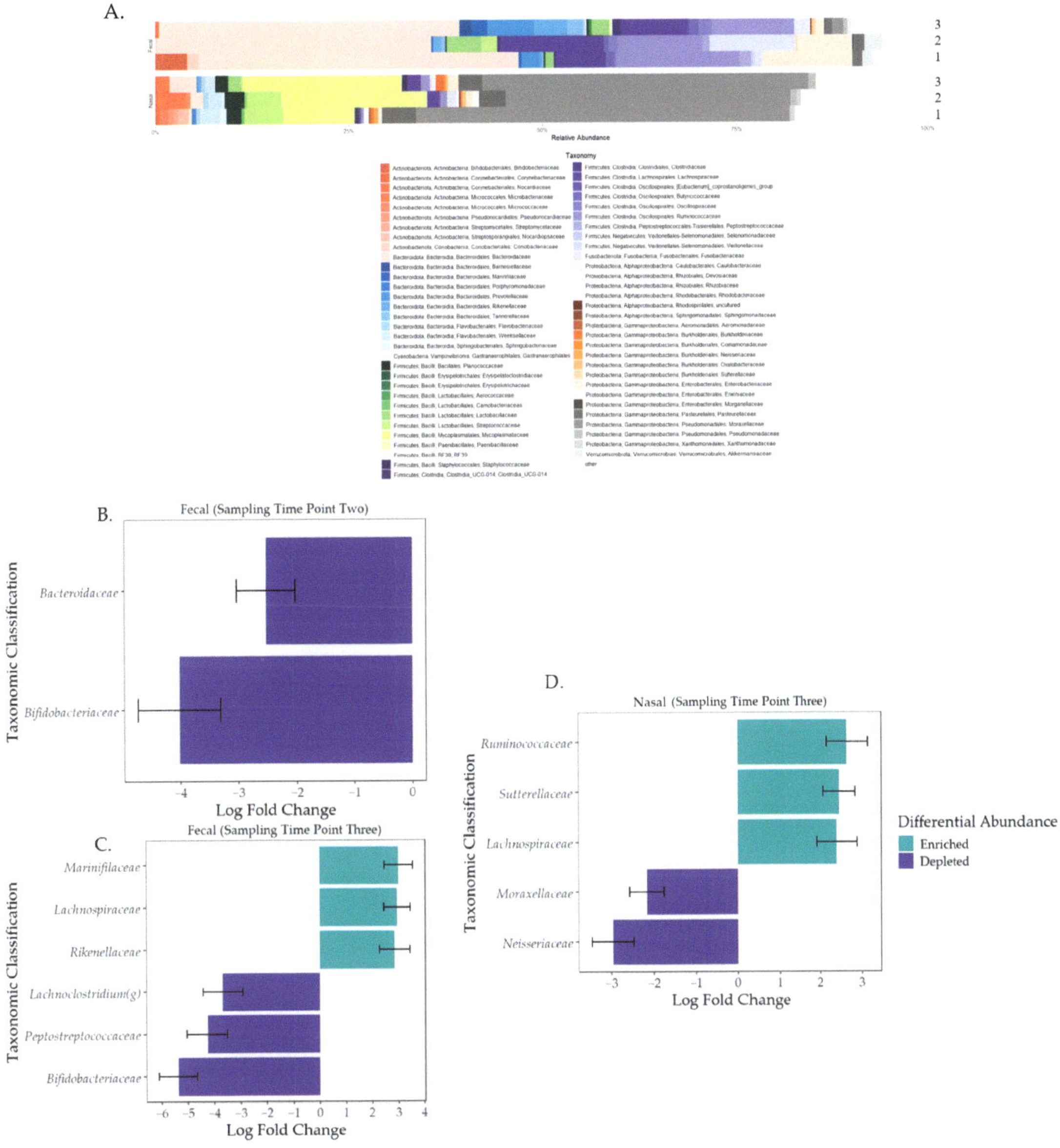

Figure 3. Taxonomy and taxonomic analysis using differential abundance testing (ANCOM-BC). (**A**) Taxonomic profiles for both fecal and nasal microbiomes using mean relative abundance at the family level across the three sampling time points (1–3) [61]. (**B**) Differentially abundant fecal taxa at sampling time point two. (**C**) Differentially abundant fecal taxa at sampling time point three. (**D**) Differentially abundant nasal taxa at sampling time point three.

3.3. Governing Forces of Microbial Community Composition

There was a difference in microbial composition in fecal samples between sampling time points one and two (unweighted UniFrac PERMANOVA, $p = 0.044$, $q = 0.044$), one and three ($p = 0.001$, $q = 0.001$), and two and three ($p = 0.001$, $q = 0.001$). The changes in community composition were analyzed for the forces driving community assembly. Fecal

NSTi, MST, SES, and β_{RC} revealed values of 0.9765, 54.91%, 0.0576, and 0.0146, respectively (Table S1). The high NSTi value indicates that stochasticity dominated community assembly, while the MST value of 54.91% indicates that community assembly favored stochastic forces. The SES and β_{RC} values confirm that community assembly favored neutrality (Figure 2C). Kruskal–Wallis testing revealed non-significant results for each of the values when tested against sampling time points ($p > 0.01$; Table 2).

Table 2. Significance results of Kruskal–Wallis with Benjamini–Hockberg multiple testing corrections (p-value, with adjusted p-value in parentheses) from cNST testing from both sampling time point and disease state.

Sample Type		NSTi	MST	SES	β_{RC}
Fecal	Sampling time point	0.3382 * (0.5411)	0.3202 (0.5411)	0.0482 (0.2185)	0.0546 (0.2185)
	Disease state	0.4595 (0.6127)	0.1415 (0.3774)	0.9219 (0.9361)	0.9361 (0.9361)
Nasal ¥	Sampling time point	0.8345 (0.8344)	0.6316 (0.8344)	0.7071 (0.8344)	0.7671 (0.8344)

* All p-values were set to significance level 0.01. ¥ Nasal samples were only tested against sampling time point due to low incidence of BRD.

Within nasal samples, microbial community composition differed between sampling time points one and two (unweighted UniFrac PERMANOVA, $p = 0.038$, $q = 0.057$) and time points one and three ($p = 0.004$, $q = 0.012$). The nasal NSTi, MST, SES, and β_{RC} values were 0.9603, 36.61%, 0.0463, and 0.0166, respectively (Table S1). The high NSTi value indicates dominance of stochastic forces driving community assembly. While the MST value of 36.61% indicates that deterministic factors were favored in nasal community assembly, the model emphasizes the observed versus the null model such that it can downplay stochasticity in datasets where deterministic processes are favored in some sample pairs. Furthermore, the SES and β_{RC} values likely confirm neutral processes in both fecal and nasal community assembly (Figure 2C). Kruskal–Wallis testing in nasal samples also returned non-significant results ($p > 0.01$) for the NST testing values when tested against sampling time points (Table 2).

3.4. Impacts of Disease on Early-Life Microbial Establishment

Early-life microbiome association with disease state was only assessed for diarrhea and fecal microbiomes given the low incidence rate of BRD-positive calves. Table S2 details the number of calves associated with fecal severity scores used in the analysis. Calves with "healthy" or "mild" fecal scores were considered healthy, while calves with "moderate" or "severe" fecal scores were considered diarrheic. In healthy calves, *Bacteroidaceae* (39.52%), *Ruminococcaceae* (13.06%), *Lachnospiraceae* (10.14%), *Enterobacteriaceae* (7.16%), and *Fusobacteriaceae* (3.53%) were the most relatively abundant bacteria at the family level. In diarrheic calves, *Bacteroidaceae* (41.42%), *Ruminococcaceae* (11.94%), *Enterobacteriaceae* (9.35%), *Lachnospiraceae* (9.12%), and *Lactobacillaceae* (5.77%) were the most relatively abundant families. Healthy calves reported a higher relative abundance in *Prevotellaceae* (3.52%) compared to diarrheic calves. Richness was higher in healthy calves compared to diarrheic calves ($p = 0.002$, $q = 0.002$), phylogenetic diversity (Faith's PD) was higher for healthy calves compared to diarrheic calves ($p = 0.0044$, $q = 0.0044$), and Shannon diversity was higher in healthy calves compared to diarrheic calves ($p = 0.043$, $q = 0.043$; Figure 4A). Multiple regression analysis indicated that fecal severity was not correlated with diversity changes (Figure S7). No compositional differences (PERMANOVA) or enriched features (ANCOM-BC) were detected (Figure 4B; $p > 0.05$). The DMM indicated that the genera

Bacteroides (two features, with one feature corresponding to the species *Bacteroides vulgatus*) and *Escherichia-Shigella* in healthy calves (Figure 4C). In diarrheic calves, the top drivers were the genera *Bacteroides*, *Faecalibacterium*, and *Escherichia-Shigella*. PERMANOVA testing using the adonis2 function revealed that the relationship between DMM clusters and composition were statistically significant (F-statistic = 8.7342; $p = 0.001$). NSTi, MST, SES, and β_{RC} values tested for significance using Kruskal–Wallis with Benjamini–Hockberg multiple-testing corrections revealed non-significant results (Table 2; $p > 0.01$), indicating the top three genera in each group were possible assembly drivers.

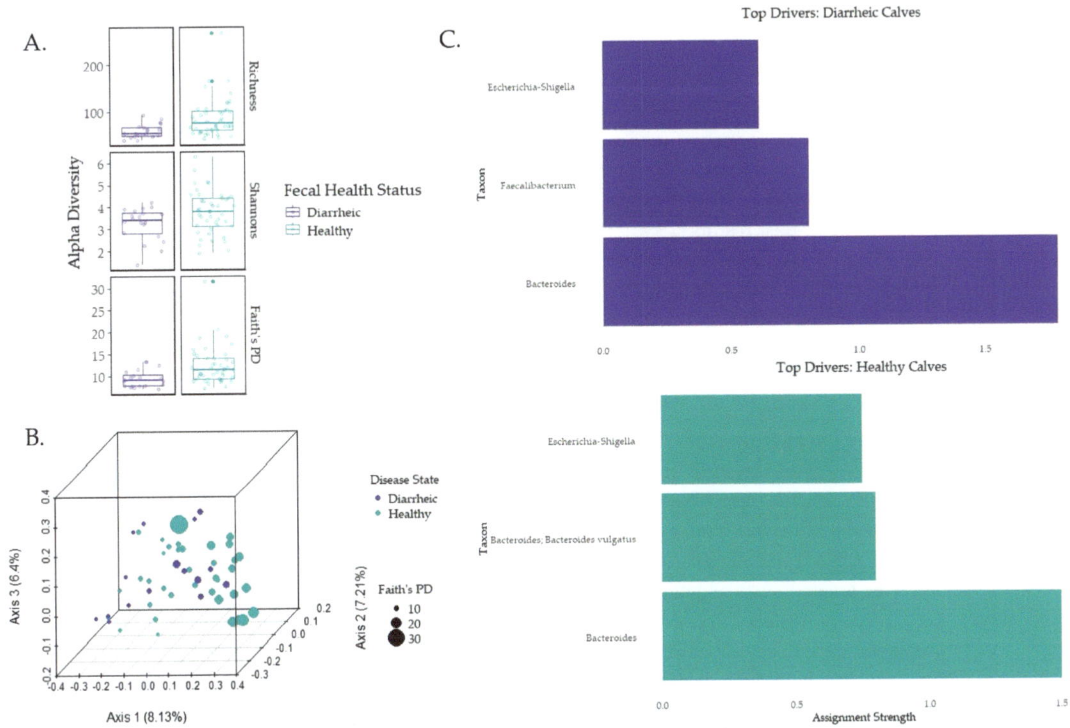

Figure 4. Alpha and beta diversity metrics for healthy and diarrheic calves with the top three community drivers at the genus level. (**A**) Alpha diversity metrics were measured for fecal samples and health status. Statistical analysis was conducted with Kruskal–Wallis pairwise testing. (**B**) Unweighted UniFrac distance was used to analyze beta diversity for fecal samples and health status and tested for significance using PERMANOVA pairwise analysis [61]. (**C**) DMM generated the top three genera for microbial community assembly using the lowest Laplace approximation score for microbial structure similarity clustering.

4. Discussion

The early-life microbiome is an important component to host maintenance and overall fitness in dairy calves. The assemblage of the dairy calf microbiome plays an important role in the outcome of host physiologies and functionalities [9–13]. Stochastic and deterministic forces can have profound effects on microbial composition, although these forces are not fully understood. Stochastic forces are those that randomly shape community structure and can be probabilistic with respect to species identity or functionality (i.e., historical contingency, drift, dispersal); on the other hand, deterministic forces are those that shape microbial assembly in a defined, predictable manner (i.e., age, diet, host genetics) [19–21]. The various factors governed by these forces can impact microbiome assembly and, ultimately, function (Figure 1). In the current study, early-life microbial ecologies varied in both fecal and nasal samples. Importantly, our findings suggest that stochastic processes, as described by neutral theory-based dynamics, may play a significant role in shaping microbial structure regardless of health status within the first three weeks of life. Neutral theory posits that all species have ecological equivalence and that community assembly is driven by random processes. Moreover, previous research supports our results indicating stochastic assembly during the first three weeks of life [22]. Our results indicate that the assembly of the microbiomes is predicated by important microbial community drivers. These microbes shape community populations in fecal and nasal microbiomes, and could be governed by stochastic, neutral theory-based dynamics.

Similar to previous work, the current study revealed that *Bacteroidaceae*, *Ruminococcaceae*, *Lachnospiraceae*, *Enterobacteriaceae*, and *Fusobacteriaceae* were the most relatively abundant families in the early-life GI microbiome regardless of health status [16,28,62]. These findings are indicative of the early, pioneering colonizers within the ruminant GI tract [16]. *Bacteroidaceae* was the most dominant family throughout the first three weeks of life, confirming past results [15,16]. It has also been reported that *Bacteroidaceae* dominate fecal microbiomes during the first two weeks of early life and decrease in abundance as the calf ages [16,28]. Our results corroborate these past findings, as *Bacteroidaceae* was depleted at sampling time point two. Importantly, our results indicate that regardless of health status, features identified in the genus *Bacteroides* were found to be top drivers for community structure assembly (DMM; p-value < 0.01). Supporting our findings, the genus *Bacteroidetes* has been well studied in its ability to maintain gut health and has shown the potential to be a biomarker in healthy cattle GI tracts [28]. Taken together, a high relative abundance of *Bacteroidaceae* indicates its importance to both GI health and shaping community structure in the calf early-life microbiome.

Additionally, the families *Ruminococcaceae* and *Lachnospiraceae* have the ability to produce butyrate and have been linked to gut health [62]. Butyrate is a short-chain fatty acid and is a primary source of energy for the rumen epithelium, which is necessary for nutrient and water absorption; butyrate can have beneficial side effects when supplemented in the diet of early-age livestock animals [12]. While not among the top three drivers of community structure, recent evidence has indicated that *Ruminococcaceae* can be predictive of microbial community assembly in both healthy and diarrheic calves [22]. Interestingly, recent findings reported an increase in the fiber-degrading family *Lachnospiraceae* as the calf aged [15]. Similarly, our results indicated that *Lachnospiraceae* was enriched at sampling time point three. Interestingly, our study also reported that *Rikenellaceae* was enriched at sampling time point three. *Rikenellaceae* is thought to be a beneficial bacterial group and has been shown to potentially improve host health and metabolism [63]. Thus, important shifts in microbial ecology are present in neonatal microbiomes, perhaps to prepare the calf for more complex diets. Like Furman et al. [15], no dietary changes were introduced to the calves, suggesting that other factors, such as niche modification by early-arriving species, were involved in microbial assembly. Interestingly, previous research has provided evidence for niche modification, as early-arriving aerobes and facultative anaerobes consume the oxygen, providing an environment for anaerobic species to thrive in the rumen [64]. Overall, our findings could indicate a shift in the microbiome to prepare the animal for solid, fiber-based diets prior to its introduction.

The early-life dairy calf microbiome is characterized by increasing diversity with age and change in diet, especially in the gut microbiome. Similar to previous findings, the current study revealed significant changes to diversity, as measured by Faith's PD and richness (p-value < 0.05) [65,66]. The change in diversity over time validates the dynamic shift in microbiota during the early life of dairy calves. Similarly, the current study revealed that *Bifidobacteriaceae* decreased in relative abundance in fecal samples over time (ANCOM-BC). Given the restrictive nature of on-farm feeding protocols to promote weaning, our results support previous findings in explaining the decrease in *Bifidobacterium* as the calf matures [67]. Bifidobacterium has a well-established role in mammalian gut health, as it provides mechanisms in gut mucosal barrier maintenance by production of beneficial metabolic substrates as well as immunological properties that prevent the attachment of pathogenic organisms [68,69]. The decrease in *Bifidobacterium* over time further highlights the shift to a microbial structure prepared for fiber-based diets. These observed shifts in microbial diversity raise critical questions about the underlying processes driving these changes, particularly the balance between stochastic and deterministic forces. The observed shifts in diversity and composition highlight the dynamic nature of microbial assembly during the first three weeks of life. To better understand the drivers of these changes, we evaluated the role of stochastic and deterministic processes in shaping microbial communities.

Importantly, our results suggest that stochastic forces dominate early-life microbial assembly in the GI microbiome, confirming previous findings [15,22]. The high NSTi and MST values (0.9765 and 54.91%, respectively) suggest that stochastic drivers shape the assembly and structure of early-life microbial populations, providing plausible explanation to observed diversity shifts over time. As an example, Kim et al. [28] reported that *Enterobacteriaceae* was significantly abundant in vaginally birthed calves aged one to four weeks, suggesting vertical transmission (i.e., stochasticity). Similarly, the elevated abundance of *Enterobacteriaceae* reported in the current study could be due to stochastic forces such as vertical transmission, given its place among the pioneer GI colonizers. Interestingly, the near-neutral values of SES and β_{RC} (0.0576 and 0.0146, respectively) indicate that the observed community assembly patterns align with predictions of neutral theory [20]. These findings suggest that the neutral theory of microbial assembly could be governing microbial assembly in the early microbiome. Only two studies have previously reported neutral-based assembly dynamics in the rumen microbiome [15,22]. Therefore, the current study could be among the first studies in suggesting the neutral-based dynamics of microbial assembly in early-life dairy calves. The consistency of diet in the current study validates earlier work of age-dependent assembly mechanisms in early-life microbial structure [15]. Further, microbial composition displayed significant differences between all sampling time points (PERMANOVA; p-value < 0.05), supporting the dynamic nature of early-life microbial assembly. Contrary to previous studies that have been confounded by diet when analyzing diversity shifts in dairy calves [65], the calves in our study were only fed line milk. Thus, the diversity and compositional changes within our study were not due to dietary alterations. While our results indicated no correlation between age and sampling time point with alpha diversity increases (p-value > 0.05, $R^2 = 0.57$), the shifts in composition suggest that age and development of the dairy calf (i.e., determinism) could be major drivers of compositional changes regardless of the influence of diet. This inference is supported by Ma et al. [62], wherein calves fed a regulated diet of milk replacer had distinct compositional shifts during the first eight weeks of life. These findings suggest that as calves age, deterministic forces increasingly shape community dynamics, signifying a gradual shift from stochastic to more deterministic assembly processes. Past studies support this claim, as a transition to more deterministic assembly dynamics was seen between weeks two and three in neonatal calves [22]. However, non-significant results from Kruskal–Wallis testing for NST, MST, SES, and β_{RC} across time points indicate that stochastic and neutral dynamics dominate consistently without temporal variability ($p > 0.01$). Therefore, while the calf might be shifting to more deterministically driven community assembly patterns due to calf age, stochastic forces remain the top drivers of population structure in the first three weeks of life. Our results strongly align with the hypothesis that neutral-based, stochastic forces govern early-life microbial assembly in the GI tract. While our results indicate that stochasticity dominates early microbial assembly and establishment, the current study does not analyze the impact of microbial genetic mutation on community assembly. Therefore, further investigation is needed to validate these claims of neutral-based stochasticity in the first three weeks of life.

While healthy calves demonstrated stochastic assembly dynamics, disease states introduced additional complexity. Diarrheic calves displayed distinct microbial patterns, suggesting that perturbations may shift the balance toward deterministic assembly. Diarrheic calves differed in the most relatively abundant taxa due to the addition of *Lactobacillacae* (5.577%). *Lactobacillus* species are dominant in neonatal calf GI microbiomes, as it is among the microbiota to utilize milk nutrients; thus, the higher relative abundance in diarrheic calves might not correlate to disease state [28]. However, alpha diversity was significantly higher in healthy calves compared to diarrheic calves (p-value < 0.05), confirming previous research [70]. Our study highlights that lower diversities are indicative of microbial changes due to perturbations, translating to Anna Karenina principles (AKP), which suggests that dysbiosis is due to stochastic effects rather than deterministic [19,71]. AKP states that "all healthy microbiomes are similar; each dysbiotic microbiome is dysbiotic in its own way." Our results support AKP, as the DMM demonstrated distinct clustering between healthy and diarrheic calves, indicating that community assembly was driven by distinct microbiota. However, the stochastic principles that create the dysbiotic community structure transition to more deterministic forces once clinical disease is observed. Previous work indicates that deterministic forces are predominant in lower-diversity microbial populations [72]. As the diversity was lower in diarrheic calves in the current study, stochastic forces may have been downplayed in diseased animals. This claim is demonstrated within the current study, as the genus *Faecallibacterium* was among the top drivers of community assembly in diarrheic calves, supporting previous studies indicating that *Faecallibacterium* was highly influential in unhealthy calves at two weeks of age [22]. This family of microbes is also among the butyrate-producing bacteria in the neonatal gut and can induce T-regulatory cells critical for host gut immunological homeostasis [22,73]. Interestingly, *Escherichia-Shigella*, an etiological agent for diarrhea, was among the top drivers in both healthy and diarrheic calves. Because pathogenic bacteria can interact with beneficial community members (e.g., *Faecallibacterium*), resulting in dysbiosis, bacteria such as *Escherichia-Shigella* are thought to be keystone taxa [22]. Thus, our results could potentially support previous findings that unhealthy animals favor deterministically driven assembly, contrasted with healthy calves, which favor stochastically driven assembly. While disease could significantly impact community assembly dynamics, the exact interaction mechanisms between *Faecallibacterium* and *Escherichia-Shigella* were not investigated in the current study. However, PERMANOVA adonis testing indicated that the top three genera in each group were likely valid community drivers ($p < 0.01$). Thus, the difference in community composition between healthy and diarrheic calves are driven by distinct microbial compositions, as supported by previous research [22]. The exact mechanisms for this distinction were not directly tested within this current study. Therefore, these results highlight how disease can shift early-life microbiomes to deterministically driven assembly patterns rather than stochasticity in the GI tract. As posited by Vellend et al. [20], niche and neutral theories can be considered together, and therefore encompass a unified neutral theory capable of explaining both the stochastic and the deterministic nature of microbial assembly. These findings emphasize a potential unified neutral theory and highlight the need for further research in the mechanisms underlying these shifts.

While the fecal microbiome provides insights into GI-related microbial dynamics, the nasal microbiome offers a complementary perspective on host-associated microbial assembly, particularly in relation to respiratory health and microbial transmission in neonatal calves. The nasal microbiome is a key component of the URT and could provide additional insights into early-life microbial assembly processes and their implications for health and disease dynamics. Similar to previous findings, the most relatively abundant bacteria in nasal samples were the families *Moraxellaceae*, *Mycoplasmataceae*, *Pasteurellaceae*, *Streptococcaceae*, and *Weeksellaceae* [30,74–76]. Our study was unable to directly test nasal samples with disease state due to the low incidence of BRD-positive animals; however, certain aspects of our findings can be speculated upon with the current literature. As with GI microbial communities, pathogenic and beneficial bacteria inhabit the nasal microbiome [27]. The route of transmission of *Moraxellaceae* and *Streptococcaceae*, commonly associated with pneumonia and otitis media in calves, might contribute to vertical transmission. Our results demonstrate that the nasal microbiome favors stochastic assembly with no temporal variation (NSTi value = 0.9603, $p > 0.05$), as potentially seen in previous studies [30,74–76]. It was demonstrated that 63% of microbial composition was shared between the neonatal calf nasal microbiome and the mother's vaginal microbiome, supporting our claim of stochastic assembly in early-life microbial assembly [74]. Moreover, similar to our findings, *Mycoplasmataceae* has been linked to both sick and healthy animals [30]. While health status could not be directly tested in the current study, as mentioned above, our findings could suggest that pathogens such as *M. dispar* interact with beneficial bacteria in a stochastic manner, creating dysbiosis sufficient for causing disease states (i.e., AKP) [30]. Interestingly, the current study identified that the families *Neisseriaceae* and *Moraxellaceae* were depleted within the nasal microbiomes. Given that the importance of *Neisseriaceae* in URT microbiomes has yet to be determined, its introduction from oral microbiomes suggests stochasticity [77]. More importantly, the depletion of *Moraxellaceae* at sampling time point three suggests that neonatal calf health was improving. Given our study saw little to no clinical evidence of BRD, these results suggest host–microbe interactions in early-life URT microbiomes supporting animal health. These dynamics indicate a shift from stochasticity to more host-derived determinism with immunological consequences.

Furthermore, no diversity changes were detected for nasal samples across time (p-value > 0.05), consistent with previous findings [30,76]. The lack of diversity change coupled with composition changes over time (PERMANOVA p- value < 0.05) supports our findings that neutral processes are present and do not vary temporally (SES = 0.0463, $\beta_{RC} = 0.0166$, $p > 0.05$ respectively). There is little evidence to support these claims within the literature, making our study among the first to report potential unified neutral-based dynamics in the nasal microbiome. Supporting this hypothesis, deterministic properties are present, as MST testing reported that only 36.61% of microbial assembly was explained by stochastic forces, with no temporal variation. Diversity can remain unchanged if stochastic events such as disease are minimal, leading to community structures defined by deterministic forces [19]. Furthermore, environmental filtering can impact microbial assembly. While the calves were individually housed, limiting animal-to-animal contact, the calves could interact with their neighbors as well as the environment. Linear regression models demonstrated that richness was correlated with time in the current study (p-value = 0.016, $R^2 = 0.72$). An increase in unique microbial members could imply that environmental filtering may have contributed to the observed compositional changes in the nasal microbiome. Environmental filtering can occur through many routes, including differences in husbandry, hygiene, and animal contact. Previous research found that nasal samples from dairy calves were dominated by *Mycoplasma* spp. due to a high prevalence of *Mycoplasma* spp. found on the farm within the study, supporting environmental filtering within the current study [78].

Moreover, *Lachnospiraceae* was enriched at sampling time point three in both fecal and nasal microbiomes. The corresponding enrichment of *Lachnospiraceae* between fecal and nasal microbiomes further supports environmental filtering of URT microbial populations. Overall, the high NSTi value (0.9603), the near-neutral SES and β_{RC} values, and the lack of associated temporal variation provide ample evidence for stochastic-based assembly consistent with unified neutral processes occurring. Both stochastic (e.g., vertical transmission) and deterministic (e.g., environmental filtering) processes may influence early microbial assembly in the URT. Therefore, further evaluation must be undertaken to untangle the balance between stochastic and deterministic processes in URT microbiome assembly.

While there are several strengths of the current study, it is not without its faults. First, multiple breeds were included. Future studies should include enough calves to control for breed statistically since commercial dairies may have more than one breed. Additionally, the number of calves enrolled in the study along with the limited number of sampling time points should be increased in future studies. Increasing the sample population and adding more sampling time points could be better suited for capturing the dynamics of microbial assembly. An increase in sample size and sampling time could also elevate the potential of disease occurrence, primarily in clinical BRD. Larger sampling populations would offer a more robust evaluation of disease state on microbial establishment and assembly.

Second, the farm's treatment protocol did not include the use of antimicrobials for treating diarrhea, although antibiotics were used for treating respiratory disease. This disparity in antimicrobial use was not assessed in the current pilot study and should be considered in future studies. Past research indicates that antimicrobials can affect microbial ecology in early life. Ma et al. [62] demonstrated that the use of antimicrobials delayed the temporal development of microbial diversity, suggesting that therapeutic use of antimicrobials negatively affects microbial development more so than diarrhea incidence in the first 21 days of life for dairy calves. While calves with two or more antibiotic treatments were removed from the current study, it is unclear which consequences arose from initial antibiotic treatment. Additionally, while past studies have reported no differences in community composition and diversity due to antibiotic treatment, others have reported significant changes in treated calves [76,79]. Thus, future studies should account for antibiotic treatment effects on microbial assembly in both GI and URT microbiomes.

Third, empirical studies face challenges when assessing stochastic effects on microbial community assembly. Primarily, mathematical calculations of within-group distance can include variance due to error and environmental factors that are not measured, resulting in an overestimation of stochasticity [21]. Furthermore, measuring microbial dispersal is often not possible in real-time datasets, making it difficult to measure the influence of generalists across studies [21]. Theoretical models such as the Decomposition Model of Enzymatic Traits (DEMENT) can be employed to better understand the complexity of microbial assembly [80]. Additionally, designing empirical studies with a more complex theoretical framework could provide more robust approaches to untangling the complexity of microbial assembly [19]. Therefore, future studies on the balance of stochastic and deterministic forces in microbial assembly should aim to incorporate many of these elements to better substantiate the hypotheses currently expressed in the pilot study.

Fourth, future studies should investigate the functional aspects of early-life dairy calf microbiomes and dysbiosis caused by disease. Multi-omics microbiome data sets that include combinations of metagenomic and metatranscriptomic sequencing, proteomics, lipidomics, and metabolomics, as well as immune measures, can provide valuable insights into the relationship between microbiomes and disease states. Past research has utilized functional analysis to study various microbial communities in dairy cattle. For instance, Pitta et al. [81] provided metagenomic evidence that metabolic functions within the rumen microbiome varied in three different treatment groups of dairy cows in response to age and physiological changes. Incorporating functional analyses could enhance our understanding of how stochasticity and deterministically driven assemblies occur in nature. These methodologies could untangle the functional potential with observed taxonomic patterns. Additionally, functional analyses could help tease apart the dynamic nature of the role of disease state in microbial assembly and composition. Future studies should aim to include functional analysis to quantify the mechanisms of microbial assembly during perturbation events. Despite these limitations, our study adds valuable insights into early-life microbial assembly and highlights critical areas for future investigation.

5. Conclusions

Early-life microbiomes in calves are important for the development of the animal and maintenance of health. The current pilot study characterizes the establishment of fecal and nasal microbiomes during the first three weeks of life of dairy calves as well as investigates the forces that drive community assembly. Our study is among the first to potentially describe stochastic, unified neutral theory-based assembly dynamics in both GI and URT microbiomes. Stochastic processes seem to dominate in both GI and URT microbial assemblies; however, evidence suggests that deterministic factors may reduce stochasticity in nasal microbiomes. Additionally, distinct microbial compositions are seen in between healthy and diarrheic calves and are driven by different microbial members. However, given the limited sample size and sampling time points, larger studies are necessary to support this hypothesis. Functional analysis could further validate the dynamics involved in early-life microbial assembly. Overall, this study provides a foundational framework for understanding microbial assembly in neonatal calves, offering novel insights into the interplay between stochastic and deterministic processes. In addition to understanding assembly dynamics, our findings on clinical diarrhea emphasize the importance of investigating perturbations and their implications on microbial ecology. Our findings suggest that healthy and diarrheic calves have distinct microbial clusters that drive community composition, indicating that perturbation could be evident of AKP. Our findings provide a roadmap for future research into neonatal calf microbial ecology, with critical implications for improving calf health and understanding microbial dynamics across host-associated environments.

Supplementary Materials: The following supporting information can be downloaded at: https://www.mdpi.com/article/10.3390/ani15030361/s1, Table S1: The results from cNST models to test stochasticity and neutrality in early-life microbiomes; Table S2: The total number of calves associated with fecal severity scores at each sampling time point; Figure S1: Mock community assessment; Figure S2. Linear regression models for breed, alpha diversity, and fecal severity; Figure S3. Linear regression models for breed, alpha diversity, and nasal samples; Figure S4. Linear regression models for sex, alpha diversity, and nasal samples; Figure S5. Linear regression models for sampling time points, alpha diversity metrics, and fecal samples; Figure S6. Multiple regression models for fecal and nasal sample types, with predictors of sampling time point, individual calf, and calf age in days for effects on alpha diversity metrics; Figure S7. Linear regression models for fecal severity scores and alpha diversity metrics.

Author Contributions: Conceptualization, J.L.M., A.D.B. and A.N.F.; methodology, A.N.F., with significant input from J.L.M. and M.C.C.; software, A.N.F. and J.L.M.; validation, A.N.F., A.D.B., K.A.-L. and J.L.M.; formal analysis, A.N.F. and K.A.-L.; investigation, A.N.F. and L.F.; resources, J.L.M. and M.C.C.; data curation, A.N.F. and L.F.; writing—original draft preparation, A.N.F.; writing—review and editing, A.N.F., L.F., A.D.B., K.A.-L., M.C.C. and J.L.M.; visualization, A.N.F.; supervision, J.L.M.; project administration, A.N.F. and J.L.M.; funding acquisition, J.L.M. All authors have read and agreed to the published version of the manuscript.

Funding: This research was funded by the Canadian Institute for Advanced Research (CIFAR) via Jessica Metcalf's Global Scholar award.

Institutional Review Board Statement: The animal study protocol was approved by the Institutional Animal Care and Use Committee at Colorado State University (IACUC #1421).

Informed Consent Statement: Dairy management provided verbal informed consent to the principal investigators prior to project commencement. The study parameters and study duration were approved by dairy management.

Data Availability Statement: The 16S rRNA gene sequencing data are publicly available in the EBI repository under study accession PRJEB76862 and on the QIITA study platform under study ID 13813 [34].

Acknowledgments: We would like to thank all authors for their contributions to this work. We appreciate the work of Kristen Otto (Colorado State University) for both her technical assistance and her laboratory expertise. The authors would like to thank Zachary Burcham (University of Tennessee) for his technical expertise and assistance in data preparation. We also would like to thank the staff and owners of the dairy for allowing us to perform this research, as well as the 2021 ANEQ505 class. This work was part of Anthony Nathan Frazier's (A.N.F.) doctoral dissertation.

Conflicts of Interest: The authors declare no conflicts of interest.

Abbreviations

The following abbreviations are used in this manuscript:

AKP	Anna Karenina Principles
ANCOM-BC	Analysis of composition of microbiomes with bias corrections
BRD	Bovine respiratory disease
DEMENT	Decomposition Model of Enzymatic Traits
DMM	Dirichlet multinomial model
FDR	False discovery rate
GI	Gastrointestinal
NMDS	Nonmetric multidimensional
NST	Normalized stochasticity testing
NSTi	Normalized stochasticity index
MST	Modified stochasticity ratio
PERMANOVA	Permutational multivariant analysis of variance
β_{RC}	Raup–Crick distance
SES	Standardized effect size index
URT	Upper respiratory tract

References

1. McFall-Ngai, M.; Hadfield, M.G.; Bosch, T.C.G.; Carey, H.V.; Domazet-Loso, T.; Douglas, A.E.; Dubilier, N.; Eberl, G.; Fukami, T.; Gilbert, S.F.; et al. Animals in a bacterial world, a new imperative for the life sciences. *Proc. Natl. Acad. Sci. USA* **2013**, *110*, 3229–3236. [CrossRef] [PubMed]
2. Tamburini, S.; Shen, N.; Wu, H.C.; Clemente, J.C. The microbiome in early-life: Implications for health outcomes. *Nat. Med.* **2016**, *22*, 713–722. [CrossRef] [PubMed]
3. Mohammadkah, A.I.; Simpson, E.B.; Patterson, S.G.; Ferguson, J.F. Development of the gut microbiome in children, and lifetime implications for obesity and cardiometabolic disease. *Children* **2018**, *5*, 160. [CrossRef] [PubMed]
4. Zhuang, L.; Chen, H.; Zhang, S.; Zhuang, J.; Li, Q.; Feng, Z. Intestinal microbiota in early-life and its implications on childhood health. *Gen. Prot. Bioinform.* **2019**, *17*, 13–25. [CrossRef] [PubMed]
5. Hooper, L.V.; Midtvedt, T.; Gordon, J.I. How host-microbial interactions shape the nutrient environment of the mammalian intestine. *Anny. Rev. Nutr.* **2002**, *22*, 283–307. [CrossRef] [PubMed]
6. Erny, D.; Hrabě de Angelis, A.L.; Jaitin, D.; Wieghofer, P.; Staszewaski, O.; David, E.; Keren-Shaul, H.; Mahlakoiv, T.; Jakobshagen, K.; Buch, T.; et al. Host microbiota constantly control maturation and function of microglia in the CNS. *Nat. Neurosci.* **2015**, *18*, 965–977. [CrossRef]

7. Dinan, T.G.; Cryan, J.F. The impact of gut microbiota on brain and behaviour: Implications for psychiatry. *Cur. Opi. Clin. Nutr. Metabl. Care* **2015**, *18*, 552–558. [CrossRef] [PubMed]
8. Johnson, K.V.-A.; Foster, K.R. Why does the microbiome affect behaviour? *Nat. Rev. Microbiol.* **2018**, *16*, 647–655. [CrossRef] [PubMed]
9. de Menezes, A.B.; Lewis, E.; O'Donovan, M.; O'Neail, B.F.; Clipson, N.; Doyle, E.M. Microbiome analysis of dairy cows fed pasture or total mixed ration diets. *FEMS Microbiol. Ecol.* **2011**, *78*, 256–265. [CrossRef] [PubMed]
10. Difford, G.F.; Plichta, D.R.; Løvendahl, P.; Lassen, J.; Noel, S.J.; Højberg, O.; Wright, A.-D.G.; Zhu, Z.; Kristensen, L.; Nielsen, H.B.; et al. Host genetics and the rumen microbiome jointly associate with methane emissions in dairy cows. *PLoS Genet.* **2018**, *14*, e10007580. [CrossRef]
11. Chen, S.; Wang, J.; Peng, D.; Li, G.; Chen, J.; Gu, X. Exposure to heat-stress environment affects the physiology, circulation levels of cytokines, and the microbiome in dairy cows. *Sci. Rep.* **2018**, *8*, 14606. [CrossRef] [PubMed]
12. O'Hara, E.; Kelly, A.; McCabe, M.S.; Kenny, D.A.; Guan, L.L.; Waters, S.M. Effect of butyrate-fortified milk replacer on gastrointestinal microbiota and products of fermentation in artificially reared dairy calves at weaning. *Sci. Rep.* **2018**, *8*, 14901. [CrossRef] [PubMed]
13. Bronzo, V.; Lopreiato, V.; Riva, F.; Amadori, M.; Curone, G.; Addis, M.F.; Cremonesi, P.; Moroni, P.; Trevisi, E.; Castiglioni, B. The role of innate immune response and microbiome in resilience of dairy cattle to disease: The mastitis model. *Animals* **2020**, *10*, 1397. [CrossRef] [PubMed]
14. Perez-Muñoz, M.E.; Arrieta, M.-C.; Ramer-Tait, A.E.; Walter, J. A critical assessment of the "sterile womb" and "in utero colonization" hypotheses: Implications for research on the pioneer infant microbiome. *Microbiome* **2017**, *5*, 48. [CrossRef] [PubMed]
15. Furman, O.; Shenhav, L.; Sasson, G.; Kokou, F.; Honig, H.; Jacoby, S.; Hertz, T.; Cordero, O.X.; Halperin, E.; Mizrahi, I. Stochasticity constrained by deterministic effects of diet and age drive rumen microbiome assembly dynamics. *Nat. Commun.* **2020**, *11*, 1904. [CrossRef] [PubMed]
16. Rey, M.; Enjalbert, F.; Combes, S.; Cauquil, L.; Bouchez, O.; Monteils, V. Establishment of ruminal bacterial community in dairy calves from birth to weaning is sequential. *J. Appl. Microbiol.* **2013**, *166*, 245–257. [CrossRef] [PubMed]
17. Takino, T.; Kato-Mori, Y.; Motooka, D.; Nakamura, S.; Iida, T.; Hagiwara, K. Postnatal changes in the relative abundance of intestinal *Lactobacillus* spp. in newborn calves. *J. Vet. Med. Sci.* **2017**, *79*, 452–455. [CrossRef]
18. Haley, B.J.; Kim, S.-W.; Salaheen, S.; Hovingh, E.; Van Kessel, J.A.S. Differences in the microbial community and resistome structures of feces from preweaned calves and lactating cows in commercial dairy herds. *Food. Path. Dis.* **2020**, *17*, 494–503. [CrossRef] [PubMed]
19. Frazier, A.N.; Beck, M.R.; Waldrip, H.; Koziel, J.A. Connecting the ruminant microbiome to climate change: Insights from current ecological and evolutionary concepts. *Front. Microbiol.* **2024**, *15*, 1503315. [CrossRef] [PubMed]
20. Vellend, M.; Srivastava, D.S.; Anderson, K.M.; Brown, C.D.; Jankowski, J.E.; Kleynhans, E.J.; Kraft, N.J.B.; Letaw, A.D.; MacDonald, A.A.M.; Myers-Smith, I.H.; et al. Assessing the relative importance of neutral stochasticity in ecological communities. *Oikos* **2014**, *123*, 1420–1430. [CrossRef]
21. Evans, S.; Martiny, J.B.H.; Allison, S.D. Effects of dispersal and selection on stochastic assembly in microbial communities. *ISME J.* **2016**, *11*, 146–185. [CrossRef] [PubMed]
22. Pan, Z.; Ma, T.; Steele, M.; Guan, L.L. Varied microbial community assembly and specialization patterns driven by early life microbiome perturbation and modulation in young ruminants. *ISME J.* **2024**, *4*, ycae044. [CrossRef]
23. USDA. Part II: Changes in the U.S. Dairy Cattle industry, 1991–2007. In *Dairy 2007*; USDA-APHIS-VS, CEAH: Fort Collins, CO, USA, 2007; pp. 57–61.
24. USDA. Part IV: Health and health management of U.S. feedlots with a capacity of 1000 or more head. In *Feedlot 2011*; APHIS, National Animal Health Monitoring System: Fort Collins, CO, USA, 2013.
25. Dunn, T.R.; Ollivett, T.L.; Renaud, D.L.; Leslie, K.E.; LeBlanc, S.J.; Duffield, T.F.; Kelton, D.F. The effect of lung consolidation, as determined by ultrasonography, on first-lactation milk production in Holstein dairy calves. *J. Dairy Sci.* **2018**, *101*, 5404–5410. [CrossRef] [PubMed]
26. Cramer, M.C.; Ollivett, T.L. Growth of preweaned, group-housed dairy calves diagnosed with respiratory disease using clinical respiratory scoring and thoracic ultrasound—A cohort study. *J. Dairy Sci.* **2019**, *102*, 4322–4331. [CrossRef] [PubMed]
27. Gomez, D.E.; Arroyo, L.G.; Costa, M.C.; Viel, L.; Weese, J.S. Characterization of the fecal bacterial microbiota of healthy and diarrheic dairy calves. *J. Vet. Intern. Med.* **2017**, *31*, 928–939. [CrossRef]
28. Kim, E.-T.; Lee, S.-J.; Kim, T.-Y.; Lee, H.-G.; Atikur, R.M.; Gu, B.-H.; Kim, D.-H.; Park, B.-Y.; Son, J.-K.; Kim, M.-H. Dynamic changes in fecal microbial communities of neonatal dairy calves by aging and diarrhea. *Animals* **2021**, *11*, 1113. [CrossRef]
29. Lima, S.F.; Teixeira, A.G.V.; Higgins, C.H.; Lima, F.S.; Bicalho, R.C. The upper respiratory tract microbiome and its potential role in bovine respiratory disease and otitis media. *Sci. Rep.* **2016**, *6*, 29050. [CrossRef] [PubMed]

30. Centeno-Martinez, R.E.; Klopp, R.N.; Koziol, J.; Boerman, J.P.; Johnson, T.A. Dynamics of the nasopharyngeal microbiome of apparently healthy calves and those with clinical symptoms of bovine respiratory disease from disease diagnosis to recovery. *Front. Vet. Sci.* **2023**, *10*, 1297158. [CrossRef]
31. Frazier, A.N.; Belk, A.D.; Beck, M.R.; Koziel, J.A. Impact of methane mitigation strategies on the native ruminant microbiome: A protocol for a systematic review and meta-analysis. *PLoS ONE* **2024**, *19*, e0308914. [CrossRef] [PubMed]
32. Ferree, L.A.; Edwards-Callaway, L.N.; Ramon-Muniz, I.N.; Coetzee, J.F.; Applegate, T.J.; Ollivett, T.L.; Cramer, M.C. Oral meloxicam given as an ancillary treatment at respiratory disease diagnosis was not associated with growth, clinical scores, or ultrasound scores in preweaned dairy calves. *J. Am. Vet. Med. Assoc.* **2022**, *261*, 1716–1723. [CrossRef] [PubMed]
33. McGuirk, S.M.; Peek, S.F. Timely diagnosis of dairy calf respiratory disease using a standardized scoring system. *Anim. Health Res. Rev.* **2014**, *15*, 145–147. [CrossRef] [PubMed]
34. Ollivett, T.L.; Buczinski, S. On-farm use of ultrasonography for bovine respiratory disease. *Vet. Clin. N. Am. Food Anim. Pract.* **2016**, *32*, 19–35. [CrossRef]
35. Belk, A.D.; Frazier, A.N.; Fuerniss, L.K.; Delmore, R.; Belk, K.; Borlee, B.; Geornaras, I.; Martin, J.N.; Metcalf, J.L. A Pilot Study: The development of a facility-associated microbiome and its association with the presence of *Listeria* spp. in one small meat processing facility. *Microbiol. Spectr.* **2022**, *10*, 5. [CrossRef] [PubMed]
36. Thompson, L.R.; Sanders, J.G.; McDonald, D.; Amir, A.; Ladau, J.; Locey, K.J.; Prill, R.J.; Tripathi, A.; Gibbons, S.M.; Ackermann, G.; et al. The Earth Microbiome Project Consortium. A communal catalogue reveals Earth's multiscale microbial diversity. *Nature* **2017**, *551*, 457–463. [CrossRef] [PubMed]
37. Gonzalez, A.; Navas-Molina, J.A.; Kosciolek, T.; McDonald, D.; Vázquez-Baeza, Y.; Ackermann, G.; DeReus, J.; Janssen, S.; Swafford, A.D.; Orchanian, S.B.; et al. Qiita: Rapid, web-enabled microbiome meta-analysis. *Nat. Methods* **2018**, *15*, 796–798. [CrossRef] [PubMed]
38. Boylen, E.; Rideout, J.R.; Dillon, M.R.; Bokulich, N.A.; Abnet, C.C.; Al-Ghalith, G.A.; Alexander, H.; Alm, E.J.; Arumugam, M.; Asnicar, F.; et al. Reproducible, interactive, scalable and extensible microbiome data science using QIIME 2. *Nat. Biotechnol.* **2019**, *87*, 852–857. [CrossRef]
39. Callahan, B.J.; McMurdie, P.J.; Rosen, M.J.; Han, A.W.; Johnson, A.J.A.; Holmes, S.P. DADA2: High-resolution sample inference from Illumina amplicon data. *Nat. Methods* **2016**, *13*, 581–583. [CrossRef]
40. Quast, C.; Pruesse, E.; Yilmaz, P.; Gerken, J.; Schweer, T.; Yarza, P.; Peplies, J.; Glöckner, F.O. The SILVA ribosomal RNA gene database project: Improved data processing and web-based tools. *Nucleic Acids Res.* **2012**, *41*, D590–D596. [CrossRef]
41. Bokulich, N.; Dillon, M.; Bolyen, E.; Kaehler, B.D.; Huttley, G.A.; Caporaso, J.G. Q2-sample-classifier: Machine-learning tools for microbiome classification and regression. *J. Open Source Softw.* **2018**, *3*, 934. [CrossRef] [PubMed]
42. Matsen, F.A.; Kodner, R.B.; Armbrust, E.V. pplacer: Linear time maximum-likelihood and Bayesian phylogenetic placement of sequences onto a fixed reference tree. *BMC Bioinform.* **2010**, *11*, 538. [CrossRef] [PubMed]
43. Janssen, S.; McDonald, D.; Gonzalez, A.; Navas-Molina, J.A.; Jiang, L.; Xu, Z.Z.; Winker, K.; Kado, D.M.; Orwoll, E.; Manary, M.; et al. Phylogenetic placement of exact amplicon sequences improves associations with clinical information. *mSystems* **2018**, *3*, 3. [CrossRef]
44. Lozupone, C.; Knight, R. UniFrac: A new phylogenetic method for comparing microbial communities. *Appl. Environ. Microbiol.* **2005**, *71*, 8228–8235. [CrossRef] [PubMed]
45. Lozupone, C.A.; Hamady, M.; Kelley, S.T.; Knight, R. Quantitative and qualitative beta diversity measures lead to different insights into factors that structure microbial communities. *Appl. Environ. Microbiol.* **2007**, *73*, 1576–1585. [CrossRef] [PubMed]
46. Lin, H.; Das Peddada, S. Analysis of compositions of microbiomes with bias correction. *Nat. Commun.* **2020**, *11*, 3514. [CrossRef] [PubMed]
47. Shannon, C.L.; Weaver, W. *The Mathematical Theory of Communication*; University of Illinois Press: Urbana, IL, USA, 1949; Volume vii, p. 117.
48. McIntosh, R.P. An Index of Diversity and the Relation of Certain Concepts to Diversity. *Ecology* **1967**, *48*, 392–404. [CrossRef]
49. Faith, D.P. Conservation evaluation and phylogenetic diversity. *Biol. Conserv.* **1992**, *61*, 1–10. [CrossRef]
50. Kruskal, W.H.; Wallis, W.A. Use of ranks in one-criterion variance analysis. *J. Am. Stat. Assoc.* **1952**, *260*, 583–621. [CrossRef]
51. R Core Team. *R: A Language and Environment for Statistical Computing*, R Foundation for Statistical Computing: Vienna, Austria. Available online: https://www.R-project.org (accessed on 10 April 2021).
52. Zhou, J.; Ning, D. Stochastic community assembly: Does it matter in microbial ecology? *Microbiol. Mol. Biol. R.* **2017**, *81*, e00002-17. [CrossRef] [PubMed]
53. Gotelli, N.J.; McCabe, D.J. Species co-occurrence: A meta-analysis of J.M. Diamond's assembly rules model. *Ecology* **2002**, *83*, 2091–2096. [CrossRef]
54. Crump, B.C.; Peterson, B.J.; Raymond, P.A.; Holmes, R.M. Circumpolar synchrony in big river bacterioplankton. *Proc. Natl. Acad. Sci. USA* **2009**, *106*, 21208–21212. [CrossRef]

55. Ning, D.; Deng, Y.; Tiedje, J.M.; Zhou, J. A general framework for quantitatively assessing ecological stochasticity. *Proc. Natl. Acad. Sci. USA* **2019**, *116*, 16892–16898. [CrossRef]
56. Holmes, I.; Harris, K.; Quince, C. Dirichlet multinomial mixtures: Generative models for microbial metagenomics. *PLoS ONE* **2012**, *7*, e30126. [CrossRef]
57. Stewart, C.J.; Ajami, N.J.; O'Brien, J.L.; Hutchinson, D.S.; Smith, D.P.; Wong, M.C.; Ross, M.C.; Lloyd, R.E.; Doddapaneni, H.; Metcalf, G.A.; et al. Temporal development of the gut microbiome in early childhood from the TEDDY study. *Nature* **2019**, *562*, 583–588. [CrossRef] [PubMed]
58. Bray, J.R.; Curtis, J.T. An ordination of the upland forest communities of southern Wisconsin. *Ecol. Monogr.* **1957**, *27*, 325–349. [CrossRef]
59. Oksanen, J.; Blanchet, F.G.; Friendly, M.; Kindt, R.; Legendre, P.; McGlinn, D.; Minchin, P.R.; O'Hara, R.B.; Simpson, G.L.; Solymos, P.; et al. *R Package*, Version 2.6-2; Vegan: Community Ecology Package; R Core Team: Vienna, Austria, 2022. Available online: https://CRAN.R-project.org/package=vegan (accessed on 9 January 2025).
60. Wickham, H. *ggplot2: Elegant Graphics for Data Analysis*; Springer: New York, NY, USA, 2009.
61. Frazier, A.N. Application of Biotechnology in Agriculture: A bioinformatic and Market Analysis of Novel Intervention Methods. Ph.D. Thesis, Colorado State University, Fort Collins, CO, USA, 2020. Available online: https://hdl.handle.net/10217/234261 (accessed on 19 January 2025).
62. Ma, T.; Villot, C.; Renaud, D.; Skidmore, A.; Chevaux, E.; Steele, M.; Guan, L.L. Linking perturbations to temporal changes in diversity, stability, and compositions of neonatal calf gut microbiota: Prediction of diarrhea. *ISME J.* **2020**, *14*, 2223–2235. [CrossRef] [PubMed]
63. Wang, L.; Wu, D.; Zhang, Y.; Li, K.; Wang, M.; Ma, J. Dynamic distribution of gut microbiota in cattle at different breeds and health states. *Front. Microbiol.* **2023**, *14*, 1113730. [CrossRef]
64. Jami, E.; Israel, A.; Kotser, A.; Mizrahi, I. Exploring the bovine rumen bacterial community from birth to adulthood. *ISME J.* **2013**, *7*, 1069–1079. [CrossRef]
65. Dill-McFarland, K.A.; Breaker, J.D.; Suen, G. Microbial succession in the gastrointestinal tract of dairy cows from 2 weeks to first lactation. *Sci. Rep.* **2017**, *7*, 40864. [CrossRef] [PubMed]
66. Malmuthuge, N.; Guan, L.L. Understanding the gut microbiome of dairy calves: Opportunities to improve early-life gut health. *J. Dairy Sci.* **2017**, *100*, 5996–6005. [CrossRef]
67. Kelly, W.J.; Cookson, A.L.; Altermann, E.; Lambie, S.C.; Perry, R.; The, K.H.; Otter, D.E.; Shapiro, N.; Woyke, T.; Leahy, S.C. Genomic analysis of three Bifidobacterium species isolated from the calf gastrointestinal tract. *Sci. Rep.* **2016**, *6*, 30768. [CrossRef]
68. Fukuda, S.; Toh, H.; Hase, K.; Oshima, K.; Nakanishi, Y.; Yoshimura, K.; Tube, T.; Clarke, J.M.; Topping, D.L.; Suzuki, T.; et al. Bifidobacteria can protect enteropathogenic infection through the production of acetate. *Nature* **2011**, *469*, 543–547. [CrossRef] [PubMed]
69. O'Connell Motherway, M.; Zomer, A.; Leahy, S.C.; Reunanen, J.; Bottacini, F.; Claesson, M.J.; O'Brien, F.; Flynn, K.; Casey, P.G.; Munoz, J.A.M.; et al. Functional genome analysis of *Bifidobacterium breve* UCC2003 reveals type IVb tight adherence (Tad) pili as an essential and conserved host-colonization factor. *Proc. Natl. Acad. Sci. USA* **2011**, *108*, 11217–11222. [CrossRef] [PubMed]
70. Nakamura, S.-I.; Kim, Y.H.; Takashima, K.; Kimura, A.; Nagai, K.; Ichijo, T.; Sato, S. Composition of the microbiota in forestomach fluids and feces of Japanese Black calves with white scours. *J. Anim. Sci.* **2017**, *95*, 3949–3960. [CrossRef] [PubMed]
71. Zanevald, J.R.; McMinds, R.; Thurber, R.V. Stress and stability: Applying the Anna Karenia principle to animal microbiomes. *Nat. Microbiol.* **2017**, *2*, 17121. [CrossRef] [PubMed]
72. Xun, W.; Li, W.; Xiong, W.; Ren, Y.; Liu, Y.; Miao, Y.; Xu, Z.; Zhang, N.; Shen, Q.; Zhang, R. Diversity-triggered deterministic bacterial assembly constraints community functions. *Nat. Commun.* **2019**, *10*, 3833. [CrossRef]
73. Appert, O.; Garcia, A.R.; Frei, R.; Roduit, C.; Constancias, F.; Neuzil-Bunesova, V.; Ferstl, R.; Zhang, J.; Akdis, C.; Lauener, R.; et al. Initial butyrate producers during infant gut microbiota development are endospore formers. *Environ. Microbiol.* **2020**, *22*, 3909–3921. [CrossRef]
74. Lima, S.F.; de Souza Bicalha, M.L.; Bicalho, R.C. The *Bos taurus* maternal microbiome: Role in determining the progeny early-life upper respiratory tract microbiome and health. *PLoS ONE* **2018**, *14*, e0208014. [CrossRef] [PubMed]
75. Centeno-Martinez, R.E.; Glidden, N.; Mohan, S.; Davidson, J.L.; Fernández-Juricic, E.; Boerman, J.P.; Schoonmaker, J.; Pillai, D.; Koziol, J.; Ault, A.; et al. Identification of bovine respiratory disease through the nasal microbiome. *Anim. Microbiome* **2022**, *4*, 15. [CrossRef] [PubMed]
76. Liu, G.; Zhang, S.; Xiang, Z.; Shirani, I.; Chen, Y.; Guo, A. Effects of vaccination and interventions on nasal microbiome and BRD-associated pathogens in calves. *Front. Microbiol.* **2024**, *15*, 1467908. [CrossRef] [PubMed]
77. Barden, M.; Richards-Rios, P.; Ganda, E.; Lenzi, L.; Eccles, R.; Neary, J.; Oultram, J.; Oikonomou, G. Maternal influences on oral and faecal microbiota maturation in neonatal calves in beef and dairy production systems. *Anim. Microbiome* **2020**, *2*, 31. [CrossRef]

78. Raabis, S.M.; Quick, A.E.; Skarlupka, V.J.H.; Suen, G.; Ollivett, T.L. The nasopharyngeal microbiota of preweaned dairy calves with and without ultrasonographic lung lesions. *J. Dairy Sci.* **2021**, *104*, 3386–3402. [CrossRef] [PubMed]
79. Amat, S.; Timsit, E.; Workentine, M.; Schwinghamer, T.; van der Meer, F.; Guo, Y.; Alexander, T.W. A single intranasal dose of bacterial therapeutics to calves confers longitudinal modulation of the nasopharyngeal microeobiota: A pilot study. *Msystems* **2023**, *8*, e01016-22. [CrossRef] [PubMed]
80. Allison, S.D. A trait-based approach for modelling microbial litter decomposition. *Ecol. Lett.* **2012**, *15*, 1058–1070. [CrossRef] [PubMed]
81. Pitta, D.W.; Indugu, N.; Kumar, S.; Vecchiarelli, B.; Sinha, R.; Baker, L.D.; Bhukya, B.; Ferguson, J.D. Metagenomic assessment of the functional potential of the rumen microbiome in Holstein dairy cows. *Anaerobe* **2016**, *38*, 50–60. [CrossRef]

animals

Article

Predicting Body Weight in Pre-Weaned Holstein–Friesian Calves Using Morphometric Measurements

Flávio G. Silva [1,2,3,4,*], Emanuel Carreira [2,3], Joana M. Ramalho [5], Tomás Correia [6], Marília Meira [3], Cristina Conceição [2,3], Severiano R. Silva [1], Alfredo M. F. Pereira [2,3] and Joaquim L. Cerqueira [1,4]

1 Veterinary and Animal Research Centre (CECAV) and Associate Laboratory for Animal and Veterinary Sciences (AL4AnimalS), Department of Animal Science, Universidade de Trás-os-Montes e Alto Douro, Quinta de Prados, 5000-801 Vila Real, Portugal; ssilva@utad.pt (S.R.S.); cerqueira@esa.ipvc.pt (J.L.C.)
2 MED-Mediterranean Institute for Agriculture, Environment and Development and CHANGE-Global Change and Sustainability Institute, Universidade de Évora, Pólo da Mitra, Ap. 94, 7006-554 Évora, Portugal; ersc@uevora.pt (E.C.); cristinaconceicao@uevora.pt (C.C.); apereira@uevora.pt (A.M.F.P.)
3 Department of Animal Science, Universidade de Évora, Pólo da Mitra, Ap. 94, 7006-554 Évora, Portugal
4 Center for Research and Development in Agrifood Systems and Sustainability (CISAS), Instituto Politécnico de Viana do Castelo, Agrarian School of Ponte de Lima, Rua D. Mendo Afonso, 4990-706 Ponte de Lima, Portugal
5 Department of Veterinary Medicine, Universidade de Évora, Pólo da Mitra, Ap. 94, 7006-554 Évora, Portugal
6 Independent Researcher, Malagueira, 7000-407 Évora, Portugal
* Correspondence: fsilva@uevora.pt

Simple Summary: Calves should be regularly weighed to monitor their growth, which can be an indication of their health and welfare. Pre-weaned calves are particularly vulnerable to health problems as their immune system is still developing. In addition, knowing the calf's weight can help to provide a more balanced diet and to be more accurate when administering medication. However, there are still many farmers who do not have a scale on their farm or who have limitations in weighing their calves. Therefore, we developed a model using morphometric traits to estimate the weight of Holstein–Friesian pre-weaned calves without the need for a scale, which should be easy to use in most circumstances. Our model used the measurement of heart girth with a simple tape measure and was able to predict the weight of the calf with a mean difference of -1.4 ± 3.24 kg from the actual weight. We did not find any differences between females and males, but we did find differences between farms, which could be due to different genetic lines associated with farm breeding protocols. In addition, a web application was developed to make it easy for farmers to use the developed model. This avoids the use of calibrated weight bands, which are usually calibrated for a wider age range or for beef breeds.

Abstract: Regularly weighing calves helps to assess the efficiency of the rearing period and contributes to animal welfare by allowing more precise feeding and medication application in dairy farming, but many farmers do not weigh their calves regularly. Improving the feasibility of this process is, therefore, important. The use of morphometric measurements has been used to estimate the weight of cattle. However, many studies have focused on adult animals or used a wide age range. As calves experience allometric tissue growth, specific models for certain ranges might be more accurate. Therefore, the aim of this work was to develop a weight estimation model specific for pre-weaned Holstein–Friesian calves using morphometric measurements and to compare the model with another equation previously validated for the same breed with young and adult animals. From four dairy farms, 237 measurements of body weight, heart girth, abdominal girth, hip height, withers height, and body length were taken from Holstein–Friesian male and female calves. Linear and non-linear regression analysis was used to test the relationship between body weight and morphometric measurements, with age, sex, and farm as possible explanatory variables. Selected models were compared with goodness of fit and agreement tests. The final model was able to accurately predict body weight ($R^2 = 0.96$) with a mean difference of -1.4 ± 3.24 kg. Differences in the relationship between body weight and morphometric traits were observed between farms, but not between males and females. The genetics of the animal population at farm level may be responsible for this variability and further studies are needed to understand this variability and improve weight prediction models. The developed model was able to perform better in the agreement tests than the previously validated model for Holstein–Friesian animals, suggesting that different equations should be used depending on the growth phase the animal is in. In addition, a web application has been developed to facilitate the use of the developed model by farmers. This avoids the use of calibrated weight bands, which are usually calibrated for a broader age range or for beef cattle.

Keywords: abdominal girth; body length; dairy calves; heart girth; hip height; weight scale; withers height

1. Introduction

Regular weighing of calves is an essential management practice for dairy farmers, contributing to animal health and welfare and the overall productivity of the dairy farm. Calf body weight (BW) information can be used for several aspects, such as feeding management practices (e.g., feeding colostrum per BW instead of a standardized volume [1]); calculating medication dosages (e.g., calculating anesthetic and analgesic dosages for disbudding [2] or reducing underdosing caused by incorrect visual weight estimation [3]); to sell animals based on BW, especially for male calves; to test changes in the environment or management practices (e.g., changes in housing system [4]), or to set weight gain targets and assess the rearing period efficiency [5,6]. All these aspects are important for improving calf welfare, financial components, and environmental aspects such as carbon footprint and antimicrobial resistance.

However, weighing calves may not be practical due to logistical or economic constraints, which may be related to the lack of a reliable scale on the farm or the lack of staff to weigh calves on a regular basis [7] or due to the misperception of visual weight estimation [8]. A survey of calf management practices in the UK found that only 55.6% (120/216) of farmers weighed their calves, and only 31.7% of these farmers used a scale, with the remainder using a weigh tape; moreover, these measurements were taken almost exclusively at birth and weaning, and only 10% weighed their animals on a weekly basis [9].

Therefore, if a scale is not available, or if it is available but difficult to use regularly because, for example, the scale is not near the calves' rearing area and is difficult to move, alternative estimation methods are valuable. Machine vision techniques have been used to reliably estimate bovine BW using automated measurements of body traits [10,11], but such technologies are not yet readily available to most livestock owners and for commercial use [12]. In this manner, linear body measurements or morphometric measurements (MSs) have been used to reliably estimate cattle BW [13–16]. However, most of these studies have focused on adult animals or in a wide age range, including both younger and older animals in the analysis, with some inconsistencies being reported for younger animals [17].

The measurement of heart girth (HG) is the most highly correlated variable with BW in cattle, and generally the addition of a second variable to these models does not compensate for the normally high correlation with HG [13]. For this reason, weight bands were developed so that livestock farmers could estimate the weight of their animals. Heinrichs et al. [18] compared the same equation used in Heinrichs et al.'s [13] study 25 years later and concluded that the previous equation remained valid for Holstein dairy heifers from 1 to 821 d of age. Nevertheless, bovine body traits can change with breed [15,19], age [20,21], and breeding programs over time [18]. Animals have allometric growth of tissues [22], so the relationship between BW and MS is likely to change with age. Sex influences the growth of body tissues [23], so this factor should also be taken into account. It is, therefore, necessary to regularly update these equations, considering breed, sex, and age, to obtain accurate prediction models. Since different phenotypic traits can be found in animals of the same breed [24], these equations should be calculated with individuals from different farms and this factor should be considered when developing predictive models for BW estimation. Furthermore, providing livestock owners with the means to implement these models on a daily basis is equally important.

Therefore, the aim of this study was to develop a new equation specifically for pre-weaned Holstein–Friesian dairy calves and to compare it with the equation proposed by [13], taking into account sex, age, and farm as possible influencing factors.

2. Materials and Methods

2.1. Data Collection

The study was approved by the Ethics Committee for Animal Welfare at Universidade Trás-os-Montes e Alto Douro under reference 2664-e-DZ-2023.

The study took place between January 2023 and November 2023 and involved 210 pre-weaned Holstein–Friesian calves, aged 1 to 90 days (160 females and 50 males), from four commercial dairy farms in the Alentejo region of Portugal. During the farm visit, calves aged between 1 and 90 days were weighed using a digital scale (Tru-Test 702, Tru-Test Datamars, Lugano, Switzerland), morphometric measurements (MS) were taken, and sex was recorded. Heart girth (HG) and abdominal girth (AG) were measured with a measuring tape (Comb MAAL, We-Bo, Denmark), and withers height (WH), hip height (HH), and body length (BL) were measured with a hipometer. Twenty-seven animals were weighed twice with a minimum interval of 30 days, as they were part of another study not yet published, resulting in a total of 237 records. All measurements were performed by the same operator on a flat surface with all limbs on the ground, according to [25].

HG was measured with a tape around the chest just behind the scapula. AG was measured similarly to HG but at the level of the last rib. WH was measured from the highest point of the withers to the ground and HH from the middle third of the sacrum to the ground. BL was measured as a straight line from the tip of the scapula (the most prominent point of the scapulohumeral joint) to the ischial tuberosity (Figure 1).

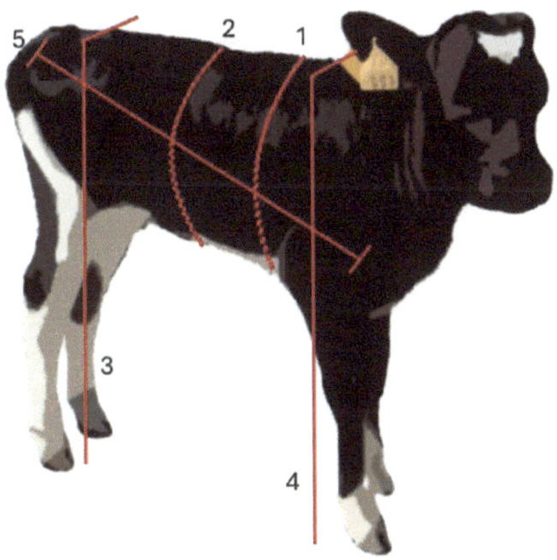

Figure 1. Schematic representation of the morphometric measurements. 1—Heart girth (HG); 2—abdominal girth (AG); 3—hip height (HH); 4—withers height (WH); 5—body length (BL).

2.2. Statistical Analysis

All statistical tests were conducted in IBM® SPSS® Statistics, v27 (SPSS).

Variables were tested for normality using graphical analysis and Kolmogorov–Smirnov test; four out of six variables did not show a normal distribution. However, the sum of all variables led to an approximated normal distribution, thus applying the Central Limit Theorem. One calf was removed due to a clear data recording error. Outliers were checked using the studentized deleted residuals and absolute values ≥ 3 were removed.

To assess initial associations between variables, Spearman's rank correlation coefficients were calculated. Subsequently, for model creation, several linear and non-linear equations (i.e., simple and multiple) were computed to analyze the relationship between BW as the dependent variable and both MS and age as independent variables. Additionally, an equation used to predict the BW in horses [26] was also tested. This equation can be described as follows:

$$\frac{BL \times HG^2}{a}$$

where a is fix parameter calculated through an iterative optimization process.

First, the Curve Estimation procedure in SPSS was used to analyze the type of relationship between variables. Linear equations were then calculated using Linear Regression, while non-linear equations were calculated using Non-Linear Regression in SPSS, which incorporates an iterative optimization process to estimate regression parameters. In this process, SPSS starts with a specified initial value for each parameter and then iteratively refines it to minimize the difference between the predicted BW values and the actual data points.

To assess the influence of farm (A, B, C, and D) and sex (female and male) on the explanatory power of the models, we employed dummy coding for these categorical variables. The dummy coded variables and their interaction terms with MS were then incorporated as independent variables in the models. If the interaction between farm/sex and MS showed a statistically effect ($p < 0.05$), indicating that the effect of MS on BW differed depending on farm or sex, separate models were built for each farm–sex combination. Separate models for farm were created using stepwise linear regression with the option to exclude cases listwise for missing values; all potential predictor variables (i.e., MS) were included in the initial model, and the variables of the final model were selected using the stepwise procedure in SPSS.

Final models were selected based on the correlation coefficient (r) and the coefficient of determination (R^2). Predicted BW from the final models was compared with the measured BW with goodness of fit and agreement parameters to select the most appropriate model. The parameters used were the R^2, the Akaike Information Criterion (AIC) calculated as $n \times \ln\left(\frac{RSS}{n}\right) + 2 \times K$, where K is the number of parameters of the model, Bland–Altman plots (i.e., plot of the difference between the measured BW and the predicted BW over the mean of these measurements, with ± 1.96 SD to define the upper and lower limits of agreement), and the Intraclass Correlation Coefficient (ICC) and their 95% CI to test the reliability between different models (two-way mixed effects, absolute agreement, single, and average measures), using the Reliability Analysis procedure, in SPSS. Student's t-tests were used to assess if the mean difference between measured BW and predicted BW differed from zero (T value = 0). In addition, the equation proposed by Heinrichs et al. (1992) [13] was applied to the database used in this study, and the predicted BW was compared with the final model using the same tests to understand if this alternative was more accurate.

2.3. *Web Application*

A web application (Figure 2) accessible at https://eloquent-begonia-6dc576.netlify.app/ (accessed on 17 July 2024) leveraging HTML, CSS, and JavaScript programming languages, was developed. The application's source code is stored on GitHub, facilitating version control and collaborative development, while Netlify serves as the platform for continuous deployment, ensuring the latest updates are readily available.

The web application provides unrestricted access and is compatible with all major web browsers. It offers a user-friendly interface designed to estimate the weight of Holstein–Friesian calves based on their heart girth measurements. It is presented in English and Portuguese languages.

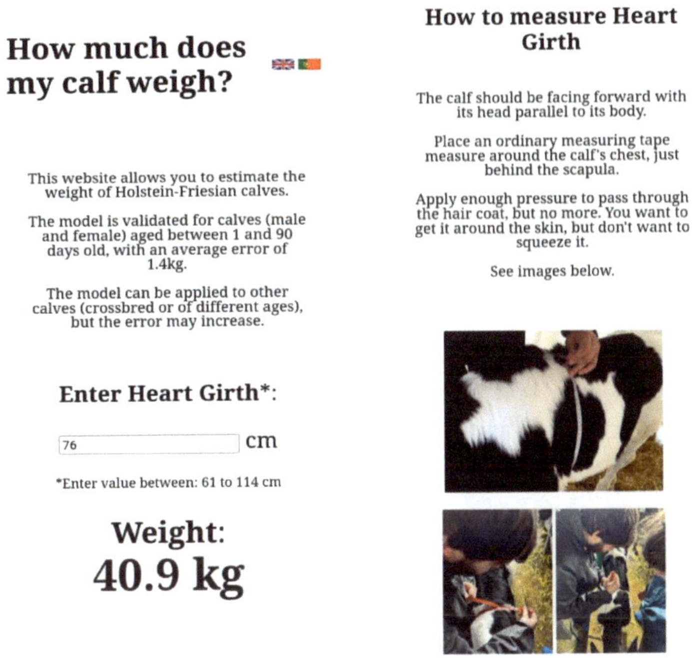

Figure 2. "How much does my calf weigh?" (mobile phone view)—web application developed to allow quick and easy application of the body weight prediction model. Source: https://eloquent-begonia-6dc576.netlify.app/ (accessed on 17 July 2024).

3. Results

The age of the calves included in this study and their BWs and MSs are shown in Table 1 and the data distribution is shown in Figure 3. The sample of calves from each farm had a similar age range and both BWs and MSs were similar between farms. From the data distribution shown in Figure 3, BW had the greatest variability between all measurements. This suggests that the relationship between BW and MS is not completely linear.

Table 1. Descriptive statistics (mean ± standard deviation) for age, body weight, and morphometric measurements across farm.

Source	N [1]	Age, Days	BW, kg	HG, cm	AG, cm	HH, cm	WH, cm	BL, cm
Farm A	43	35.6 ± 15.98	56.3 ± 11.23	85.7 ± 6.04	91.3 ± 8.90	86.8 ± 4.66	82.1 ± 4.28	77.7 ± 5.67
Farm B	50	27.8 ± 19.34	56.9 ± 16.02	84.6 ± 7.52	93.4 ± 10.44	88.7 ± 5.93	82.6 ± 4.91	77.4 ± 6.74
Farm C	90	32.7 ± 22.32	50.8 ± 11.14	82.1 ± 6.63	84.1 ± 7.02	86.3 ± 5.13	78.0 ± 7.15	77.1 ± 4.74
Farm D	54	24.4 ± 27.77	59.7 ± 22.38	85.7 ± 10.1	89.9 ± 9.14	88.8 ± 7.03	83.2 ± 5.94	77.3 ± 8.52

[1] Number of weight measurements, other measurements besides HG were not measured in all calves. BW—Body weight; HG—heart girth; AG—abdominal girth; HH—hip height; WH—withers height; BL—body length.

All MSs were correlated with BW and with each other ($p < 0.001$; Table 2). HG had the best correlation with BW, followed by HH. Plotting BW against BL showed that there were two different relationships (i.e., BL(1) and BL(2)). The BL(1) distribution was best fit by a quadratic curve (n = 129), while the BL(2) distribution was best fit by a linear line ($n = 58$); therefore, two separate correlations were calculated for BL. Age was also positively correlated with BW (0.81, $p < 0.001$).

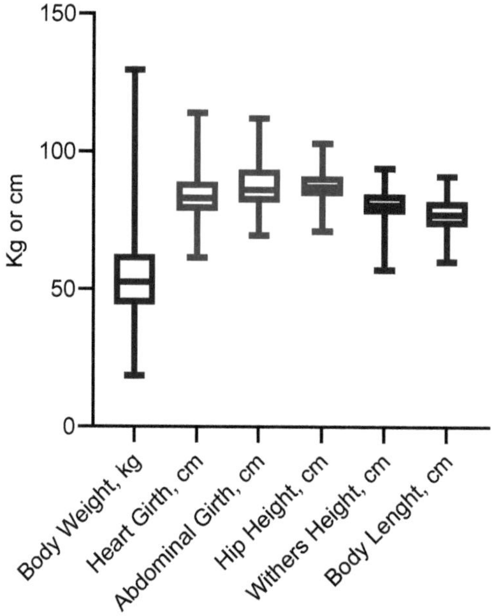

Figure 3. Data distribution of body weight and morphometric measurements.

Table 2. Spearman's correlations between body weight and morphometric measures in pre-weaned Holstein–Friesian calves.

	BW	HG	AG	HH	WH
HG	0.97				
AG	0.87	0.86			
HH	0.88	0.85	0.72		
WH	0.81	0.79	0.77	0.81	
BL(1) *	0.88	0.86	0.77	0.79	0.78
BL(2) *	0.94	0.93	0.78	0.82	0.84

All correlations where significant at $p < 0.001$. BW—Body weight; HG—hearth girth; AG—abdominal girth; HH—hip height; WH—wither height; BL—body length. * BL was divided into two, since two very distinct relationships were found in this measurement.

Quadratic equations provided higher R^2 values than linear equations with all MSs (Table 3). In simple linear and quadratic regression analysis, HG had the best fit. Plotting BW against BL showed that there were two different relationships, so two separate models were calculated for BL, which produced different results.

Sex and farm effects were tested only with the HG model as this was the variable most closely associated with BW. Regressing the HG linear equation using sex as a dummy variable showed no significant effect of sex on the BW prediction ($p = 0.141$). However, the interaction between farm and HG was statistically significant ($p < 0.001$), indicating that the predicted values from the HG equation might depend on the farm where the measurements took place. Consequently, separate linear models were created to analyze the relationship between the MSs and BWs for each farm (Table 4). While HG remained the best predictor of BW in every farm, the inclusion of other MS factors improved the model fit depending on the specific farm.

Table 3. Regression coefficients of linear and quadratic equations for pre-weaned Holstein–Friesian calves body weight estimated from morphometric measures.

Measurement	Intercept	Linear	Quadratic	*r*	R^2	N
HG	−108.660	1.943		0.97	0.94	234
	58.459	−1.979	0.023	0.98	0.96	
AG	−55.298	1.248		0.89	0.79	168
	−86.236	1.943	−0.004	0.89	0.80	
HH	−143.701	2.273		0.90	0.81	214
	149.210	−4.469	0.039	0.91	0.83	
WH	−98.606	1.901		0.84	0.70	219
	193.423	−5.555	0.047	0.86	0.75	
BL(1) *	−97.034	2.009		0.88	0.78	116
	181.688	−5.274	0.047	0.90	0.81	
BL(2) *	−50.216	1.175		0.96	0.93	61
	−95.090	2.304	−0.007	0.97	0.93	

All parameters where significant at $p < 0.001$. HG—Hearth girth; AG—abdominal girth; HH—hip height; WH—wither height; BL—body length; *r*—correlation coefficient; R^2—coefficient of determination; N—number of measurements * BL was divided into two models, since two distinct relationships were found with this measurement.

Table 4. Multiple linear regression models for pre-weaned Holstein–Friesian calves body weight estimated from morphometric measurements for each farm.

β Coefficient	N	Intercept	HG	AG	WH	HH	BL	*r*	R^2
Farm A	42	−112.739	1.472 ***		0.524 *			0.96	0.95
Farm B	28	−197.34	1.115 ***	0.207 *	1.701 ***			0.99	0.97
Farm C	90	−94.214	1.209 ***	0.283***		0.253 *		0.97	0.95
Farm D	43	−141.146	1.842 ***			0.485 **		0.99	0.97

*** $p < 0.001$, ** $p < 0.01$, * $p < 0.05$. HG—Hearth girth; AG—abdominal girth; HH—hip height; WH—wither height; BL—body length; r—correlation coefficient; R^2—coefficient of determination; N—number of measurements.

3.1. Model Selection

The model using the horse equation [26] resulted in a considerably lower coefficient of determination (R^2 = 0.69) and was therefore excluded. Four final models were selected and compared using goodness of fit tests and measurements of agreement to identify the most appropriate one.

The first model (BW1) consisted of a quadratic equation with HG. The second model (BW2) consisted of a quadratic model with HG and HH. Sex did not significantly affect the predictability of BW using HG; however, females and males were not equally distributed in our sample, so separated models were created as well (i.e., a model for female—BW(f) and a model for male calves—BW(m)). The addition of age to a model with any other MS did not improve the fit and the beta coefficient was always non-significant. Therefore, besides the farm-dependent models, four additional quadratic models were created and presented in Table 5.

Table 5. Selected models for estimating body weight from morphometric measurements. Model identification and their respective equation and description are provided.

Model ID	Equation	Description
BW1	$BW = 58.459 - 1.979 \times HG + 0.023 \times HG^2$	simple quadratic regression with HG (n = 234)
BW2	$BW = 62.449 - 1.979 \times HG + 0.026 \times HG^2 - 0.284 \times HH + 0.01 \times HG^2 - 0.011 \times (HG \times HH)$	multiple quadratic regression with HG and HH (n = 214)
BW(f)	$BW = 71.128 - 2.28 \times HG + 0.025 \times HG^2$	simple quadratic regression with HG, females only (n = 185)
BW(m)	$BW = 14.756 - 0.946 \times HG + 0.017 \times HG^2$	simple quadratic regression with HG, males only (n = 49)

3.2. Models' Goodness of Fit and Agreement Tests

In all models except BW(m), the mean difference was statistically different from zero ($p < 0.001$; Student's *t*-test), meaning that they were not in perfect agreement with the measured BW and that there is a statistically relevant difference (Table 6). However, when the difference between BW1 and BW was regressed on the mean of BW1 and BW, the β coefficient was close to zero ($\beta = 0.006$) and $p = 0.683$, indicating that there is no deterministic bias towards lower or higher BWs for the BW1 model, as can be seen in the Bland–Altman plot (Figure 4). The vast majority (93.2%) of the differences between the two methods fall within the limits of agreement, indicating a good level of agreement between the methods despite the statistical difference. Furthermore, as shown by the graphical distribution of the data (Figure 4), there appears to be no trend in the difference between the BWs at any particular range of BW.

Table 6. Characteristics of the measured body weight (BW) and the predicted body weight (BWp) for each corresponding model.

Model	BW [1]	BWp [1]	BW-BWp [2]	*p* Value	N
BW1	55.0 ± 15.65	56.3 ± 15.56	−1.4 ± 3.24	<0.001	234
BW2	55.8 ± 15.77	52.9 ± 14.99	2.9 ± 2.97	<0.001	218
BW(f)	55.8 ± 16.43	59.1 ± 16.80	−3.2 ± 2.95	<0.001	185
BW(m)	51.8 ± 11.90	52.7 ± 11.27	−0.9 ± 4.19	0.137	49

[1] Mean ± standard deviation; [2] mean difference; N—number of measurements.

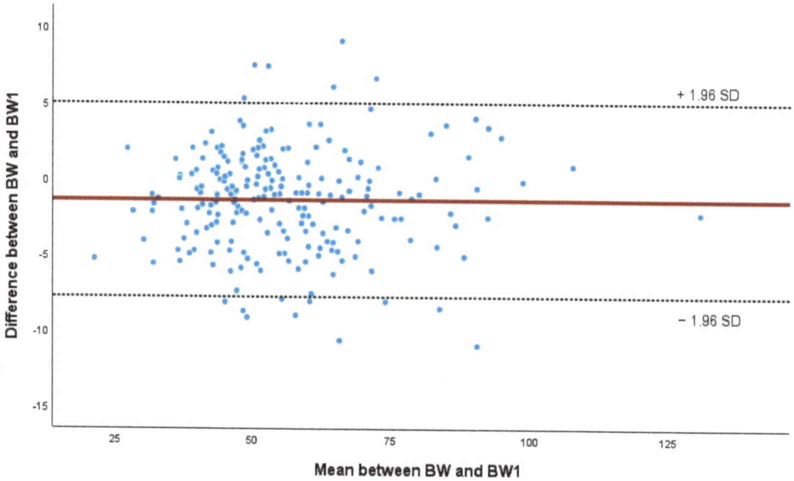

Figure 4. Bland–Altman's plot of estimated weight (BW1) against measured body weight (BW). Red line represents the mean difference between BW and BW1 and doted lines the upper and lower limits of agreement.

The R^2 values may not be an adequate measure of the goodness of fit in non-linear models [27]; thus, models were also compared with AIC, Bland–Altman's plots, and ICC (Table 7). Despite having a lower R^2 and a higher AIC than BW2, BW1 had a better ICC in both single and average correlations. Additionally, BW2 is more cumbersome since it must use two MS and the increase in R^2 (1%) was not considered significant.

The models separated by sex showed a better AIC but a worst ICC than BW1. Therefore, the equation from model BW1 was compared with BW(f) and BW(m) in each respective subsample (i.e., only female or only male calves). In both cases, BW1 was preferred as it had a higher ICC and a lower mean difference between the observed and predicted values (Table 8), which is consistent with the non-significant effect previously reported for sex.

Therefore, BW1 was considered the most accurate model. Graphical analysis of this model is shown in Figures 5 and 6.

Table 7. Goodness of fit and agreement tests between models.

Model	*r*	R^2	AIC	ICC (Single)	ICC (Average) *
BW1	0.98	0.96	575.2	0.98 [0.96–0.98]	0.99 [0.98–0.99]
BW2	0.98	0.97	495.6	0.97 [0.78–0.99]	0.92 [0.87–0.99]
BW(f)	0.98	0.97	397.8	0.97 [0.70–0.99]	0.98 [0.82–0.99]
BW(m)	0.93	0.88	149.1	0.93 [0.88–0.96]	0.97 [0.94–0.98]

r—Correlation coefficient; R^2—coefficient of determination; AIC—Akaike information criterion; ICC—intraclass correlation coefficient. AIC calculated as: $n \times \ln\left(\frac{RSS}{n}\right) + 2 \times K$, where K is the number of parameters. * ICC of the average between measured body weight and predicted body weight. The ICC 95% CI are shown between square brackets.

Table 8. Characteristics of the measured body weight (BW) and the predicted body weight (BWp) and Intraclass Correlation coefficients for female and male data using the BW1 model.

Database	Model	BW [1]	BWp [1]	BW-BWp [2]	*p*-Value	ICC (Single)	ICC (Average) [3]	N
Female	BW1	55.8 ± 16.43	57.4 ± 16.41	−1.6 ± 2.90	<0.001	0.99 [0.96–0.99]	0.99 [0.98–0.99]	185
Male	BW1	51.8 ± 11.9	52.2 ± 11.05	−0.4 ± 4.22	<0.46	0.93 [0.89–0.96]	0.97 [0.94–0.98]	49

[1] Mean ± standard deviation; [2] mean difference; [3] ICC of the average between measured body weight and predicted body weight. ICC—Intraclass correlation coefficient; N—number of measurements.

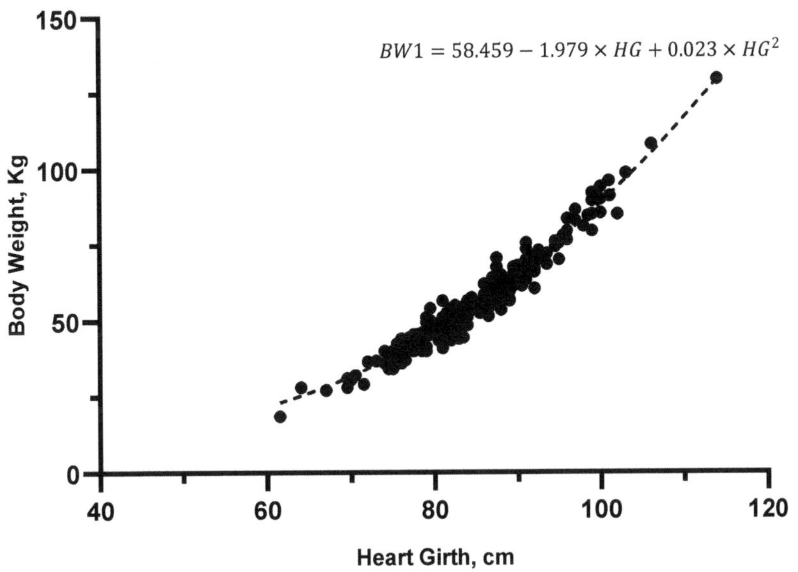

Figure 5. Scatter plot of model BW1, with hearth girth (cm) and body weight (kg), and its respective equation, R^2 = 0.96.

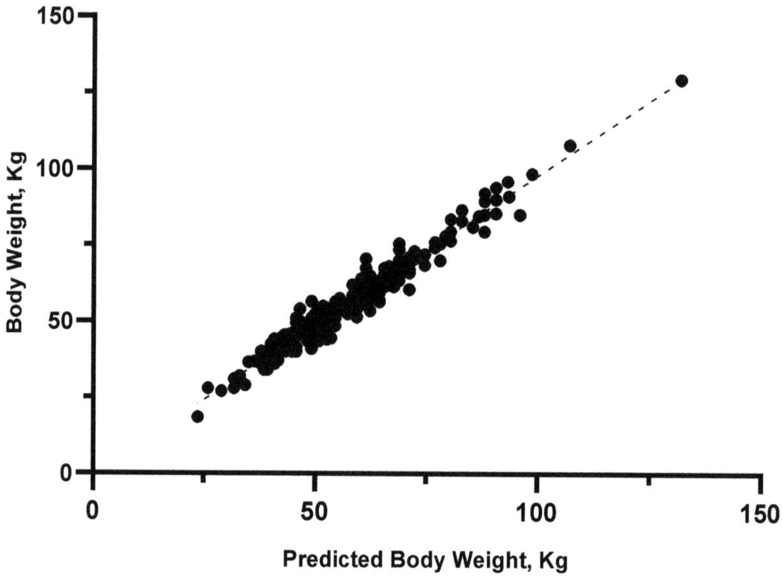

Figure 6. Scatter plot of predicted body weight using model BW1 and measured body weight, $R^2 = 0.96$.

3.3. Comparison with Equation from Heinrichs et al. (1992)

Using the same tests as in the previous analysis, the equation from [13], hereafter referred to as BW(H92), was applied to our sample and compared with BW1 (Table 9). The results are similar with both models, but while BW1 tended to overestimate the measured BW by 1.4 ± 3.24 kg, BW(H92) tended to underestimate it by 4.7 ± 3.47 kg. The ICC showed a large variability within the 95% CI for BW(H92).

Table 9. Comparison of the measured body weight (BW) with the predicted body weight (BWp) using BW1 and BW(H92) [13] equations.

Model	BW [1]	BWp [1]	BW-BWp [2]	ICC (Single)	ICC (Average) [3]	N
BW1	55.0 ± 15.65	56.3 ± 15.56	−1.4 ± 3.24 *	0.98 [0.96–0.98]	0.99 [0.98–0.99]	234
BW(H92)	55.0 ± 15.65	50.3 ± 14.01	4.7 ± 3.47 *	0.93 [0.30–0.98]	0.96 [0.45–0.99]	234

[1] Mean ± standard deviation; [2] mean difference; [3] ICC of the average between measured body weight and predicted body weight. N—number of measurements; * $p < 0.001$.

The Bland–Altman plot (Figure 7) compares the measured BW with the predicted BW using BW(H92) equation, with 93.6% of values being correctly measured within the 95% confidence interval. However, there seems to be a bias towards underestimating BW as it increases, as can be seen in the Bland–Altman plot.

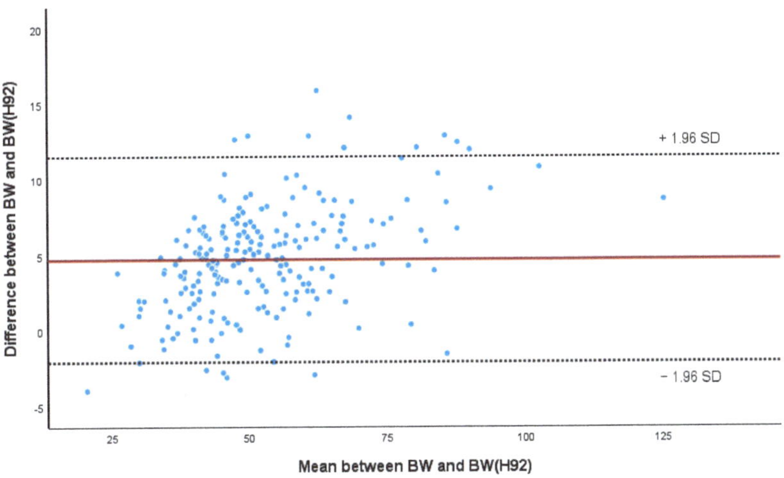

Figure 7. Bland–Altman's plot of predicted weight (BW(H92)) against measured body weight (BW). Red line represents the mean difference between BW and BW(H92) and doted lines the upper and lower limits of agreement.

4. Discussion

The use of MSs can be used to predict BW in animals; therefore, several models have been developed over the years. However, due to genetic selection, this relationship can change [18]. In addition, breed and age can also affect this relationship [15,19]; therefore, breed- and age-dependent models are needed [28]. A very strong relationship is needed to reliably predict calf weight to increase precision and production efficiency in calf rearing. Besides reliability, feasibility and robustness are important features for producers to systematically measure calf BW.

As in previous studies [13,19,29], HG was found to be the variable with the best relationship with BW. In our study, besides HG, other MSs showed good results as well, so if HG could not be measured (e.g., in a chute), HH could be measured instead, with an increased error. However, due to the general calmness of this breed and the way animal management is set up, it is usually easy to regularly measure HG in pre-weaned calves. Height may also be more difficult to measure systematically if the ground is uneven, which is not unusual on most farms. In the study by Henrichs et al. [13], the R^2 values obtained are higher than in this paper, but there could be some collinearity in this case as the same animals were measured seven to thirty-three times. Nevertheless, studies with other breeds, had similar results to our study [8,19,30,31]. García et al. [29] found higher correlations with HG (99.1%) and WH (96.8%) in 104 Holstein heifers aged 4 days to 2 years.

Few studies have focused on predictive models for MSs in calves and they seem to focus on hoof circumference at the level of the coronary band [32–34]. It has been shown that weight estimation with HG has greater variability for heifers with a BW less than 150 kg compared to heavier heifers, with a variability of less than 10% [35]. In our case, the mean variation was 2.55%, similar to the results obtained in [36]. Dingwell et al. [21] also showed that HG measurements in heifers between 3 and 15 months of age were in good agreement with BW, but not in heifers under 3 months of age. In one study, the hoof circumference of newborn beef calves was linearly related to BW (R^2 = 0.69) [33]. In another study with Holstein and Jersey calves, MSs were related to BW by linear (R^2 = 0.91) and quadratic (R^2 = 0.93) regression [34]. Both studies found significant differences between actual and estimated weight as a function of BW range. In the study of Long et al. [34], these differences occurred for calves with a BW lower than 31.3 kg and higher than 44.9 kg. Different goodness of fit were found between different ages (i.e., 2, 8, and 16 weeks), with similar R^2 values to the present study [17]. It is clear that age plays a significant role in how the BW can be predicted by MS. This finding helps to explain the slightly lower, although still high, correlations in this study, which used calves from a narrower age range, compared to studies using heifers from a wider age range. In addition, some of this variation may be due to abomasum and rumen filling, which should be proportionally more representative in younger than in older animals.

There is no evidence to suggest the benefits of using sex-specific models in calves of the age in our sample. Although other authors have shown different models according to sex and obtained different results [8,14,37], this does not mean that it is necessarily caused by sex and could be due to the nature of the sample, as we have suggested in this paper, or other factors like age and breed [20,38]. Therefore, the influence of sex on the predictive models for BW should be tested in the presence of other factors (e.g., age and breed). In addition to the differences in MSs between farms, a different relationship between BL and BW was found in one of the farms (farm C). This farm had been visited previously (i.e., 7 months apart) and the distribution of data was similar to that of the other farms. The variability found between farms may be the result of different genetic selection on each farm [13,15]. A difference in BW between farms was found in other work [34], but it was not tested whether 'farm' as a factor influenced the relationship between BW and MSs, as we did in the present study, since BW may be different, but the relationship may be identical. However likely this may be, we cannot be sure, and a larger sample size from more farms would provide more information on this topic. Besides genetic selection, considering farm as a factor could imply other influencing factors, like housing conditions, animal handling, and percentage of sick animals, thus further studies are needed. Considering that the farm-specific models developed are not necessarily better than the general models, we decided to use the BW1 for the web application, as it was intended to be used in different farms.

The predicted weights using our model BW1, or BW(H92) were similar, but our model seems to be more appropriate for our sample. BW(H92) had a wider age range, and since the growth tissues of calves are allometric, it seems more advantageous to have a different predictive model for younger ages. Using an analysis that includes age as a factor of variation is difficult to implement in a weigh tape, so with a simple tape measure and a mobile application we can use more appropriate equations that include the age of the animal. There are several calibrated HG bands, but they can vary in the equation used and are not specific to dairy calves. Thus, the web application developed can overcome this limitation, since a regular tape measure can be used to take the HG and then the result is given using a digital platform (e.g., mobile phone and laptop). We believe that this method can increase the accuracy and feasibility of the measurements and improvements to the model can be made in further studies.

5. Conclusions

In this study, we proposed a model for estimating body weight from morphometric measurements in pre-weaned Holstein–Friesian calves aged 1 to 90 days. In addition, we developed a web application to increase uniformity and feasibility at farm level. The results indicate that calf body weight can be reliably estimated using heart girth measurements. Dairy farmers without access to a weighing scale can use this model to reliably assess the efficiency of the pre-weaning period or to provide colostrum, milk, starter, or medication based on weight. The relationship between morphometric measurements and body weight may depend on the genetic make-up of the population. Therefore, further studies to validate this model in different farms would be beneficial to increase its robustness.

Author Contributions: Conceptualization, F.G.S., C.C., S.R.S., A.M.F.P., and J.L.C.; methodology, F.G.S., E.C., J.M.R., T.C., M.M., and J.L.C.; software, T.C.; validation, S.R.S., A.M.F.P., and J.L.C.; formal analysis, F.G.S.; investigation, F.G.S., E.C., C.C., S.R.S., and J.L.C.; writing—original draft preparation, F.G.S.; writing—review and editing, F.G.S., C.C., S.R.S., A.M.F.P., and J.L.C.; supervision, C.C., S.R.S.,

A.M.F.P., and J.L.C.; funding acquisition, S.R.S., C.C., A.M.F.P., and J.L.C. All authors have read and agreed to the published version of the manuscript.

Funding: This work is funded by National Funds through FCT—Foundation for Science and Technology under the Projects CECAV UIDB/00772/2020 (https://doi.org/10.54499/UIDB/00772/2020), MED UIDB/05183/2020 (https://doi.org/10.54499/UIDB/05183/2020; https://doi.org/10.54499/UIDP/05183/2020) and CHANGE (https://doi.org/10.54499/LA/P/0121/2020) and F.G.S. PhD grant UI/BD/150834/2021.

Institutional Review Board Statement: The study was approved by the ORBEA (Ethics Committee for Animal Welfare) at Universidade Trás-os-Montes e Alto Douro (UTAD) under reference 2664-e-DZ-2023.

Informed Consent Statement: Informed consent was obtained from all subjects involved in the study.

Data Availability Statement: Data is contained within the article. The original contributions presented in the study are included in the article, further inquiries can be directed to the corresponding author.

Conflicts of Interest: The authors declare no conflicts of interest.

References

1. Conneely, M.; Berry, D.P.; Murphy, J.P.; Lorenz, I.; Doherty, M.L.; Kennedy, E. Effect of feeding colostrum at different volumes and subsequent number of transition milk feeds on the serum immunoglobulin G concentration and health status of dairy calves. *J. Dairy Sci.* **2014**, *97*, 6991–7000. [CrossRef]
2. Glynn, H.D.; Coetzee, J.F.; Edwards-Callaway, L.N.; Dockweiler, J.C.; Allen, K.A.; Lubbers, B.; Jones, M.; Fraccaro, E.; Bergamasco, L.L.; Kukanich, B. The pharmacokinetics and effects of meloxicam, gabapentin, and flunixin in postweaning dairy calves following dehorning with local anesthesia. *J. Vet. Pharmacol. Ther.* **2013**, *36*, 550–561. [CrossRef]
3. van Dijk, J.; Eagle, S.J.; Gillespie, A.V.; Smith, R.F.; Holman, A.N.; Williams, H.J. Visual weight estimation and the risk of underdosing dairy cattle. *Vet. Rec.* **2015**, *177*, 75. [CrossRef]
4. Costa, J.H.C.; Meagher, R.K.; von Keyserlingk, M.A.G.; Weary, D.M. Early pair housing increases solid feed intake and weight gains in dairy calves. *J. Dairy Sci.* **2015**, *98*, 6381–6386. [CrossRef]
5. Brickell, J.S.; Mcgowan, M.M.; Pfeiffer, D.U.; Wathes, D.C. Mortality in holstein-friesian calves and replacement heifers, in relation to body weight and IGF-I concentration, on 19 farms in England. *Animal* **2009**, *3*, 1175–1182. [CrossRef]
6. Tautenhahn, A.; Merle, R.; Müller, K.E. Factors associated with calf mortality and poor growth of dairy heifer calves in northeast Germany. *Prev. Vet. Med.* **2020**, *184*, 105154. [CrossRef]
7. de Villiers, J.F.; Gcumisa, S.T.; Gumede, S.A.; Thusi, S.P.; Dugmore, T.J.; Cole, M.; Du Toit, J.F.; Vatta, A.F.; Stevens, C. Estimation of live body weight from the heart girth measurement in KwaZulu-Natal goats. *Appl. Anim. Husb. Rural Dev.* **2009**, *1*, 1–8.
8. Tebug, S.F.; Missohou, A.; Sabi, S.S.; Juga, J.; Poole, E.J.; Tapio, M.; Marshall, K. Using body measurements to estimate live weight of dairy cattle in low-input systems in Senegal. *J. Appl. Anim. Res.* **2018**, *46*, 87–93. [CrossRef]
9. Mahendran, S.A.; Wathes, D.C.; Booth, R.E.; Blackie, N. A survey of calf management practices and farmer perceptions of calf housing in UK dairy herds. *J. Dairy Sci.* **2022**, *105*, 409–423. [CrossRef]
10. Nir, O.; Parmet, Y.; Werner, D.; Adin, G.; Halachmi, I. 3D Computer-vision system for automatically estimating heifer height and body mass. *Biosyst. Eng.* **2018**, *173*, 4–10. [CrossRef]
11. Weales, D.; Moussa, M.; Tarry, C. A robust machine vision system for body measurements of beef calves. *Smart Agric. Technol.* **2021**, *1*, 100024. [CrossRef]
12. Silva, F.G.; Conceição, C.; Pereira, A.M.F.; Cerqueira, J.L.; Silva, S.R. Literature Review on Technological Applications to Monitor and Evaluate Calves' Health and Welfare. *Animals* **2023**, *13*, 1148. [CrossRef] [PubMed]
13. Heinrichs, A.J.; Rogers, G.W.; Cooper, J.B. Predicting Body Weight and Wither Height in Holstein Heifers Using Body Measurements. *J. Dairy Sci.* **1992**, *75*, 3576–3581. [CrossRef] [PubMed]
14. Bahashwan, S. Application of Morphometric Traits for Live Body Weight Estimation in Dhofari Calves. *Int. J. Sci. Res. Agric. Sci.* **2014**, *1*, 90–96. [CrossRef]
15. Menezes, G.L.; Bresolin, T.; Halfman, W.; Sterry, R.; Cauffman, A.; Stuttgen, S.; Schlesser, H.; Nelson, M.A.; Bjurstrom, A.; Rosa, G.J.M.; et al. Exploring associations among morphometric measurements, genetic group of sire, and performance of beef on dairy calves. *Transl. Anim. Sci.* **2023**, *7*, txad064. [CrossRef] [PubMed]
16. Rocha-Silva, M.; Britto, F.B.; Silva, D.L.S.; Oliveira do O, A.; Silva, L.A.S.; Oliveira, M.B.; Araújo, C.V.; Carvalho, G.M.C.; Sarmento, J.L.R. Prediction of body weight of Curraleiro Pé-Duro cattle based on morphometric measurements. *Trop. Anim. Health Prod.* **2024**, *56*, 1–11. [CrossRef] [PubMed]
17. Wilson, L.L.; Egan, C.L.; Terosky, T.L. Body Measurements and Body Weights of Special-Fed Holstein Veal Calves. *J. Dairy Sci.* **1997**, *80*, 3077–3082. [CrossRef] [PubMed]

18. Heinrichs, A.J.; Heinrichs, B.S.; Jones, C.M.; Erickson, P.S.; Kalscheur, K.F.; Nennich, T.D.; Heins, B.J.; Cardoso, F.C. Short communication: Verifying Holstein heifer heart girth to body weight prediction equations. *J. Dairy Sci.* **2017**, *100*, 8451–8454. [CrossRef] [PubMed]
19. Udoh, J.E.; David, E.G.; Unah, U.L. Prediction of body weight from linear body measurement in two breeds of cattle. *Open Access Res. J. Biol. Pharm.* **2021**, *3*, 041–046. [CrossRef]
20. Sow, R.S.; Denis, J.-P.; Trail, J.C.M.; Thiongane, P.I.; Mbaye, M. Note sur la barymétrie comme moyen de sélection indirecte du poids vif chez le zébu Gobra au Sénégal. *Rev. D'élevage Médecine Vétérinaire Pays Trop.* **1991**, *44*, 97–100. [CrossRef]
21. Dingwell, R.T.; Wallace, M.M.; McLaren, C.J.; Leslie, C.F.; Leslie, K.E. An evaluation of two indirect methods of estimating body weight in Holstein calves and heifers. *J. Dairy Sci.* **2006**, *89*, 3992–3998. [CrossRef] [PubMed]
22. Bruns, K.W. The Allometric Growth of Adipose Tissues in Beef Cattle. Ph.D. Thesis, South Dakota State University, Brookings, SD, USA, 2001.
23. Berg, R.T.; Butterfield, R.M. *New Concepts of Cattle Growth*, 1st ed.; Sydney University Press: Sydney, Australia, 1976; ISBN 0424000024.
24. Bohlouli, M.; Alijani, S.; Varposhti, M.R. Genetic relationships among linear type traits and milk production traits of Holstein dairy cattle. *Ann. Anim. Sci.* **2015**, *15*, 903–917. [CrossRef]
25. Cerqueira, J.O.L.; Araújo, J.P.P.; Vaz, P.S.; Cantalapiedra, J.; Blanco-Penedo, I.; Niza-Ribeiro, J.J.R. Relationship between Zoometric Measurements in Holstein-Friesian Cow and Cubicle Size in Dairy Farms. *Int. J. Morphol.* **2013**, *31*, 55–63. [CrossRef]
26. Wagner, E.L.; Tyler, P.J. A Comparison of Weight Estimation Methods in Adult Horses. *J. Equine Vet. Sci.* **2011**, *31*, 706–710. [CrossRef]
27. Spiess, A.-N.; Neumeyer, N. An evaluation of R2 as an inadequate measure for nonlinear models in pharmacological and biochemical research: A Monte Carlo approach. *BMC Pharmacol.* **2010**, *10*, 6. [CrossRef]
28. Lesosky, M.; Dumas, S.; Conradie, I.; Handel, I.G.; Jennings, A.; Thumbi, S.; Toye, P.; de Clare Bronsvoort, B.M. A live weight–heart girth relationship for accurate dosing of east African Shorthorn Zebu Cattle. *Trop. Anim. Health Prod.* **2012**, *45*, 311–316. [CrossRef]
29. García, L.; Ferreño, V.; Calviño, F.; Galego, V.; García, L. Previsión del Peso Vivo en animales Holstein mediante barimetria: Equaciones de predicción simples y múltiples. In Proceedings of the AIDA, XIII Jornadas sobre Producción Animal, Tomo II, Zaragoza, Spain, 12–13 May 2009; pp. 469–471.
30. Siddiqui, M.; Lateef, M.; Khalid Bashir, M.; Qamar Bilal, M.; Muhammad, G.; Mustafa, M.; ur Rehman, S. Pakistan Journal of Life and Social Sciences Estimation of Live Weight Using Different Body Measurements in Sahiwal Cattle. *Pak. J. Life Soc. Sci.* **2015**, *13*, 12–15.
31. Reis, G.L.; Albuquerque, F.H.M.A.R.; Valente, B.D.; Martins, G.A.; Teodoro, R.L.; Ferreira, M.B.D.; Monteiro, J.B.N.; de Almeida E Silva, M.; Madalena, F.E. Predição do peso vivo a partir de medidas corporais em animais mestiços Holandês/Gir. *Cienc. Rural* **2008**, *38*, 778–783. [CrossRef]
32. Pruitt, R.J.; Vanderwal, K.; Britt, M. Evaluation of Hoof Circumference to Predict Birth Weight. *S. D. Beef Rep.* **1995**, *17*, 17–19.
33. Parish, J.A.; Smith, T.; Parish, J.R.; Best, T.F.; Boland, H.T. Evaluation of Four Different Methods of Calf Birth Weight Data Collection. *Prof. Anim. Sci.* **2009**, *25*, 716–721. [CrossRef]
34. Long, N.M.; Collier, R.J.; Smith, J.F. Short communication: Comparison of 2 methods of assessing calf birth weights in dairy calves. *J. Dairy Sci.* **2012**, *95*, 7206–7209. [CrossRef] [PubMed]
35. Heinrichs, A.J.; Erb, H.N.; Rogers, G.W.; Cooper, J.B.; Jones, C.M. Variability in Holstein heifer heart-girth measurements and comparison of prediction equations for live weight. *Prev. Vet. Med.* **2007**, *78*, 333–338. [CrossRef] [PubMed]
36. Sherwin, V.; Hyde, R.; Green, M.; Remnant, J.; Payne, E.; Down, P. Accuracy of heart girth tapes in the estimation of weights of pre-weaned calves. *Vet. Rec. Open* **2021**, *8*, e16. [CrossRef]
37. Tüzemen, N.; Yanar, M.; Akbulut, Ö.; Ugur, F.; Aydin, R. Prediction of Body Weights from Body Measurements in Holstein-Friesian Calves. *Atatürk Üniv. Ziraat Fak. Derg.* **1995**, *26*, 245–252.
38. Magnabosco, C.U.; Lopes, F.B.; Brunes, L.C.; de Souza, F.M.; Mamede, M.M.S.; dos Santos, M.F.; de Queiroz, L.C.R.; Junior, A.R.B.; Carvalho, M.A.A.; Oliveira, M.F.C.E.; et al. *Estimativa do Peso ao Nascimento em Bovinos das Raças Brahman e Nelore Mocho por meio do Perímetro Torácico*; Embrapa Cerrados: Fortaleza, Brasília, 2017; Volume 341.

animals

MDPI

Article

Calf Management: Individual or Paired Housing Affects Dairy Calf Health and Welfare

David S. McFarland [1,*], Lorna M. McFarland [2], Darren J. Shaw [1] and Alastair I. Macrae [1]

[1] Division of Farm Animal, Veterinary Public Health and Epidemiology, Royal Dick School of Veterinary Studies and the Roslin Institute, Easter Bush Campus, Midlothian EH25 9RG, UK; darren.shaw@ed.ac.uk (D.J.S.); a.i.macrae@ed.ac.uk (A.I.M.)

[2] Independent Researcher, Newtownstewart BT78 4LG, UK

* Correspondence: davidcoolaghy@gmail.com; Tel.: +44-(0)131-445-4468

Simple Summary: This study involved 138 Holstein dairy calves from one UK farm, who were either reared individually or as pairs until 2–4 weeks of age, when they were subsequently batched together. Although there were no differences in growth rate, disease treatments or mortality, the paired calves demonstrated improved behavioural indicators of calf welfare. In addition, the heaviest calf in the pair grew more quickly and displayed less fear and more exploratory behaviour than the lightest calf in the pair. Although recent advice to farmers has been to rear calves in pairs, this study suggests that this may be potentially detrimental to the smaller calf.

Abstract: Previous research has indicated that preweaned dairy calves reared in pairs compared with individually have improved performance and indicators of animal welfare. One hundred and thirty Holstein female calves completed the trial, with eighty-five being allocated to paired housing and forty-five calves being allocated to individual housing. Daily live weight gain (DLWG), treatments and mortality were recorded throughout the preweaning period. Salivary cortisol, latency to feed and latency to approach a novel object were assessed at batching. There were no significant differences in DLWG, mortality and disease treatments between the average of the pair and the individually housed calves, although the pair-reared calves were quicker to approach the milk feed after batching and interacted more quickly with a novel object. The heaviest born calves within the pair had the highest DLWG from birth to weaning, with a higher percentage of calves approaching the novel object, compared with the lightest born calf within the pair. This study shows that calves within a pair may have significantly different performance and welfare during the preweaning period, with the heavier calf outperforming and displaying less fear and more exploratory behaviour than the lighter calf within a pair.

Keywords: calf; social housing; daily live weight gain; salivary cortisol; pair; novel object; latency to feed

1. Introduction

The traditional rearing of dairy calves individually in pens or hutches allows for easier observation and management whilst preventing the transmission of infectious diseases such as diarrhoea and pneumonia. However, recent studies have shown benefits to dairy calves reared in social housing (either in groups or pairs), in comparison to housing calves individually. Improved performance has been observed in group housing systems, with improved daily live weight gain (DLWG) and solid feed intake being seen during the preweaning stage [1–3]. Previous work has shown that individually reared calves are more fearful in comparison to calves reared in social housing [4,5], with paired calves showing less vocalisation at weaning in comparison to individually reared calves [6,7]. These studies suggest that rearing calves in pairs or groups may equip calves to better adapt to changes in the environment or management during the preweaning period [3].

Whilst various behavioural measurements have been used to assess calf welfare, few studies have used cortisol as a measurement of the stress response in calf social housing studies [8–10]. As an alternative to blood sampling, salivary cortisol is a non-invasive approach, having approximately a 10% yield in cortisol in comparison to plasma [11]. Salivary cortisol samples have been shown to peak approximately sixty minutes after a stressful event in previous calf studies [11–13]; so, salivary cortisol is a useful tool to help assess the stress response in calves during the preweaning period.

The majority of studies to date have compared paired and individually reared calves for all of the milk rearing phase, for between six and sixteen weeks of age [2,5,7]. However, a recent study found that only 40% of UK farms made no change in housing groups during the preweaning period, with the most common change being a move from individual into group pens [14]. There is little research to date focussing on the effects of calf rearing in pairs for a limited period in early life, with early grouping during the milk rearing phase [15].

With the majority of published research using the average of the pair to assess the performance of calves during calf social housing experiments [1,16,17], no research has been published on whether both calves within the pair have the same outcome, compared with averaging the data from the pair.

The objective of this study was to assess the effect of the single and paired housing of calves on dairy calf performance and welfare during the preweaning period. The aim was to assess this by using two methods: the average of the paired calves in the pen and looking at each calf within the pair separately to assess if there was a difference between calves within the pair. With little research having been carried out on calf social housing based on group movement and batching during the milk rearing phase, as commonly carried out on UK dairy farms [14,18], this study also aimed to provide data directly relevant to UK commercial dairy farms about the pair rearing of calves. Using behavioural measurements such as latency to feed and latency to interact with novel objects would enable assessment of calf fearfulness, exploratory behaviour and welfare, especially after stressful events such as batching. Our hypothesis was that the social relationship developed from pairing calves from birth would equip them with positive attributes to help improve their welfare, with the ability to adapt and manage potentially stressful situations such as batching.

2. Materials and Methods

This study was carried out on a 450-cow Holstein herd based in Northern Ireland, UK. The study was approved by University of Edinburgh, Royal (Dick) School of Veterinary Studies, Veterinary Ethical Review Committee (VERC; reference number 5.21).

The required sample sizes were calculated by using Minitab 20.3 (Minitab, LLC, State College, PA, USA), using a 95% confidence interval and 0.8 power. Calculations were based on an average DLWG of 0.7 kg with a 0.2 kg difference between groups (sample size of 64 animals) and a 10% difference in morbidity based on a 30% treatment rate (sample size of 62 animals). By using an 8% average UK calf mortality rate, the required sample size of over 1000 calves meant that it was not going to be possible to have sufficient calves to assess any differences in mortality rate [19,20].

One hundred and thirty-eight female Holstein dairy calves were enrolled in the trial from 14 September 2021 to 22 January 2022. Of the 138 calves enrolled, 92 were allocated to paired housing at birth, and 46 calves were allocated to individual housing. Mortality and morbidity were the primary reason for removal of calves from the study. Calves in pairs that did not complete the full pre-batching period as a pair were both removed from the trial. The six calves removed from the study for these reasons were included in the morbidity and mortality treatment only. One calf was incorrectly sexed at birth and was thus not included in any of the data analysis, along with the other paired calf in its pen from birth. Of the one hundred and thirty-two calves that were batched in groups of twelve, ten groups contained eight paired calves and four individual calves, with one group containing seven paired calves and five individual calves. Of the one hundred and thirty Holstein female calves with full datasets that completed the trial finishing on 9 March 2022, eighty-five were pair-housed at birth and forty-five were individually housed at birth. These 130 calves were included in the analysis for DLWG, salivary cortisol, latency to feed and latency to approach a novel object. A total of 136 calves were included in the morbidity and mortality data. This included calves removed from the trial due to illness or death.

2.1. Calf Management

Calves were born in a group maternity pen on deep bedded straw and were removed from their dam within twelve hours of birth. Calves were fed 3.5 L of their dam's colostrum within twelve hours of birth, with a further 3.5 L being provided within twenty-four hours of life. Two feeds of 3.5 L of the dam's colostrum was fed on day two. Calves were then fed powdered milk replacer on day three after birth (Triple AAA Golden Maverick; Volac, Orwell, UK). This was fed at 125 g per litre, with calves being built up to 4 L fed twice daily by seven days of age (8 L in total per day). Feeding times were from 6 am to 8 am and from 4 pm to 6 pm, with milk being delivered by using the same mobile milk tank. Calves were reared on both teat feeders and buckets during the pre-batching stage, initially starting on teat feeders before being moved onto buckets in the week prior to batching. The milk feed equipment, teat feeders and milk buckets were washed on a daily basis by using warm water. Calves also had access to fresh water and starter concentrate (Thompson Calf Pride; Thompsons, Belfast, UK) from entering the pen on day one.

Holstein heifer calves were placed into either paired or individual pens based on their order of birth. For example, calf one was placed in an individual pen, calves two and three were placed in a paired pen, and calf four was placed in an individual pen (Figure 1).

Figure 1. An example of a paired and an individual pen used in the study.

All calves were housed in the same shed at birth, made up of twelve individual pens and twelve paired pens (24 paired calves in twelve pairs, 12 individually penned calves), giving 36 calves in total in the shed. The paired pens were placed throughout the shed to minimise the potential variation in the environment within the shed (Figure 2).

Pen divisions were made with metal bar panels (183 cm in length) allowing for auditory, visual and physical contact between neighbouring pens. Paired pens were 183 cm in length by 183 cm in width, and individual pens were 183 cm by 91.5 cm in width, giving both paired and individually housed calves the same floor space of 1.67 m^2. Calves were bedded on deep barley straw, which was replenished with fresh straw daily.

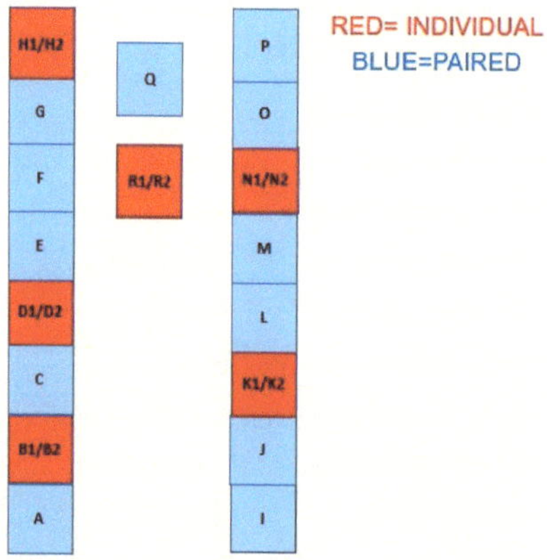

Figure 2. Layout of the calf shed used from birth to batching. Individual pens are in red and paired pens in blue.

At batching, at around 2–4 weeks of age, calves were moved into a separate batching shed in groups of twelve. Where possible, batches were made up of 4 individually housed and 8 pair-housed calves, giving 12 calves in total per pen. The batching shed consisted of eight pens 4.57 m in length and 4.27 m in width, providing 1.62 m^2 of floor space per calf. Pen barriers between batches were made of solid boarding; therefore, nose-to-nose contact between pens was limited (Figure 3). Once calves were in their batch, there were no group changes until after weaning. Water was available in each pen via a drinker, and milk was fed from a trough with sufficient space for all twelve calves in the pen to drink at once. The trough had no dividers, with all calves in the batch having access to the total quantity of milk for the group. The milk replacer used (Triple AAA Golden Maverick; Volac, UK) was fed at one hundred and twenty-five grams per litre, with calves drinking 4 L twice daily. Calves had ad libitum access to starter concentrate and straw, which were available throughout the day and refreshed on a daily basis. Calves were bedded on deep barley straw, which was replenished with fresh straw daily. Calves were routinely disbudded post-weaning; so, no disbudding of calves occurred during the period in which the calves were on the study.

2.2. Body Weight and Calf Performance

Calves were weighed at three timepoints throughout the trial: at birth, the day prior to batching and one week prior to weaning. Calves were fed four litres of milk replacer twice daily whilst in the batches and completed the trial once they had their final weight recorded. Milk feeding was then reduced for one week until weaning after calves completed the study. The weighing of calves was carried out by using a Ritchie weigh crate with a fitted Tru-test digital load cell and load bars.

Calves were diagnosed with clinical disease according to standard farm protocols. These were recorded and treated accordingly, in relation to farm treatment protocols. For example, calves with respiratory disease (diagnosed primarily based on clinical signs such as laboured breathing, coughing, nasal discharge or lethargy with a rectal temperature of over 39.6 °C) were treated with 2.5 mg tulathromycin/kg bodyweight subcutaneously (Draxxin, Zoetis, Leatherhead, UK) and 0.5 mg meloxicam/kg body weight subcutaneously (Metacam 20 mg/mL solution for injection; Boehringer Ingelheim Animal Health UK Ltd., Bracknell, UK). Calves with diarrhoea (diagnosed by fluid faeces and evidence of dehydration) were treated with oral electrolytes (Life Aid Xtra; one sachet per two litres of water), 0.2 mg butylscopolamine bromide/kg and 25 mg metamizole/kg bodyweight intramuscularly (Buscopan compositum solution for injection; Boehringer Ingelheim Animal

Health UK Ltd.). In addition, 8 mg procaine penicillin/kg and 10 mg dihydrostreptomycin sulphate/kg bodyweight (Pen & Strep suspension for injection; Newry, Northern Ireland) was given by intramuscular injection for three consecutive days. Mortality and reason for calves leaving the trial were also recorded.

Figure 3. Example of calves in a batch with their unique colour identification marking during the latency to approach a novel object test.

2.3. Salivary Cortisol

Saliva samples were collected at three timepoints from each calf: twenty-four hours prior to the day of batching, forty-five minutes after batching and twenty-four hours after the day of batching. Calves were sampled in the same order each day, with the aim to keep the time of sampling during the day consistent between each sample, with a twenty-four-hour time-period difference between each saliva sample collected per calf. Two synthetic swabs (Salivette®® Cortisol Swabs; Sarstedt, Germany) were held in the calf's mouth by using forceps for ninety seconds. Saliva sampling took approximately forty-five minutes per group. Restraint of calves was kept to a minimum to avoid unnecessary stress, with the majority of calves willing to chew on the forceps and Salivette swabs. If calves needed further restraint, they were guided to a corner of the pen to help facilitate handling. The routine was identical for all three saliva collection timepoints. Plastic gloves and the forceps were washed and dried with disposable paper towels between each calf to avoid cross-contamination of samples. Following collection, the swabs were immediately frozen at −20 °C for subsequent analysis. Following thawing, the cotton swabs were centrifuged at 3000× *g* for 15 min to extract the saliva, and concentrations of salivary cortisol were measured by using a commercial kit (Cortisol ELISA (Saliva); ALPCO®®, Salem, NH, USA) by SRUC Biomarkers Laboratory, SRUC, Roslin Institute Building, Easter Bush, Midlothian, EH25 9RG, UK.

2.4. Behavioural Measurements

On the day prior to batching, calves were given an individual colour marking using two coloured stripes marked on each side of the thorax (Figure 3). Saliva samples were taken first on the day prior to batching, before weighing and marking with their individual colour marking. This gave each calf a unique colour marking within its batching pen, and allowed for the identification of the calf on the video recordings. The video recording of the pen was taken by using a Reveal video camera positioned to view the whole pen and subsequently recorded onto digital media (Reveal DEMS360 version 5.4 software) to allow for subsequent analysis.

Video analysis was undertaken to assess latency to feed and latency to approach a novel object. Latency to feed was assessed at both the first and second feed after batching and consisted of a five-minute video recording from the time that milk entered the trough at feeding. Calves were deemed to be drinking milk if they were in a feeding position at the trough with their mouth entering the milk, and the time during which this drinking behaviour occurred following the arrival of milk in the trough was then determined.

Latency to approach a novel object was assessed on the day of batching and consisted of a fifteen-minute video recording from the time in which a novel object was placed in the centre of the pen. This test was performed at various times throughout the day, ranging between 11 am and 10 pm. The novel object was a 55 cm diameter purple exercise ball placed within a blue open top container to help keep the novel object's position central within the pen (Figure 3).

One person ensured that all calves within the pen were standing and then given a minute to settle, prior to placement of the novel object in the centre of the pen. Calves were deemed to have approached the novel object if they came within 30 cm of the novel object whilst showing clear interest in the novel object, and the time during which this behaviour occurred following the arrival of the novel object in the pen was then determined. During the latency to approach a novel object test, if any calves lay down, this was also recorded, and the time at which calves lay down was also recorded following the arrival of the novel object in the pen.

2.5. Data Analysis

All data were entered into an Excel file, with subsequent statistical analysis in Minitab 20.3 (Minitab, LLC) and R (version 3.4.4; R Foundation for Statistical Computing, Vienna, Austria). Standard errors were calculated for normally distributed data and interquartile ranges for those not normally distributed.

Four sets of mixed-effect models were established: (a) average of a pair versus single penned; (b) individuals within a pair; (c) heaviest calf at birth within a pair versus single penned; (d) lightest calf at birth within a pair versus single penned. Which batch the calves were part of was considered the random effect in all models. Linear mixed-effect models were established to assess differences in daily live weight gain and cortisol. Normality of residuals from these models was assessed prior to reporting analytical results, with cortisol levels requiring $\log_{10}$ transformation. General linear mixed-effect models with binomial errors were established to assess differences in the 4 sets of comparison for treatment of diarrhoea or pneumonia; total treatments; calves approaching their feed within 10 s or a novel object within 30 s; or calves did not approach their feed within five minutes or the novel object within 15 min. Statistical significance was taken as $p \leq 0.05$.

3. Results

3.1. Descriptive Analyses

Of the 130 Holstein female calves that completed the trial finishing on 9 March 2022, 85 were pair-housed at birth, and 45 were individually housed at birth. Calves had a mean birthweight of 39.81 ± SEM 0.37 kg. The median time from birth to the day prior to batching was 24 (IQR: 15–31) days. The mean weight on the day prior to batching was 51.40 ± 0.62 kg. Calves were weighed prior to reducing the milk for weaning at a median age of 64 (58–69) days, with a mean weight of 80.61 ± 1.24 kg.

3.2. Daily Live Weight Gain

When using the average DLWG for the paired calves in the pen, there was no significant difference in the mean DLWG between the individually reared calves and the paired calves from birth to batching (individual versus pair: 0.50 ± 0.02 kg per day versus 0.48 ± 0.02 kg per day (mean ± SEM); p = 0.248; Figure 4), from batching to weaning (individual versus pair: 0.70 ± 0.04 kg per day versus 0.74 ± 0.04 kg per day; p = 0.390; Figure 4) or from birth to weaning (individual versus pair: 0.63 ± 0.03 kg per day versus 0.64 ± 0.03 kg per day; p = 0.545; Figure 4). There was no significant difference in calf birthweight between the mean of the paired calves and the individually housed calves (individual versus pair: 39.7 ± 0.58 kg versus 39.9 ± 0.60 kg; p = 0.731).

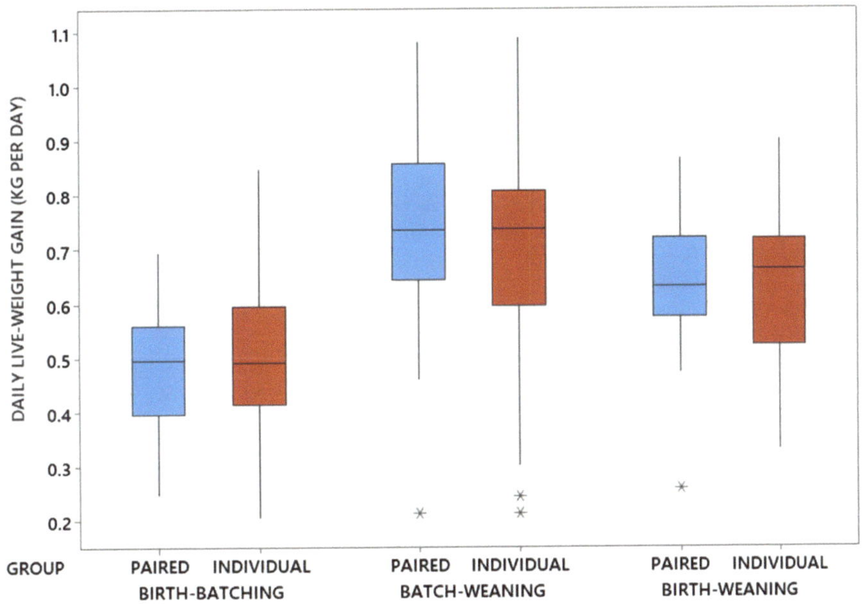

Figure 4. Mean daily live weight gain (DLWG) for calves either individually reared or pair-reared. The line in the box represents the median value, and the box represents the interquartile range. Outliers are identified by asterisks (*).

To assess whether there was any difference in DLWG between the paired calves, the pair within each pen were split according to their birthweight into "heavy" and "light" calves at birth. As expected, there was a significant difference in calf birthweight between the heavy and light calves within a pair (heavy versus light: 42.5 ± 0.59 kg versus 37.4 ± 0.59 kg; $p < 0.001$; Table 1). There was no significant difference in DLWG between the heavy and light calves within a pair from birth to batching (heavy versus light: 0.50 ± 0.03 kg per day versus 0.45 ± 0.03 kg per day; p = 0.196; Table 1) and from batching to weaning (heavy versus light: 0.77 ± 0.04 kg per day versus 0.71 ± 0.04 kg per day; p = 0.132; Table 1). However, there was a significant difference in DLWG from birth to weaning (heavy versus light: 0.68 ± 0.03 kg per day versus 0.61 ± 0.03 kg per day; p = 0.046; Table 1).

There was no significant difference in DLWG between the heavy calf in a pair at birth and the individually reared calves from birth to batching (heavy versus individual: 0.50 ± 0.03 kg per day versus 0.50 ± 0.03 kg per day; p = 0.986; Table 1), from batching to weaning (heavy versus individual: 0.78 ± 0.04 kg per day versus 0.70 ± 0.04 kg per day; p = 0.072; Table 1) or from birth to weaning (heavy versus individual: 0.68 ± 0.03 kg per day versus 0.63 ± 0.02 kg per day; p = 0.070; Table 1).

Table 1. DLWG and weight for different stages during pre-weaning between individual and pair-reared calves.

	Birthweight (kg) Mean ± SEM	Birth to Batching DLWG (kg per day) Mean ± SEM	Batching Weight (kg) Mean ± SEM	Batching to Weaning DLWG (kg per day) Mean ± SEM	Weaning Weight (kg) Mean ± SEM	Birth to Weaning DLWG (kg per day) Mean ± SEM	Time in Days from Birth to Weaning Median (IQR)
Individual (I) vs. Paired Average (PA)	I = 39.7 ± 0.58 PA = 39.9 ± 0.60 p = 0.731	I = 0.50 ± 0.02 PA = 0.48 ± 0.02 p = 0.248	I = 51.95 ± 0.97 PA = 51.18 ± 0.89	I = 0.70 ± 0.04 PA = 0.74 ± 0.04 p = 0.390	I = 80.09 ± 2.07 PA = 80.89 ± 1.88	I = 0.63 ± 0.03 PA = 0.64 ± 0.03 p = 0.545	I = 65 (58–70) PA = 64 (59–69)
Heavy (H) vs. light (L)	H = 42.5 ± 0.59 L = 37.4 ± 0.59 p < 0.001	H = 0.50 ± 0.03 L = 0.45 ± 0.03 p = 0.196	H = 54.66 ± 1.09 L = 47.79 ± 0.97	H = 0.77 ± 0.04 L = 0.71 ± 0.04 p = 0.132	H = 85.94 ± 2.03 L = 76.30 ± 2.17	H = 0.68 ± 0.03 L = 0.61 ± 0.03 p = 0.046	H = 64 (59–69) L = 64 (59–69)
Heavy (H) vs. individual (I)	H = 42.5 ± 0.59 I = 39.7 ± 0.58 p = 0.000	H = 0.50 ± 0.03 I = 0.50 ± 0.03 p = 0.986	H = 54.66 ± 1.09 I = 51.95 ± 0.97	H = 0.78 ± 0.04 I = 0.70 ± 0.04 p = 0.072	H = 85.94 ± 2.03 I = 80.09 ± 2.07	H = 0.68 ± 0.03 I = 0.63 ± 0.02 p = 0.070	H = 64 (59–69) I = 65 (58–70)
Light (L) vs. individual (I)	L = 37.4 ± 0.59 I = 39.7 ± 0.58 p = 0.007	L = 0.45 ± 0.03 I = 0.50 ± 0.03 p = 0.067	L = 47.79 ± 0.97 I = 51.95 ± 0.97	L = 0.71 ± 0.04 I = 0.70 ± 0.04 p = 0.968	L = 76.30 ± 2.17 I = 80.09 ± 2.07	L = 0.61 ± 0.03 I = 0.63 ± 0.03 p = 0.550	L = 64 (59–69) I = 65 (58–70)

There was no significant difference in DLWG between the light calf in a pair at birth and the individually reared calves from birth to batching (light versus individual: 0.45 ± 0.03 kg per day versus 0.50 ± 0.03 kg per day; p = 0.067; Table 1), from batching to weaning (light versus individual: 0.71 ± 0.04 kg per day versus 0.70 ± 0.04 kg per day; p = 0.968; Table 1) or from birth to weaning (light versus individual: 0.61 ± 0.03 kg per day versus 0.63 ± 0.03 kg per day; p = 0.550; Table 1).

To assess whether the difference in DLWG between the paired calves was due to the difference in their birthweights, the individual penned calves were ranked according to their birthweights. The bottom 50% of the individual penned calves based on birthweight were then compared to the light calf in a pair at birth. Despite their similar median birthweights (light pair 37.4 (IQR: 35.7–39.5) kg versus bottom 50% of individual calves 37.8 (IQR: 34.9–38.9) kg; p = 0.880), the mean DLWG from birth to batching was significantly lower for the light calf in a pair at birth (light pair 0.45 ± 0.024 kg per day versus bottom 50% of individual calves 0.54 ± 0.031 kg per day; p = 0.020).

3.3. Mortality and Morbidity

Only 2 of the 92 paired calves and 1 of the individually housed calves died during the study. These three deaths were related to diarrhoea, with one death also showing signs of respiratory disease. There was no significant difference in treatments between the individually reared and pair-reared calves for diarrhoea (individual 70.9 ± 8.0% versus paired 71.6 ± 6.3%; p = 0.932), with both calves in 27 of the 43 pairs being treated for diarrhoea, and 9 pairs with neither individual treated. Diarrhoea cases were reported predominantly from birth to batching. Overall higher levels of pneumonia were reported in individual calves (23.9 ± 4.1%) versus paired (12.2 ± 8.2%), although this was not statistically significant (p = 0.078). Only in 1 pair did both calves have reported pneumonia, compared with 35 pairs where no pneumonia was reported. If total disease treatments were considered, there were no differences between groups (individual 81.2 ± 6.2% versus paired 77.5 ± 5.1%; p = 0.623), although in 29 of the 43 pairs, both calves required a treatment, compared with only 5 pairs where no treatment was applied.

No significant difference was found between the heavy and light calves within a pair for diarrhoea (heavy 65.3 ± 8.9% versus light 77.8 ± 7.4%; p = 0.220), pneumonia (heavy 4.8 ± 4.2 versus light 8.1 ± 6.2%; p = 0.465) and total disease treatments (heavy 70.4 ± 8.2 versus light 84.8 ± 6.2; p = 0.120).

No significant difference was found between the heavy calves in a pair at birth and the individually reared calves for diarrhoea (heavy 64.4 ± 7.6% versus individual 69.0 ± 7.1%; p = 0.649), pneumonia (heavy 8.0 ± 4.6% versus individual 21.9 ± 7.8%; p = 0.072) and total disease treatments (heavy 69.0 ± 7.1% versus individual 80.0 ± 6.0%; p = 0.243).

No significant difference was found between the light calves in a pair at birth and the individually reared calves for diarrhoea (light 76.2 ± 6.7% versus individual 68.9 ± 7.0%; p = 0.448), pneumonia (light 8.8 ± 5.7% versus individual 17.4 ± 8.9%; p = 0.215) and total disease treatments (light 83.3 ± 5.8% versus individual 80.0 ± 6.0%; p = 0.689).

3.4. Salivary Cortisol

When the individually housed calves were compared to the average of the paired calves in a pen, there was no significant difference in $\log_{10}$-transformed salivary cortisol at Timepoint 1 (individual back-transformed mean = 11.4 ± 1.19 ng/mL versus paired 12.0 ± 1.19 ng/mL; p = 0.712; Figure 5), Timepoint 2 (individual 13.6 ± 1.16 ng/mL versus paired 18.0 ± 1.16 ng/mL; p = 0.070; Figure 5) and Timepoint 3 (individual 10.5 ± 1.19 ng/mL versus paired 12.3 ± 1.19 ng/mL; p = 0.342; Figure 5).

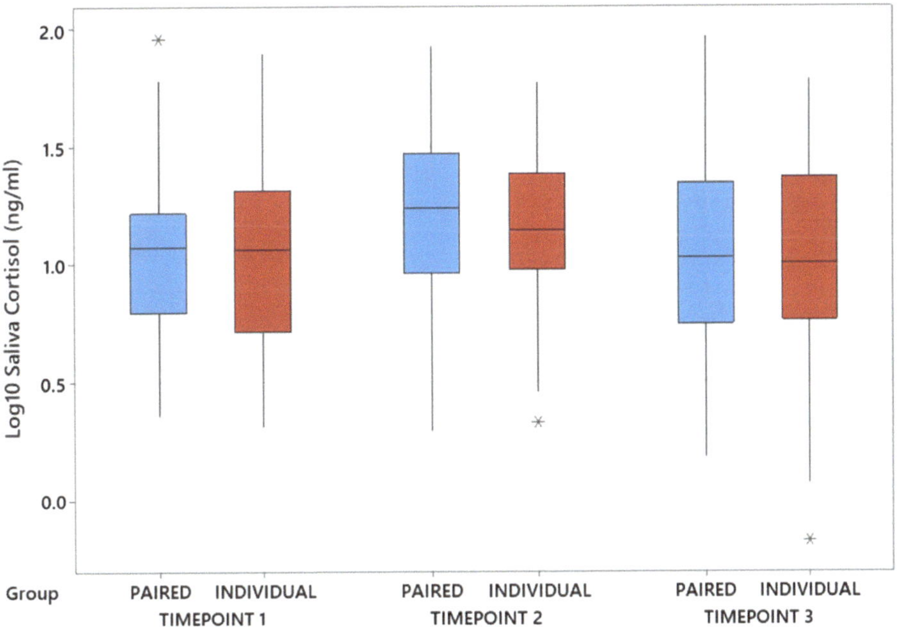

Figure 5. Salivary cortisol measurements in calves reared individually or in pairs. Timepoint 1: twenty-four hours prior to the day of batching. Timepoint 2: forty-five minutes after batching. Timepoint 3: twenty-four hours after the day of batching. The line in the box represents the median value, and the box represents the interquartile range. Outliers are identified by asterisks (*).

No significant difference was found between the heavy and light calves within a pair at birth for salivary cortisol at Timepoint 1 (heavy 12.1 ± 1.19 ng/mL versus light 12.5 ± 1.19 ng/mL; p = 0.857), Timepoint 2 (heavy 14. 5 ± 1.18 ng/mL versus light 21.6 ± 1.19 ng/mL; p = 0.085) or Timepoint 3 (heavy 11.6 ± 1.22 ng/mL versus light 13.6 ± 1.22 ng/mL; p = 0.505).

No significant difference was found between the heavy calves in a pair at birth and the individually reared calves for salivary cortisol at Timepoint 1 (heavy 13.5 ± 1.25 ng/mL versus individual 11.4 ± 1.22 ng/mL; p = 0.351), Timepoint 2 (heavy 15.0 ± 1.21 ng/mL versus individual 13.6 ± 1.16 ng/mL; p = 0.642) and Timepoint 3 (heavy 11.0 ± 1.27 ng/mL versus individual 10.5 ± 1.22 ng/mL; p = 0.807).

No significant difference was found between the light calves in a pair at birth and the individually reared calves for salivary cortisol at Timepoint 1 (light 11.5 ± 1.25 ng/mL versus individual 11.4 ± 1.21 ng/mL; $p = 0.988$) and Timepoint 3 (light 13.9 ± 1.26 ng/mL versus individual 10.5 ± 1.19 ng/mL; $p = 0.237$), but there were differences at Timepoint 2 (light 21.3 ± 1.20 ng/mL versus individual 13.7 ± 1.15 ng/mL; $p = 0.015$).

3.5. Latency to Feed

There were 6 individual calves and 14 paired (8 heavy and 6 light) calves that did not approach the first feed after batching during the observation period. Of the 39 individual calves that did approach the first feed after batching, they did so in a median time of 15.0 s [IQR: 0.0–73.0], whereas the 71 paired calves took a median time of 26.0 s [0.0–149.5].

There was one individual calf and six paired (four heavy and two light) calves that did not approach the second feed after batching during the observation period. Of the 44 individual calves that did approach the second feed after batching, they did so in a median time of 6.0 s [IQR: 0.0–21.0]. The 79 paired calves took a median time of 6.0 s [0.0–16.0].

A significantly higher proportion of the paired calves (either one or both of the calves in the pair) approached the milk within 10 s of feeding at both the first milk feed after batching (individual calves versus paired; 44.4% versus 69.5%) and the second milk feed after batching (60.0% versus 90.5%) (Table 2). However, there were no significant differences in the proportion of calves that did not approach the milk feed within 5 min, or between the heavy and light calves in the pair.

Table 2. Latency to feed from milk trough following batching between individual and pair-reared calves.

	First Milk Feed after Batching		Second Milk Feed after Batching	
	Percentage of Calves Approaching Milk within 10 s	**Percentage of Calves That Did Not Approach within 5 min**	**Percentage of Calves Approaching Milk within 10 s**	**Percentage of Calves That Did Not Approach within 5 min**
Individual (I) vs. Paired Average (PA)	I = 44.4 ± 11.4% PA = 69.5 ± 10.4% $p = 0.032$	I = 13.3 ± 5.1% PA = 9.5 ± 4.5% $p = 0.579$	I = 60.0 ± 7.3% PA = 90.5 ± 4.5% $p = 0.002$	I = 2.2 ± 2.2% PA = 2.4 ± 2.4% $p = 0.961$
Heavy (H) vs. light (L)	H = 51.1 ± 8.3% L = 43.6 ± 8.3% $p = 0.502$	H = 15.1 ± 7.8% L = 10.1 ± 7.8% $p = 0.533$	H = 65.9 ± 7.4% L = 78.0 ± 6.5% $p = 0.222$	H = 9.8 ± 4.6% L = 4.9 ± 3.4% $p = 0.405$
Heavy (H) vs. individual (I)	H = 51.0 ± 9.2% I = 44.4 ± 8.8% $p = 0.556$	H = 18.6 ± 6.8% I = 13.3 ± 5.4% $p = 0.441$	H = 65.9 ± 7.4% I = 60.0 ± 7.3% $p = 0.575$	H = 9.8 ± 4.6% I = 2.2 ± 2.2% $p = 0.171$
Light (L) vs. individual (I)	L = 43.3 ± 8.9% I = 44.4 ± 8.5% $p = 0.922$	L = 14.6 ± 5.5% I = 13.3 ± 5.1% $p = 0.862$	L = 78.0 ± 6.5% I = 60.0 ± 7.3% $p = 0.075$	L = 4.9 ± 3.4% I = 2.2 ± 2.2% $p = 0.513$

3.6. Latency to Approach a Novel Object

There were 6 individual calves and 14 paired (3 heavy and 11 light) calves that did not approach the novel object during the observation period. Of the 39 individual calves that did approach the novel object, they did so in a median time of 69.0 s [IQR: 23.0–255.0] whereas the 71 paired calves took a median time of 90 s [35.0–470.0].

A significantly higher proportion of the paired calves (either one or both of the calves in the pair) approached the novel object within 30 s (individual calves versus paired: 26.5% versus 64.2%) (Table 3). Whilst there were no differences in the speed at which the heavy and light calves approached the novel object, more of the lighter calves did not approach the novel object within 15 min compared with the heavier calves in the pair (heavier calves 7.3% versus lighter calves 26.8%) (Table 3).

Table 3. Latency to approach a novel object between individual and pair-reared calves.

	Percentage of Calves Approaching Object within 30 s	Percentage of Calves That Did Not Approach within 15 min
Individual (I) vs. Paired Average (PA)	I = 26.5 ± 9.6% PA = 64.2 ± 11.0% p = 0.003	I = 13.3 ± 5.1% PA = 4.2 ± 3.3% p = 0.182
Heavy (H) vs. light (L)	H = 46.3 ± 8.1% L = 26.5 ± 7.2% p = 0.069	H = 7.3 ± 4.1% L = 26.8 ± 6.9% p = 0.027
Heavy (H) vs. individual (I)	H = 44.6 ± 12.1% I = 26.5 ± 9.9% p = 0.112	H = 7.3 ± 4.1% I = 13.3 ± 5.1% p = 0.369
Light (L) vs. individual (I)	L = 26.5 ± 7.8% I = 26.5 ± 7.9% p = 0.646	L = 26.8 ± 6.9% I = 13.3 ± 5.1% p = 0.123

4. Discussion

To our knowledge, this is the first study to investigate the effects of paired housing by both using the average of the pair and looking at each calf within a pair separately. The majority of previous studies on the effects of pair-reared calves have used the average of the pair in their analysis [21–23]. When using the average measurement for the two calves in the pair, this study showed no significant differences between individual and pair-reared calves for the first 2–4 weeks of life in terms of average DLWG, morbidity, mortality and salivary cortisol measurements. However, pair-reared calves were quicker to approach the milk feed after batching and to interact with the novel object. When assessing the individual calves within the pair, there were significant differences indicating that the higher-birthweight calf consistently outperformed and displayed less fear and more exploratory behaviour than the lower-birthweight calf within the pair.

In this study, no significant difference was found between the average of the paired and individually housed calves for DLWG, similar to previous work [6,24]. However, a significant difference was found in DLWG when the paired calves were assessed separately. DLWG during the pre-weaning period can have long-term effects on age at first calving, milk production, fertility and longevity within the herd [25–27], highlighting the importance of understanding the social relationship occurring between calves in a pair.

Little previous research has used salivary cortisol as an indicator of stress in calves during the pre-weaning period. One study [28] assessed salivary cortisol around weaning, with some studies using salivary cortisol when assessing stress associated with castration, dystocia and heat stress in calves [29–31]. Salivary cortisol has been used to show that regrouping can be a stressful event for six-month-old dairy heifers [32]. This is similar to our study, where batching was found to have a significant relationship with salivary cortisol. Little research has been carried out to date using salivary cortisol to assess stress at batching at 2–4 weeks of age, as was carried out in this study. We found a significant difference in salivary cortisol between the light and individual calves at Timepoint 2. Given that salivary cortisol is mainly affected by severe acute stress [33], it may be difficult to determine if the significant differences found were due to stress or related to the exercise or excitability of the calves being grouped in a new environment with unfamiliar calves. Research in stallions and geldings has found that cortisol levels increase with exercise, and in stallions, cortisol levels also increase with excitement due to the presence of a mare [34], indicating that there may be other factors that influence salivary cortisol levels in calves. This is a limitation of measuring cortisol, as it can be affected by other factors, such as exercise. However, given that the lighter calf in the pair at birth also showed more fear and less exploratory behaviour in the latency to approach a novel object, the salivary cortisol results represent further evidence suggesting a significant difference in the journey of two calves within a pair from birth to weaning.

Latency to feed can be used to assess how well calves adapt to new situations, such as batching, learning where the feed is in a new environment and showing the social capability to approach the feed in a group setting [35,36]. One study [37] found that visual contact alone is not sufficient for calves to develop social bonds and that physical contact among calves is necessary. Physical contact through the pen boundaries with neighbouring calves in our study meant that even individually housed calves had a degree of physical contact with another calf. Our study found a significant difference in latency to feed between paired and individually housed calves for the first and second feed after batching. This is different from other studies [35,38], where no difference was found in latency to feed in paired and individually housed calves. This may be explained by all calves on our study having auditory, visual and physical contact with other calves; however, calves within a pair had significantly more physical contact within their pen with another calf. One study [5] found that paired or individually housed calves with physical contact through the pen boundary with other calves had a lower heart rate than individual calves with no physical contact with other calves during a novel-environment test. This shows that physical contact through the pen boundaries may allow individually housed calves to develop some social skills similar to paired calves, but not to the same level. Competition for feed within the pair may also help explain why more of the paired reared calves approached the milk feed quicker after batching.

Latency to approach a novel object can be a useful indicator of fear and exploratory behaviour in calves [39]. In our study, a significant difference was found between the heavy and light calves within a pair for latency to approach a novel object, with more of the lighter calves not interacting with the novel object. A significant difference was also found with paired calves approaching the novel object more quickly in comparison to individually housed calves. Few studies have assessed calves approaching a novel object within a group [40], as carried out in this study. Social housing has been found to affect play and exploratory behaviour, promoting positive calf welfare [41]. One study [42] found that the age of the calf can change their behaviour with a novel object, potentially becoming more responsive to new situations as they get older during the pre-weaning period.

Some studies have shown the benefits that paired housing can have on improving a calf's competitiveness to feed at three weeks of age [43] and after weaning [38], which is in agreement with our results. No studies to our knowledge have assessed if calves being pair-housed from birth develop a relationship that allows one of the paired calves to benefit in performance and welfare to the detriment of the other paired calf, potentially related to competitiveness for feed within the pen. Some of the literature [44] has found pairing calves from birth to affect their personality, with calves becoming bolder than individually housed calves, while personality has been linked with feeding behaviours in calves around weaning [45]. Some studies have shown personality traits among calves to be different [46]. One study showed that some calves can have a more active and exploratory personality than others [47]. This could play a role in how calves interact and adapt to new situations such as batching, with personality traits being suggested to be considered when forming social pairings or groups [48]. In this study, calves were fed through a teat before changing onto buckets, to allow for a smoother transition to the milk troughs used when group-housed. Although cross-suckling was not reported to be an issue in this study, competition at the teats was reported within the paired pens in the first 2–4 weeks of life prior to batching, particularly once one calf had finished the milk in its teat feeder. Milk ingestion stimulates the calf's sucking motivation and declines 10–15 min after a milk meal [49], with the competition seen at the teat feeders likely linked to calves wanting to satisfy their motivation to suck. Some studies have shown that competition is increased with teat feeders in comparison to buckets [50] and that calves are more likely to suck teats when milk is involved [51], potentially increasing competition at the teat feeders in paired housing. Ad libitum milk feeding has also been shown not to remove the motivation for non-nutritive sucking in bucket-fed calves [52] but high milk allowances do help reduce cross-suckling [51].

Potentially, feeding calves on a high milk allowance through a bucket early in life may help reduce competition in pair-housed calves while allowing access to a teat for each calf to help satisfy non-nutritive sucking in the 10–15 min period after milk feeding. This may help provide a more uniform experience for both pair-housed calves within a pen to obtain the benefits social housing can offer, such as improved solid feed intake and calves becoming less fearful [3,5]. Competition for feed may be affecting the lighter calf at birth in the pair, as it is more likely to be dominated by the heavier calf. This negative experience for the lighter calf could, therefore, be linked with their lower DLWG, increased fear and less exploratory behaviour shown through the novel object approach test.

Competition at feeding, feeding method, delaying age at pairing [53] and calf personality are areas for future research in paired housing and may explain some of the differences seen in DLWG, feeding behaviour and approaching a novel object when calves within a pair were looked at separately in this study.

This study was carried out on a working dairy farm utilising commonly used calf management practices within the UK. One of the limitations within this study is that only one farm with a set calf management protocol was used in the trial, to minimize management variation. Further research work would be beneficial on farms with different calf management practices, such as spring calving, automatic calf feeders or use of milk bars with dividers allowing for an allocated amount of milk per calf, to assess if similar results are found.

5. Conclusions

This study showed the benefits of paired housing in comparison to individual housing, with less fear, more exploratory behaviour and a better ability to adapt to a new environment being found during the behavioural tests. This is combined with no negative effects on DLWG, morbidity and mortality. The area of concern found in this study is within a pair, where one calf will thrive in both performance and welfare to the detriment of the other calf when paired from birth. This was shown by a significant difference in DLWG and approaching a novel object between the calves within a pair. A significant difference was also found in salivary cortisol between the light and individual calves. More research is required to determine if paired housing can be used for allowing calves within a pair to have more uniform growth and positive welfare experience. This study shows the risk of using the average of the pen when assessing calf performance and welfare in paired housing, with looking at the paired calves separately being recommended.

Author Contributions: The study was designed by D.S.M. and A.I.M. The experimental work and collection of data was carried out by D.S.M. and L.M.M. Data analysis was conducted by D.S.M., D.J.S. and A.I.M. The paper was written by D.S.M., L.M.M., D.J.S. and A.I.M. All authors have read and agreed to the published version of the manuscript.

Funding: This research was supported through funding from Hannah Dairy Research Foundation (MMI/JS/HA0514-0081) and University of Edinburgh (Farm Animal Practice and Dairy Herd Health and Productivity Service).

Institutional Review Board Statement: The study was conducted according to the guidelines of the Royal Dick School of Veterinary Studies and was approved by the Veterinary Ethical Review Committee of the Royal Dick School of Veterinary Studies (VERC reference number 5.21).

Informed Consent Statement: Not applicable.

Data Availability Statement: The raw data supporting the conclusions of this article will be made available by the authors upon request.

Acknowledgments: Special thanks to the McFarland family and the team at Moo Milk.

Conflicts of Interest: The authors declare no conflicts of interest.

References

1. Whalin, L.; Weary, D.M.; von Keyserlingk, M.A.G. Short communication: Pair housing dairy calves in modified calf hutches. *J. Dairy. Sci.* **2018**, *101*, 5428–5433. [CrossRef] [PubMed]
2. Knauer, W.A.; Godden, S.M.; Rendahl, A.K.; Endres, M.I.; Crooker, B.A. The effect of individual versus pair housing of dairy heifer calves during the preweaning period on measures of health, performance, and behavior up to 16 weeks of age. *J. Dairy. Sci.* **2021**, *104*, 3495–3507. [CrossRef] [PubMed]
3. Costa, J.H.C.; von Keyserlingk, M.A.G.; Weary, D.M. Invited review: Effects of group housing of dairy calves on behavior, cognition, performance, and health. *J. Dairy. Sci.* **2016**, *99*, 2453–2467. [CrossRef] [PubMed]
4. Jensen, M.B.; Vestergaard, K.S.; Krohn, C.C.; Munksgaard, L. Effect of single versus group housing and space allowance on responses of calves during open-field tests. *Appl. Anim. Behav. Sci.* **1997**, *54*, 109–121. [CrossRef]
5. Jensen, M.B.; Larsen, L.E. Effects of level of social contact on dairy calf behavior and health. *J. Dairy Sci.* **2014**, *97*, 5035–5044. [CrossRef]
6. Bolt, S.L.; Boyland, N.K.; Mlynski, D.T.; James, R.; Croft, D.P. Pair Housing of Dairy Calves and Age at Pairing: Effects on Weaning Stress, Health, Production and Social Networks. *PLoS ONE* **2017**, *12*, e0166926. [CrossRef]
7. De Paula Vieira, A.; von Keyserlingk, M.A.; Weary, D.M. Effects of pair versus single housing on performance and behavior of dairy calves before and after weaning from milk. *J. Dairy Sci.* **2010**, *93*, 3079–3085. [CrossRef]
8. Stewart, M.; Stafford, K.J.; Dowling, S.K.; Schaefer, A.L.; Webster, J.R. Eye temperature and heart rate variability of calves disbudded with or without local anaesthetic. *Physiol. Behav.* **2008**, *93*, 789–797. [CrossRef]
9. Stewart, M.; Verkerk, G.A.; Stafford, K.J.; Schaefer, A.L.; Webster, J.R. Noninvasive assessment of autonomic activity for evaluation of pain in calves, using surgical castration as a model. *J. Dairy Sci.* **2010**, *93*, 3602–3609. [CrossRef]
10. Pagani, E.; Valle, E.; Barbero, R.; Russo, N.; Osella, M.C.; Schiavone, A.; Prola, L. Effects of different housing systems on haematological profile, salivary cortisol concentration, and behavioural stress responses in calves of different ages. *Large Anim. Rev.* **2017**, *23*, 11–19.
11. Kovacs, L.; Kezer, F.L.; Bodo, S.; Ruff, F.; Palme, R.; Szenci, O. Salivary cortisol as a non-invasive approach to assess stress in dystocic dairy calves. *Sci. Rep.* **2021**, *11*, 6200. [CrossRef] [PubMed]
12. Fujiwara, M.; Haskell, M.J.; Macrae, A.I.; Rutherford, K.M.D. Impact of Maternal High Stocking Density during the Dry Period on Dairy Calf Health, Behaviour, and Welfare. *Animals* **2020**, *10*, 922. [CrossRef] [PubMed]
13. Nagel, C.; Aurich, J.; Trenk, L.; Ille, N.; Drillich, M.; Pohl, W.; Aurich, C. Stress response and cardiac activity of term and preterm calves in the perinatal period. *Theriogenology* **2016**, *86*, 1498–1505. [CrossRef] [PubMed]
14. Mahendran, S.A.; Wathes, D.C.; Booth, R.E.; Blackie, N. A survey of calf management practices and farmer perceptions of calf housing in UK dairy herds. *J. Dairy Sci.* **2022**, *105*, 409–423. [CrossRef] [PubMed]
15. Lindner, E.E.; Gingerich, K.N.; Miller-Cushon, E.K. Effects of early social contact on dairy calf response to initial social grouping and regrouping. *J. Dairy Sci.* **2021**, *104*, 10090–10099. [CrossRef] [PubMed]
16. Jensen, M.B.; Duve, L.R.; Weary, D.M. Pair housing and enhanced milk allowance increase play behavior and improve performance in dairy calves. *J. Dairy Sci.* **2015**, *98*, 2568–2575. [CrossRef] [PubMed]
17. Wormsbecher, L.; Bergeron, R.; Haley, D.; de Passille, A.M.; Rushen, J.; Vasseur, E. A method of outdoor housing dairy calves in pairs using individual calf hutches. *J. Dairy Sci.* **2017**, *100*, 7493–7506. [CrossRef] [PubMed]
18. Tapki, İ. Effects of individual or combined housing systems on behavioural and growth responses of dairy calves. *Acta Agric. Scand. Sect. A Anim. Sci.* **2007**, *57*, 55–60. [CrossRef]
19. Brickell, J.S.; Bourne, N.; McGowan, M.M.; Wathes, D.C. Effect of growth and development during the rearing period on the subsequent fertility of nulliparous Holstein-Friesian heifers. *Theriogenology* **2009**, *72*, 408–416. [CrossRef]
20. Brickell, J.S.; McGowan, M.M.; Pfeiffer, D.U.; Wathes, D.C. Mortality in Holstein-Friesian calves and replacement heifers, in relation to body weight and IGF-I concentration, on 19 farms in England. *Animal* **2009**, *3*, 1175–1182. [CrossRef]
21. Mahendran, S.A.; Wathes, D.C.; Booth, R.E.; Blackie, N. The Health and Behavioural Effects of Individual versus Pair Housing of Calves at Different Ages on a UK Commercial Dairy Farm. *Animals* **2021**, *11*, 612. [CrossRef] [PubMed]
22. Miller-Cushon, E.K.; DeVries, T.J. Effect of social housing on the development of feeding behavior and social feeding preferences of dairy calves. *J. Dairy Sci.* **2016**, *99*, 1406–1417. [CrossRef] [PubMed]
23. Costa, J.H.; Meagher, R.K.; von Keyserlingk, M.A.; Weary, D.M. Early pair housing increases solid feed intake and weight gains in dairy calves. *J. Dairy Sci.* **2015**, *98*, 6381–6386. [CrossRef] [PubMed]
24. Liu, S.; Ma, J.; Li, J.; Alugongo, G.M.; Wu, Z.; Wang, Y.; Li, S.; Cao, Z. Effects of Pair Versus Individual Housing on Performance, Health, and Behavior of Dairy Calves. *Animals* **2019**, *10*, 50. [CrossRef] [PubMed]
25. Ettema, J.F.; Santos, J.E. Impact of age at calving on lactation, reproduction, health, and income in first-parity Holsteins on commercial farms. *J. Dairy Sci.* **2004**, *87*, 2730–2742. [CrossRef] [PubMed]
26. Wathes, D.C.; Pollott, G.E.; Johnson, K.F.; Richardson, H.; Cooke, J.S. Heifer fertility and carry over consequences for life time production in dairy and beef cattle. *Animal* **2014**, *8* (Suppl. 1), 91–104. [CrossRef] [PubMed]
27. Bach, A. Associations between several aspects of heifer development and dairy cow survivability to second lactation. *J. Dairy Sci.* **2011**, *94*, 1052–1057. [CrossRef] [PubMed]

28. Rubio, C.P.; Escribano, D.; Mainau, E.; Ceron, J.J.; Navarro, E.; Manteca, X. Changes in salivary biomarkers of oxidative status in calves at weaning and grouping. *BMC Vet. Res.* **2021**, *17*, 373. [CrossRef] [PubMed]
29. Gonzalez, L.A.; Schwartzkopf-Genswein, K.S.; Caulkett, N.A.; Janzen, E.; McAllister, T.A.; Fierheller, E.; Schaefer, A.L.; Haley, D.B.; Stookey, J.M.; Hendrick, S. Pain mitigation after band castration of beef calves and its effects on performance, behavior, Escherichia coli, and salivary cortisol. *J. Anim. Sci.* **2010**, *88*, 802–810. [CrossRef]
30. Kovacs, L.; Kezer, F.L.; Ruff, F.; Szenci, O.; Bakony, M.; Jurkovich, V. Effect of artificial shade on saliva cortisol concentrations of heat-stressed dairy calves. *Domest. Anim. Endocrinol.* **2019**, *66*, 43–47. [CrossRef]
31. Barrier, A.C.; Haskell, M.J.; Birch, S.; Bagnall, A.; Bell, D.J.; Dickinson, J.; Macrae, A.I.; Dwyer, C.M. The impact of dystocia on dairy calf health, welfare, performance and survival. *Vet. J.* **2013**, *195*, 86–90. [CrossRef] [PubMed]
32. Denham, J.; Adams Progar, A. Changes in Holstein Heifer Salivary Cortisol Concentrations and Behavior after Regrouping. *Ruminants* **2023**, *3*, 255–265. [CrossRef]
33. Bozovic, D.; Racic, M.; Ivkovic, N. Salivary Cortisol Levels as a Biological Marker of Stress Reaction. *Med. Arh.* **2013**, *67*, 374–377. [CrossRef] [PubMed]
34. Colborn, D.R.; Thompson, D.L.; Roth, T.L.; Capehart, J.S.; White, K.L. Responses of Cortisol and Prolactin to Sexual Excitement and Stress in Stallions and Geldings. *J. Anim. Sci.* **1991**, *69*, 2556–2562. [CrossRef] [PubMed]
35. Fujiwara, M.; Rushen, J.; de Passillé, A.M. Dairy calves' adaptation to group housing with automated feeders. *Appl. Anim. Behav. Sci.* **2014**, *158*, 1–7. [CrossRef]
36. Costa, J.H.; Daros, R.R.; von Keyserlingk, M.A.; Weary, D.M. Complex social housing reduces food neophobia in dairy calves. *J. Dairy Sci.* **2014**, *97*, 7804–7810. [CrossRef] [PubMed]
37. Lindner, E.E.; Gingerich, K.N.; Burke, K.C.; Doyle, S.B.; Miller-Cushon, E.K. Effects of Social Housing on Dairy Calf Social Bonding. *Animals* **2022**, *12*, 821. [CrossRef] [PubMed]
38. Duve, L.R.; Weary, D.M.; Halekoh, U.; Jensen, M.B. The effects of social contact and milk allowance on responses to handling, play, and social behavior in young dairy calves. *J. Dairy Sci.* **2012**, *95*, 6571–6581. [CrossRef]
39. Lindner, E.E.; Gingerich, K.N.; Utzig, S.; Miller-Cushon, E.K. Effects of social housing on dairy calf response to novelty shortly after weaning and grouping. *JDS Commun.* **2023**. [CrossRef]
40. Cramer, M.C.; Stanton, A.L. Associations between health status and the probability of approaching a novel object or stationary human in preweaned group-housed dairy calves. *J. Dairy Sci.* **2015**, *98*, 7298–7308. [CrossRef]
41. Zhang, C.; Juniper, D.T.; Meagher, R.K. Effects of physical enrichment items and social housing on calves' growth, behaviour and response to novelty. *Appl. Anim. Behav. Sci.* **2021**, *237*, 105295. [CrossRef]
42. Lauber, M.C.Y.; Hemsworth, P.H.; Barnett, J.L. The effects of age and experience on behavioural development in dairy calves. *Appl. Anim. Behav. Sci.* **2006**, *99*, 41–52. [CrossRef]
43. Suchon, M.; Ede, T.; Vandresen, B.; von Keyserlingk, M.A.G. Social housing improves dairy calves' performance in a competition test. *JDS Commun.* **2023**, *4*, 479–483. [CrossRef] [PubMed]
44. Gingerich, K.N.; Lindner, E.E.; Kalman, S.; Miller-Cushon, E.K. Social contact from birth influences personality traits of group-housed dairy calves. *JDS Commun.* **2023**, *4*, 484–488. [CrossRef] [PubMed]
45. Neave, H.W.; Costa, J.H.C.; Weary, D.M.; von Keyserlingk, M.A.G. Personality is associated with feeding behavior and performance in dairy calves. *J. Dairy. Sci.* **2018**, *101*, 7437–7449. [CrossRef] [PubMed]
46. Carslake, C.; Occhiuto, F.; Vazquez-Diosdado, J.A.; Kaler, J. Repeatability and Predictability of Calf Feeding Behaviors-Quantifying Between- and Within-Individual Variation for Precision Livestock Farming. *Front. Vet. Sci.* **2022**, *9*, 827124. [CrossRef] [PubMed]
47. Occhiuto, F.; Vazquez-Diosdado, J.A.; Carslake, C.; Kaler, J. Personality and predictability in farmed calves using movement and space-use behaviours quantified by ultra-wideband sensors. *R. Soc. Open Sci.* **2022**, *9*, 212019. [CrossRef] [PubMed]
48. Occhiuto, F.; Vazquez-Diosdado, J.A.; King, A.J.; Kaler, J. Evidence of personality-dependent plasticity in dairy calf movement behaviours derived from automated data collection. *Sci. Rep.* **2023**, *13*, 18243. [CrossRef]
49. de Passille, A.M. Sucking motivation and related problems in calves. *Appl. Anim. Behav. Sci.* **2001**, *72*, 175–187. [CrossRef] [PubMed]
50. Jensen, M.B.; Budde, M. The effects of milk feeding method and group size on feeding behavior and cross-sucking in group-housed dairy calves. *J. Dairy. Sci.* **2006**, *89*, 4778–4783. [CrossRef]
51. Jung, J.; Lidfors, L. Effects of amount of milk, milk flow and access to a rubber teat on cross-sucking and non-nutritive sucking in dairy calves. *Appl. Anim. Behav. Sci.* **2001**, *72*, 201–213. [CrossRef] [PubMed]
52. Jensen, M.B. The effects of feeding method, milk allowance and social factors on milk feeding behaviour and cross-sucking in group housed dairy calves. *Appl. Anim. Behav. Sci.* **2003**, *80*, 191–206. [CrossRef]
53. Duve, L.R.; Jensen, M.B. Social behavior of young dairy calves housed with limited or full social contact with a peer. *J. Dairy Sci.* **2012**, *95*, 5936–5945. [CrossRef] [PubMed]

animals MDPI

Article

Effect of a Lactobacilli-Based Direct-Fed Microbial Product on Gut Microbiota and Gastrointestinal Morphological Changes

John I. Alawneh [1,*], Hena Ramay [2], Timothy Olchowy [2], Rachel Allavena [1], Martin Soust [3] and Rafat Al Jassim [4]

1 School of Veterinary Science, University of Queensland, Gatton, QLD 4343, Australia; r.allavena@uq.edu.au
2 Faculty of Veterinary Medicine, University of Calgary, Calgary, AB T3R 1J3, Canada; henaramay@gmail.com (H.R.)
3 Terragen Biotech Pty Ltd., Coolum Beach, QLD 4573, Australia; martins@terragen.com.au
4 Queensland Alliance for Agriculture and Food Innovation, St Lucia, QLD 4072, Australia; r.aljassim@uq.edu.au
* Correspondence: jalawneh@gmail.com

Simple Summary: This study aimed to characterise and compare the development of gut microbiota in dairy calves from birth to weaning, focusing on the impact of a direct-fed microbial (DFM) product containing three strains of lactic acid bacteria (LAB). Forty-four Holstein-Friesian calves were randomly assigned to Treatment (TRT) and Control (CON) groups. TRT calves received a daily dose of the DFM, while CON calves received a placebo and served as the control. Faecal samples and necropsies were collected for analysis. TRT calves exhibited higher live weights at weaning and comparable average daily live weight gain and feed intake to CON calves. TRT calves also demonstrated greater weights of specific gut segments (duodenum, abomasum, reticulum) and enhanced rumen and intestinal development. The microbial diversity was more pronounced in the TRT group, with differences in the relative abundances of eight genera. This study suggests that supplementing with the LAB-based DFM positively influenced calfs' weight, gut development, and microbial diversity. Further research is recommended to explore potential associations between DFM products and gut mucosa-associated microbiota.

Abstract: The calf's gastrointestinal tract (GIT) microbiome undergoes rapid shifts during early post-natal life, which can directly affect calf performance. The objectives of this study were to characterise and compare differences in the establishment and succession of GIT microbiota, GIT morphological changes, and the growth of dairy calves from birth until weaned. Forty-four newborn Holstein-Friesian calves were randomly selected and assigned to Treatment (TRT) and Control (CON) groups. The TRT group calves received a once-daily dose of a direct-fed microbial (DFM) liquid product containing *Lacticaseibacillus paracasei*, *Lentilactobacillus buchneri*, and *Lacticaseibacillus casei*, all formerly known as *Lactobacillus*. Fresh faecal samples were manually taken from the rectum of all calves, and gross necropsy was performed on the forestomachs and gastrointestinal tracts. Bacterial DNA was extracted from frozen faecal samples for 16S rRNA gene amplicon sequencing. Calves in the TRT group had greater live weights ($p = 0.02$) at weaning compared with calves in the CON group (mean = 69.18 kg, SD = 13.37 kg). The average daily live weight gain (ADG) and total feed intake were similar between the two groups. Calves in the TRT group had greater duodenum, abomasum, and reticulum weights ($p = 0.05$). Rumen and intestinal development ($p < 0.05$) and faecal microbial diversity ($p < 0.05$) were more pronounced in the TRT group. The relative abundances of eight genera differed ($p < 0.001$) between the groups. Supplementing calves with the LAB-based DFM increased live weight at weaning and had a more pronounced effect on the development of rumen and the gastrointestinal tract and on microbiota diversity and evenness. Future work is needed to better understand the potential association of LAB-DFM products on gut mucosa-associated microbiota.

Keywords: direct-fed microbials; GIT morphology; calves; live weight; diversity; microbiota

1. Introduction

The colonisation of the gut microbiota of calves and the operation of the microbiota has been recognised as significant factors influencing their growth and development [1]. During early post-natal life, the colonisation of the gut by various microbial species plays a pivotal role in nutrient metabolism, immune system maturation [2–4], and overall health [5,6]. The microbial composition and diversity in the calf's gastrointestinal tract undergoes dynamic changes, with initial colonisation by facultative anaerobes (i.e., *Lactobacillus*, *E. Coli*, and other genera of the *Enterobacteriaceae* family) [6,7], followed by the establishment of anaerobic bacterial populations (e.g., *Clostridium* spp., *Bifidobacterium* spp.) [8,9]. These microbiota alterations have been associated with shifts in energy extraction from feed, which can impact growth efficiency [10]. Furthermore, a balanced gut microbiome contributes to immune competence and defence against pathogens, ultimately influencing calves' health [5,6,11] and growth rates [10]. Further insights into the interplay between calves' gut microbiota during critical growth stages have the potential for devising targeted interventions aimed at enhancing animal health and welfare, as well as production efficiency.

The weaning age of calves holds a significant influence over subsequent productivity [12,13]. The timing for weaning [6,14] has been shown to influence the gut microbiota's structure and the maturation of the calf's gastrointestinal system as calves transition from a liquid milk diet to solid feed [6,15,16]. During this critical period, it has been shown that modification of calf feeding management could have prolonged effects on the gastrointestinal microbiota [17,18].

This dominance of *Lactobacilli* in the GIT ecosystem is of particular significance, as research has highlighted their pivotal role in modulating host defences [19,20]. Solid feed introduction triggers ruminal fermentative processes and enriches the indigenous microbiota, with a significant shift during weaning due to the alteration of nutrition sources that impact ruminal and intestinal microbiomes [1,21–23]. Rumen microbial fermentation products are pivotal for the development of rumen wall papillae, facilitating microbial colonisation as animals age. Recent studies indicate that a shift from aerobic to anaerobic microbes occurs around six weeks of age [14,24]. Within this context, the concept of utilising direct-fed microbials (DFMs) gains significance [25]. An effective DFM should ideally be tailored to support the proliferation of the indigenous microbiota that naturally inhabits the calf's GIT [1,26]. This microbial intervention can potentially influence the trajectory of GIT development and microbial colonisation, thus influencing the calf's overall health and growth [27]. Considering this, we hypothesised that the dairy calf's enteric microbiota structure and function would change in response to DFM administered as a supplement in their milk replacer diet over the entire pre-weaning period. We also investigated a secondary hypothesis: that changes in the GIT morphology and calves' growth would also occur in response to the earlier maturation of the GIT and the improved nutrient intake. Thus, the objective of this study was to characterise and compare differences in the establishment and evolution of GIT microbiota, GIT morphological changes, and the growth of dairy calves from birth to weaning in response to feeding a LAB-based DFM product as part of their daily milk diet.

2. Materials and Methods

2.1. Study Animals

This longitudinal study was conducted between June and October 2018 at the University of Queensland Gatton Commercial Dairy, Queensland, Australia (27.5636° S, 152.2800° E). The experimental protocol was reviewed and approved by the University of Queensland's Animal Ethics committee (animal ethics approval No. SVS/128/18). A detailed description of the study's methodology is described elsewhere [28]. Briefly, forty-four healthy newborn (male = 16, female = 28) Holstein-Friesian (HF; n = 26) and HF cross breed (n = 18) calves were randomly selected and enrolled. Following separation from their dams, calves were randomly assigned to individual pens in a well-ventilated, draft-free calf housing facility. Each calf was fed 2 L of high-quality colostrum (specific gravity of >1.050 [29] and IgG concentrations of >50 g/L [30]) via oral intubation within 8 to 12 h of birth [31]. The calves were provided with water (ad libitum), a calf-starter pellet (Calf Starter Crumble, Norco®, Brisbane, QLD, Australia; 20% crude protein, 8% crude fibre, 0.5% salt, 12.5 MJ/kg dry matter, 12% moisture/88% dry matter) and short chopped pasture hay from birth. The study animals were offered milk replacer (Norcovite, Norco®, Brisbane, QLD, Australia) mixed with water at the rate of 125 g/L at the same rate as the rest of the calves in the commercial dairy herd. Composition of the powdered calf milk replacer was 28% protein, 22% fat, 5% moisture, 40% lactose, and 95% dry matter. Calves were fed 4 L/d (divided over two meals) of milk replacer until three days of age. Thereafter, calves were fed milk replacer at 15% of live weight (LW), divided into two daily portions. The milk replacer was fed at 37 °C. The study animals were offered the same grain pellets as the rest of the calves in the commercial dairy herd. There were no feed additives or antibiotics added to the feed. Refused chaff and pellets were weighed and discarded every three days for each calf, and water was changed daily. Milk replacer feeding rate remained unchanged until calves were consuming at least 700 g of pellets with evidence of hay consumption for three consecutive days. At this time (approximately Day 42), calves were weaned off milk by being offered 2 L of milk only in the morning for two days, then offering 1 L each morning for three days. Weaning was considered successful if calves maintained feed intake and LW at the next weighing interval (by approximately Day 56). Calf live weights were recorded on Days 0, 14, 28, 42, and 56, and twice-daily calf health monitoring was performed by skilled workers (n = 6) throughout the study period. Using single-use disposable gloves, fresh faecal samples were collected directly from the rectum of all calves on Days 0 (before consumption of colostrum or milk replacer), 14, 28, 42, and 56, within 20 min of the morning feeding. Each sample was placed in a sterile plastic container and immediately placed and kept on ice until frozen at −80 °C.

2.2. Study Design and Animal Enrolment

This was a double-blinded, placebo-controlled study, with randomised design blocked on treatment groups (Control [CON] and Treatment [TRT] groups) to minimise the confounding effect of breed and age at enrolment. Sample size calculation was based on the following assumptions: an average live weight gain in the CON group of 0.8 kg/day with an average standard deviation of 0.2 kg/day. A total of forty-four calves were required to demonstrate a 25% difference (alpha = 0.05, power = 80%) in live weight change between the experimental groups. Prior to the start of the study, a list of all pregnant cows (n = 60) in the source herd was obtained. A simple randomisation without replacement technique was used to randomly assign treatment (CON or TRT) to the calves that were subsequently born to these cows. Treatments were administered by mixing 1 mL of the DFM supplement (liquid product containing *L. paracasei*, *L. buchneri*, and *L. casei* at a minimum of 10^9 cfu of each) or placebo (as appropriate) with approximately 25% of a calf's allotted milk replacer meal. After this treated milk was consumed, the remaining 75% of the meal was provided to the calf. All study personnel were blinded to the specific identity of the individual treatments (i.e., DFM vs. placebo) being administered to the study animals and to the treatment group assignments.

2.3. Bacterial DNA Extraction from Faecal Samples

Bacterial DNA was extracted from frozen faecal samples using the Quick-DNA™ Faecal/Soil Microbe Miniprep Kit (QIAGEN Chadstone, VI, Australia) following the manufacturer's guidelines. DNA was evaluated based on optical densities at 230, 260, and 280 nm wavelength using a NanoDrop ND-1000 spectrophotometer (Thermo Scientific NanoDrop™, Brisbane, QLD, Australia).

2.4. 16S rRNA Gene Amplicon Sequencing

16S rRNA gene amplicon sequencing of the gDNA extracted from each sample was conducted by the Australian Centre for Ecogenomics (ACE; St Lucia, QLD, Australia).

Samples were submitted in 20 μL aliquots, with a minimum concentration of 5 ng/μL, accompanied by the quantification data. The V6–V8 variable region of the 16S rRNA gene was amplified using the following forward and reverse primers (with Illumina adaptors): 926F (5′-TCG TCG GCA GCG TCA GAT GTG TAT AAG AGA CAG AAA CTY AAA KGA ATT GRC GG-3′) and 1392wR (5′-GTC TCG TGG GCT CGG AGA TGT GTA TAA GAG ACA GAC GGG CGG TGW GTR C-3′) [32]. Paired-end sequencing (2 × 300 bp) was carried out by ACE using the Illumina MiSeq platform (Illumina, San Diego, CA, USA), and each sample was sequenced to a depth of 3 GB. All fastq files were trimmed to remove primer sequence using Cutadapt (Cutadapt, version 1.2.1). Filtered de-multiplexed reads were then analysed using the DADA2 pipeline plugin to resolve reads to high-resolution amplicon sequence variants (ASVs), which represent, as closely as possible, the original biological sequence of the sequenced amplicon [33]. Trimming was performed to remove poor-quality sequences using a sliding window of 4 bases, with an average base quality above 15 for sequence quality plots as guidance. All reads were then hard-trimmed to 220 bases, and any with less than 210 bases were excluded.

Multiple sequence alignment of ASV representative sequences was carried out using DADA2. FastTree V2.1 [34] software was then used to infer unrooted and subsequently rooted maximum likelihood phylogenetic trees representing the phylogenetic relatedness of ASVs. ASVs were taxonomically classified using a downloaded Naïve Bayes classifier, pre-trained on Silva database V13.1. Following taxonomic classification, ASVs comprising <3% of all reads, found in only one sample, or classified as mitochondria or chloroplast, were removed.

2.5. Gross Pathology and Histopathology Examination

After weaning (age 56 days), three Friesian male calves from the CON (n = 3) and TRT (n = 3) groups were randomly selected for postmortem. All postmortem examinations were carried out by the same pathologist from the Veterinary Laboratory Services, University of Queensland. At postmortem, the individual compartments of forestomachs (rumen, reticulum, and omasum) and the gastrointestinal tract (abomasum, duodenum, jejunum, ileum, caecum, and colon) were isolated and ligated. The oesophagus and rectum were tied proximally and distally, respectively, to avoid leakage of ingesta. The forestomachs and abomasum were isolated, and the visceral fat was removed with surgical scissors. The total weights of the forestomachs, abomasum, and intestine (duodenum, jejunum, ileum, caecum, and colon) were weighed when they were full of digesta and again when emptied. The rumen was dissected into its component parts, as was the gastrointestinal tract. Samples (2 cm × 2 cm, approximately) of the rumen (antrum, ventral, caudal, and dorsal sacs) were collected. Within 10 cm of the pylorus, 2 cm × 2 cm of duodenal (entire length) and jejunal segments (proximal, mid, and distal) were collected. Ileum was the last small intestine collected, and it was collected within 10 cm of the caecum. The entirety of the small intestine, caecum, and large intestine were weighed, and their individual lengths were measured with ingesta. Ingesta was removed and the structures were re-weighed. The entire rumen and intestinal tissues were fixed in ~4% formaldehyde solution for morphometric measurement. Concurrently, a gross necropsy examination was performed to evaluate sections of the gastrointestinal tract (forestomach, abomasum, duodenum, jejunum, ileum, caecum, and colon). If observed, a sample of any pathology or abnormality was collected. All tissue samples were immediately placed into cassettes, identified, and then placed in a ~10× volume of 10% neutral buffered formalin for a minimum of 48 h. The fixed samples were then trimmed, pressed, and processed in paraffin wax for serial recuts of 40-micron thickness. The slides were mounted and scanned to obtain an electronic image of the histological cuts. Samples of the rumen epithelium of the antrum, ventral sac, and ventral blind sac were used for determination of papilla mitotic rate, length, width, density, and surface area. In the intestine, villus length, width, and surface area were measured in the mucosa of the duodenum, proximal, mid- and distal jejunum, and ileum. Using Leica Image-Scope, histological samples were chosen based on systematic uniform random sampling (SURS) in a correct plane of a sample. The average length and width of the tissue were measured, and a sampling interval was chosen that captured approximately 20 measurements per slide per animal. With a pooled variance of 0.1, a power of 80%, and an alpha of 5%, the sample size for detecting a difference of 0.2 in villi measurements was 160 measurements from each group.

For the histological linear measurements of the sampled structures, the area of interest was assessed based on fractionator sampling, where rectangular sampling fields were chosen using systematic random sampling. For intestinal villi (caecum, jejunum, duodenum, ileum, colon) length and for abomasal thickness measurements, a section of 20–30 grids were chosen randomly over the area of the slide where the most measurable area was positioned. Every second square was selected, and those squares with full-thickness measurable villi were then selected for measurement. Any given villus was measured from the basement membrane to the villus tip, regardless of whether that basement membrane was outside the square (Figure S1). Forestomach (reticulum, rumen, omasum) linear measurements of the villi were obtained through the measurement of all intact villi to ensure the minimum sample number needed was attained (Figure S2). When required (n = 10), histological recuts were acquired to obtain an adequate number of measurements to meet sample size requirements.

Surface area measurements were acquired through the use of a Nikon™ Eclipse® microscope linked to a Nikon™ viewing platform and Nis-Elements® software to measure histology surface area. Using the 40× objective on both microscope and software, each villus or mucosal thickness was measured by adjusting the microscope slide on the microscope to the correct sample field. As the grid fractionator was not available as part of the Nikon software and microscope layout, the sample field was chosen at random, along with samples of intact tissue, for each sample on each slide. Each villus was meticulously outlined to ensure that the correct surface area measurement was obtained (Figure S2).

For larger villi or the fronds of the forestomach, multiple adjustments with the microscope were required to measure the entire surface area of the villi. For these larger measurements, the surface area was measured consecutively and added up to give the entire surface area of the frond/villi.

2.6. Alpha/Beta Diversity Analyses

Microbial diversity was assessed by calculating the following alpha diversity metrics: Shannon's (Shannon's DI) and Simpson's (Simpson DI) Diversity Indices and observed ASVs. Evenness (a comparison of the relative abundance of each species in different samples) was calculated using Pielou's Evenness (Pielou's E). Compositional similarity/dissimilarity between samples (beta diversity) was estimated by generating weighted and unweighted UniFrac, Jaccard, and Bray–Curtis dissimilarity matrices for all pairwise sample comparisons. Compositional dissimilarity of samples was visualised using principal co-ordinates analysis (PCoA) of beta diversity distance matrices. Significant associations between alpha diversity metrics and metadata variables were tested using Kruskal–Wallis with Benjamin–Hochberg multiple test correction. Pairwise comparison of beta diversity distances between categorical metadata groups was analysed employing permutational multivariate analysis of variance (PERMANOVA), whilst significant correlations between numerical metadata categories and beta diversity distances were investigated using Mantel tests with 1000 permutations. To test for associations between longitudinal changes in alpha and beta diversity over time and for the different treatment groups we performed linear mixed-effects (LMEs) regression analysis. This accounted for subject-specific variation by using calf ID as a random effect, whilst allowing for identification of longitudinal differences in alpha/beta diversity due to treatment group by using that category as a fixed effect. The LME models were fitted with a first-order autoregressive correlation. Fitted residuals were assumed to follow a normal distribution with a mean of zero and a variance of σ^2.

Differential ASV abundances at the genus level were compared among groups (CON vs. TRT) and calf age (Days 0, 14, 28, 42, and 56). This analysis was performed in DESeq2 using CON and TRT groups as a covariate and Benjamini–Hochberg (BH) adjustment for multiple tests [35]. Open-source software PICRUSt2 (phylogenetic investigation of communities by reconstruction of unobserved states) [35] and aldex2 package [36,37] in R [38] were used on the 16S rRNA gene sequencing data to predict functional genes of the classified members of the rumen and faecal microbiota resulting from reference-based OTU picking against the Greengenes database. Predicted genes were then hierarchically clustered and categorised under Kyoto Encyclopaedia of Genes and Genomes (KEGG) [39] orthologs (KOs) and pathways [40] (level 3).

2.7. Histopathology and Live Weight Data Analyses

First-order descriptive statistics were generated from continuous variables, while categorical variables were presented as counts and percentages. For histopathology data, Student's *t*-test or Kruskal–Wallis test were used (as appropriate) to test if a given set of histopathology measurements were associated with treatment group. *p* values were adjusted for multiple comparison using Bonferroni correction method [41]. Applying Bonferroni correction, we strengthened the result reliability by rigorously adjusting *p*-values for multiple comparisons to minimise false positives [41]. The analysis was carried out using the epi.R package [42] in R [38].

A mixed-effects linear mode was fitted to the data to estimate calf's live weight as a function of calf's age (days), breed, and gender. The model was fitted with calf as a random intercept and age as a random slope. The error terms of the residuals were assumed to follow a normal distribution (mean of zero, variance of σ^2) and to follow an autoregressive correlation structure of the first order. First-order interaction terms were tested and were retained if the interaction term was significant at a likelihood ratio test *p* value of 0.05 or less. Explanatory variables were retained in the final model if they achieved statistical significance at a likelihood ratio test *p* value of 0.05 or less. Overall model fit was based on the Akaike information criterion (AIC) and visual assessment of *Pearson's* residuals against fitted values, and *Q-Q* standardised residuals against standardised normal quantiles were used to assess normality assumption. All analyses were conducted using nlme and lme4 [43] statistical packages in R.

3. Results

3.1. Live Weight Comparisons

The live weight measurements for the CON and TRT groups followed a similar trend (Table S1 and Figure S3). Calves in the TRT group had greater live weights (mean = 75 kg; standard deviation [SD] = 10 kg, $p = 0.02$) at weaning (Day 56) compared with calves in the CON group (mean = 69 kg, SD = 13 kg). Calves in the TRT group showed a tendency to be heavier on Day 42 (mean = 61 kg; SD = 8 kg; $p = 0.09$) compared with calves in the CON group (mean = 57 kg; SD = 11 kg). The average live weight gain, average daily live weight gain, average and total feed intake, and average feed efficiency were not statistically different between the groups.

3.2. Sequencing Results

The sequencing of faecal samples generated a total of 15,585,062 reads, of which 4,502,360 effective reads were ultimately analysed by the SILVA classifier after exclusion due to trimming and quality control. The number of effective merged reads per sample ranged from 7278 to 13,221 (median 9608; mean 11,032). Of these, 58% of the sequences could be assigned to the levels of phyla, class, order, and family, and 38% of the sequences could be assigned to the level of genus.

The phylogenetic classification demonstrated that bacterial communities were composed of 10 phyla, which were dominated by *Bacteroidetes* (average relative abundance = 32%), *Firmicutes* (32%), *Proteobacteria* (21%), *Euryarchaeota* (5%), *Actinobacteria* (4%), and *Tenericutes* (3%) (Figure 1A). *Seventeen* bacterial genera were most abundant over time (Figure 1C). The different bacterial genera were dominated by *Methanobrevibacter* (16%), *Prevotella_9* and *Succinivibrio* (11% each), *Escherichia/Shigella* (10%), *Faecalibacterium* (7%), *Collinsella* (7%), and *Salmonella* (7%). The phyla distribution for the CON and TRT calves followed a similar pattern, with the relative abundances for the two groups (CON, TRT) being as follows: *Bacteroidetes* (33%, 32%), *Firmicutes* (33%, 32%), *Proteobacteria* (18%, 22%), *Euryarchaeota* (5%, 4%), *Actinobacteria* (4%, 5%), and *Tenericutes* (3%, 3%).

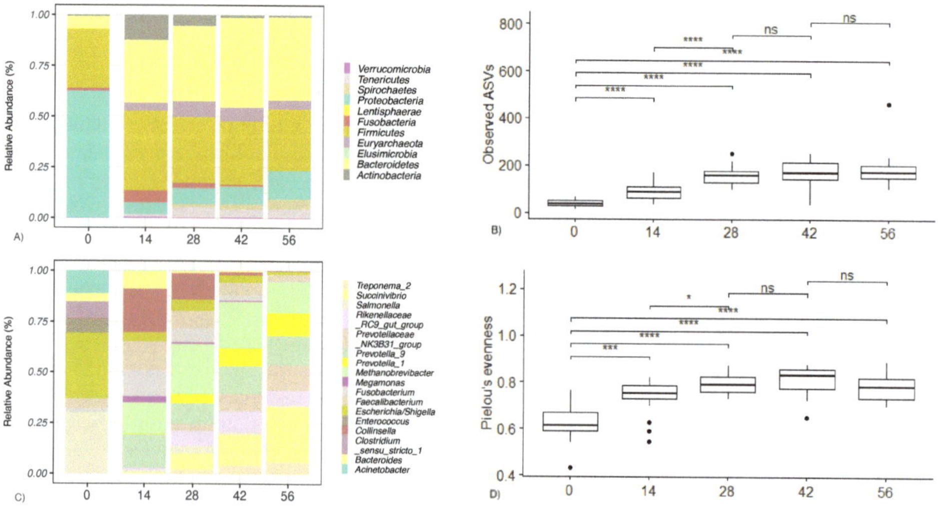

Figure 1. Changes in relative abundance of ASVs at the phyla (**A**) and genus level (**C**) in faecal samples from CON and TRT Groups at Days 0 (Birth), 14, 28, 42, and 56. Box and whisker plots showing level of microbial diversity (**B**) and evenness (**D**) of the calf faecal microbiome over time. Boxplots display the median as the middle line, whilst the perimeters of the box display the 1st and 3rd quantiles of the data. The whiskers extend to the highest and lowest values. ns not significant * $p < 0.05$ *** $p < 0.01$ **** $p < 0.001$. Solid black circles are potential outliers.

At Day 0, the faecal microbiome was dominated by *Gammaproteobacteria* (mean 62%) and *Clostridia* (mean 19%). The abundances of *Bacteroidia*, *Fusobacteriia*, and *Coriobacteriia* were substantially lower (mean 6%, 2%, and 0.2%, respectively). By Day 14, however, the levels of *Gammaproteobacteria*, exclusively represented by *Enterobacteriaceae*, had declined to a steady-state abundance from Day 14 onwards (mean 7–13%). *Bacteroidia* (composed predominantly of *Bacteriodaceae*) displayed an opposite profile, exhibiting a lower abundance at Day 0 (6%), before rapidly increasing to mean abundances of 31%, 37%, 44%, and 41% at the subsequent sampling points. This was attributable to the emergence of *Prevotellaceae*, *Bacteroidaceae*, *Rikenellaceae*, and *Paraprevotellaceae*. The *Bacilli* abundance decreased to <1% by Day 42, due to decline in *Streptococcaceae*. *Actinobacteria*, with a mean abundance of <1% at Day 0, peaked at 3% on Day 14 (due to increased *Coriobacteriaceae*), before returning to a mean of 1% on Day 28. The mean levels of *Clostridia* remained stable between 3% and 5% of the total throughout the experiment. However, during successive time points, there was a marked decline in the abundance of *Clostridiaceae* and an increased abundance of *Prevotellaceae*, *Ruminococcaceae*, *Lachnospiraceae*, and *Muribaculaceae*.

In all calves, compared to Day 0, the microbial diversity and evenness showed a curvilinear increase between each sampling time point. The increase was linear between Day 0 and Day 28 ($p < 0.05$; Figure 1B; Data S1). The levels of microbial diversity did not differ after 28 days, whilst the level of microbial evenness plateaued between Day 28 and Day 42 and significantly decreased at Day 56 ($p < 0.05$). Both microbial diversity and evenness showed a significant curvilinear association with time in all calves, indicating an increase in the overall microbial diversity, with a gradual decrease after Day 28. This enhancement and changes in diversity were also demonstrated by the increasing numbers of distinct taxa at each time point. The faecal microbiome on Day 0 comprised an average of 131 ASVs, and it contained an average of 147 ASVs on Day 56 (Figure 1B; Data S1).

The absolute and relative change in the observed microbial alpha diversity and microbial evenness followed distinct curvilinear trends ($p < 0.001$). The alpha diversity and microbial evenness increased exponentially between Day 0 and Day 42 and decreased post Day 42. On average, when compared to the CON calves, the alpha diversity in the TRT calves was higher by Day 56 ($p = 0.02$), while no significant difference in microbial evenness was observed (Figure 2A). The rate of accumulation of the microbial diversity of the CON or TRT calves did not differ (Figure 2B). However, a difference in the rate of accumulation of evenness between the CON or TRT calves was observed. Compared to Day 0, the levels of microbial evenness ($p = 0.02$) increased at a faster rate in the CON calves compared with the TRT calves (Figure 1D). By comparing the change in microbial composition from Day 0 to each subsequent time point, on average, the TRT calves had a greater change in composition relative to Day 0 than the CON calves did ($p = 0.02$). This indicates that by Day 56, the composition of the microbiome in the TRT calves was significantly more dissimilar to Day 0 than it was in the CON calves (Figure 2B). This corresponds to the differences observed in alpha diversity, which demonstrated that DFM-fed calves accumulated a greater degree of diversity when compared to Day 0 than the CON calves did (Figure 2A,B).

Moreover, the level of microbial dissimilarity was positively correlated with time, indicating that the microbial composition became significantly more divergent throughout the study period (PERMANOVA; $p < 0.001$). A strong correlation in microbial compositional abundance was explained with qualitative (unweighted UniFrac; rho = 0.62; $p < 0.001$) and quantitative dissimilarity measures (weighted UniFrac; rho = 0.51; $p < 0.001$), showing that compositional changes over time were related to both changes in the abundance of existing community members and the introduction of new species into the community. Furthermore, principal co-ordinates analysis (PCoA) of the *Jaccard* dissimilarity revealed that the faecal microbiota of all calves grouped into three distinct clusters: Day 0 (cluster 1), Days 14 and 28 (cluster 2), and Days 42 and 56 (cluster 3). This demonstrated that samples within each of these three clusters showed similar microbial community composition, but also that there was divergence in the microbial communities from the previous time period (Figure 3).

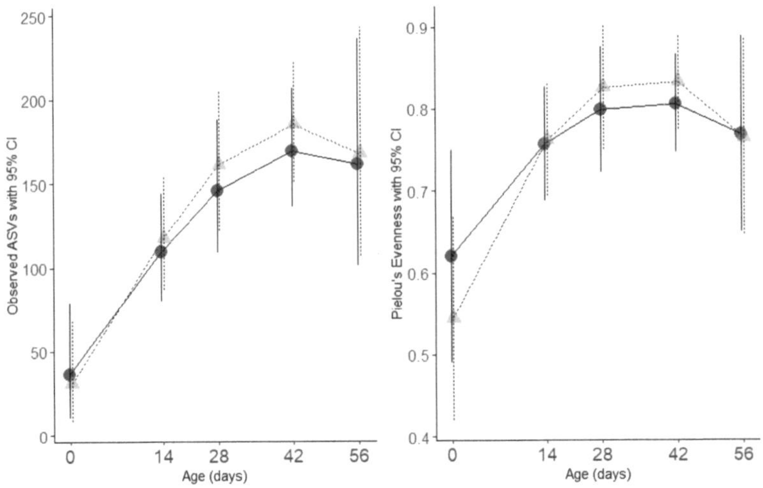

(A)

Figure 2. *Cont.*

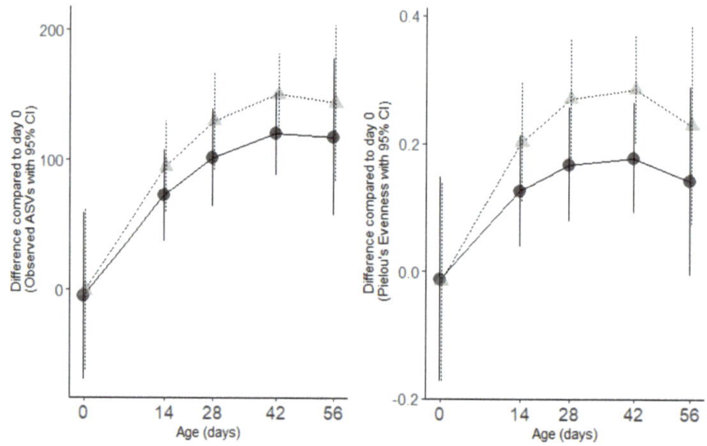

(B)

Figure 2. Line plots showing level of microbial diversity and evenness (**A**) and the difference in the level of microbial diversity (**B**) of the calf faecal microbiota over time for CON (solid grey triangles and dashed line) and TRT (solid black circle and solid black line) groups. Dashed and solid vertical lines display their 95% confidence intervals.

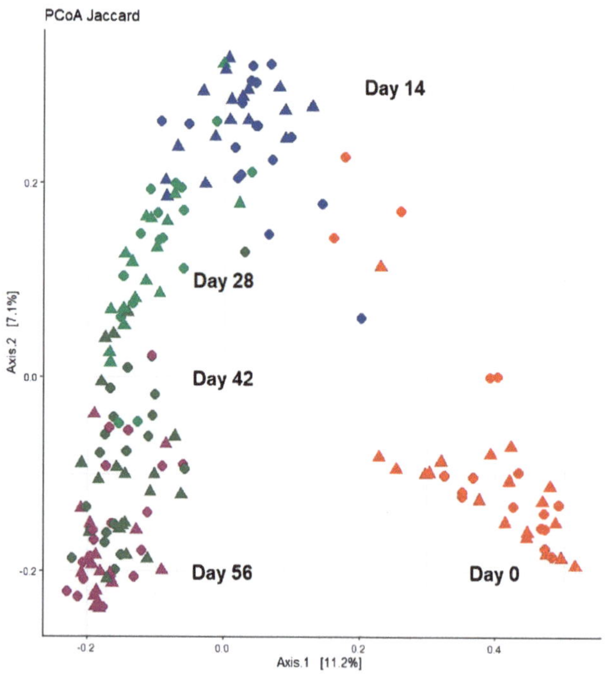

Figure 3. Clustering of calf faecal microbiome over time based on principal co-ordinates analysis (PCoA) of weighted unifrac using Bray–Curtis dissimilarity measures. Plotted using the first two principal co-ordinates, accounting for ~25% of the observed variation. Samples are coloured by individual sampling days, where solid triangles represent CON group and solid circles represent TRT group.

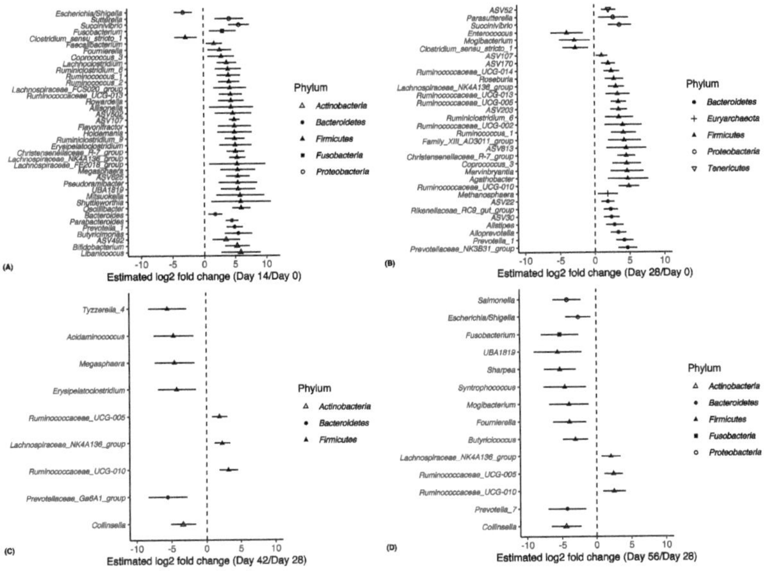

Figure 4. Significant log2-fold differences in bacterial abundance between calves' age groups are presented. Bacterial abundances on Day 14 (**A**) and Day 28 (**B**) were compared to Day 0, and those at Day 42 (**C**) and Day 56 (**D**) were compared to Day 28. Each point is shaped according to its assigned phyla and represents a bacterial genus that showed a significant difference in abundance between the age groups compared. For example, in (**A**), a negative value denotes a log2-fold reduction in that genus on Day 14 compared to Day 0. Vertical dashed lines represent the null. Adjusted *p*-values of 0.01 or less were considered significant.

3.3. Compositional Differences between Different Groups

To identify differences in the microbial composition of faecal samples between different groups, LEfSe was used to provide biomarkers at the genus level. Overall, there were 105 genera that were variably enriched in the CON and TRT groups on Days 0, 14, 28, 42, and 56 (Figure 4; Data S1). The relative abundance of eight genera differed ($p < 0.001$) between the CON and TRT groups. *Fournierella*, *Barnesiella*, and *Dialister* were significantly enriched in the CON group. *Carnobacterium*, *Pseudomonas*, *Prevotellaceae* UCG 003, *Prevotellaceae* Ga6A1 group, and *Candidatus_soleaferrea* were significantly enriched in the TRT group. The comparison between groups showed that *Bacteroidetes* were more prevalent in the TRT group, while *Firmicutes* were more prevalent in the CON group (Figure 5).

3.4. Metabolic Functions and Capacity of Ruminal and Faecal Microbiomes

To assess the metabolic potentials of faecal microbiomes, taxa were entered into PICRUSt2, and the inferred gene families were annotated against KOs and then collapsed into KEGG pathways. The gene families on Day 0 and Day 14 were collapsed into one category and used as the baseline. The gene families that were annotated to bacterial metabolism (Figure 6A) and carbohydrate metabolism (Figure 6B) declined ($p < 0.01$) post Day 28 (Figure 6A). The relative abundance of KEGG L2 pathways associated with carbohydrate metabolism did not differ ($p > 0.05$) at Day 42 and 56 but were both lower ($p < 0.01$) compared with the baseline and Day 28 (Figure 6B). A total of 288 KEGG level 3 pathways were identified, and 135 were related to carbohydrate metabolism. Of the microbiota-inferred gene families that were assigned to KEGG pathways, the four most abundant related to carbohydrate metabolism that showed variability ($p < 0.01$) between the CON and TRT groups were citrate cycle (TCA cycle; hsa00020), glycolysis/gluconeogenesis (hsa00010), galactose metabolism (hsa00052), and the pentose phosphate pathway (hsa00040; Figure 6C–F). For both the CON and TRT groups, galactose metabolism and the pentose phosphate pathway declined ($p < 0.001$) at a steady rate until Day 42. The glycolysis/gluconeogenesis pathway followed a similar trend for the CON but was stable in the TRT group ($p > 0.05$) for the first 28 days on DFM. The citrate cycle also followed the same trend. The functional pathway-level (*MetaCyc* pathway) predictions based on the metagenomics of gene families, weighted by the relative abundance of taxa, were explored for each time point and compared between the CON and TRT groups (Figure 6G). A total of 11 functional pathways were identified on Day 14 as contributing to the microbiota-wide pathway abundance (Benjamini–Hochberg [wi.eBH]-corrected $p < 0.05$). The expressions of eight functional pathways—predominantly menaquinol biosynthesis (PWY-5838, PWY-5840, PWY-5897/8/9), followed by demethylmenaquinol-8 biosynthesis (PWY-5861) and histidine, purine, and pyrimidine biosynthesis (PRPP-PWY)—were suppressed in the TRT group compared with the CON group. Conversely, fatty acid salvage (PWY-7094), tRNA processing (PWY0-1479), and ortho-cleavage pathway (Protocatechuate-ortho-cleavage-pwy) expressions were more pronounced in the CON compared to the TRT group (Figure 6F).

3.5. Gross Pathology and Histopathology Examination

The calves' attributes and study measurement characteristics at postmortem are summarised in Tables S2–S7. Within the study calves, the LW and ADG did not differ (Table S2), and no gross pathological abnormalities were observed. Compared to calves in the CON group, calves in the TRT group had greater duodenum (with digesta) weights (mean = 87 g, SD = 64 g), greater abomasum (without digesta) weights (mean = 450 g, SD = 53 g), and greater reticulum (without digesta) weights (mean = 357 g, SD = 47 g) compared with calves in the CON group (mean = 33 g, SD = 5 g, $p = 0.04$; mean = 390 g, SD = 17 g, $p = 0.05$; mean = 257 g, SD = 31 g, [Bonferroni-adjusted] $p = 0.05$, respectively; Table S3). Rumen and intestinal development adaptation were more pronounced in the TRT calves (Tables S4–S7). On average, rumen and intestinal organs' folding and crypts in the TRT calves were greater in length (mean = 1.8 mm, $p = 0.03$) and denser (2.1 unit, $p = 0.09$) than those of the CON calves (1.3 mm vs. 1.5 mm; Table S5). The calves in the TRT group had longer villi on the rumen ventral sac (TRT = 1.75 mm vs. CON = 1.31 mm, $p = 0.04$) and ileum (DFM = 0.80 mm vs. CON 0.62 mm, $p < 0.01$; Table S6). The TRT group's calves' omasum (DFM = 0.42 vs. CON = 0.66, $p = 0.02$) and colon (TRT = 0.09 vs. CON = 0.07, $p = 0.02$) villi widths were greater than those of the CON group (Table S6). The rumen's ventral sac villi width (TRT = 0.37 mm vs. CON = 0.22 mm, $p = 0.06$), caecum (TRT = 0.07 vs. CON = 0.05, $p = 0.07$) villi width, surface area of the duodenum (TRT = 0.50 μm vs. CON = 0.38 μm, $p = 0.08$), and middle jejunum (TRT = 0.42 vs. CON = 0.50, $p = 0.09$) tended to be greater in the TRT group compared with the CON (Table S7).

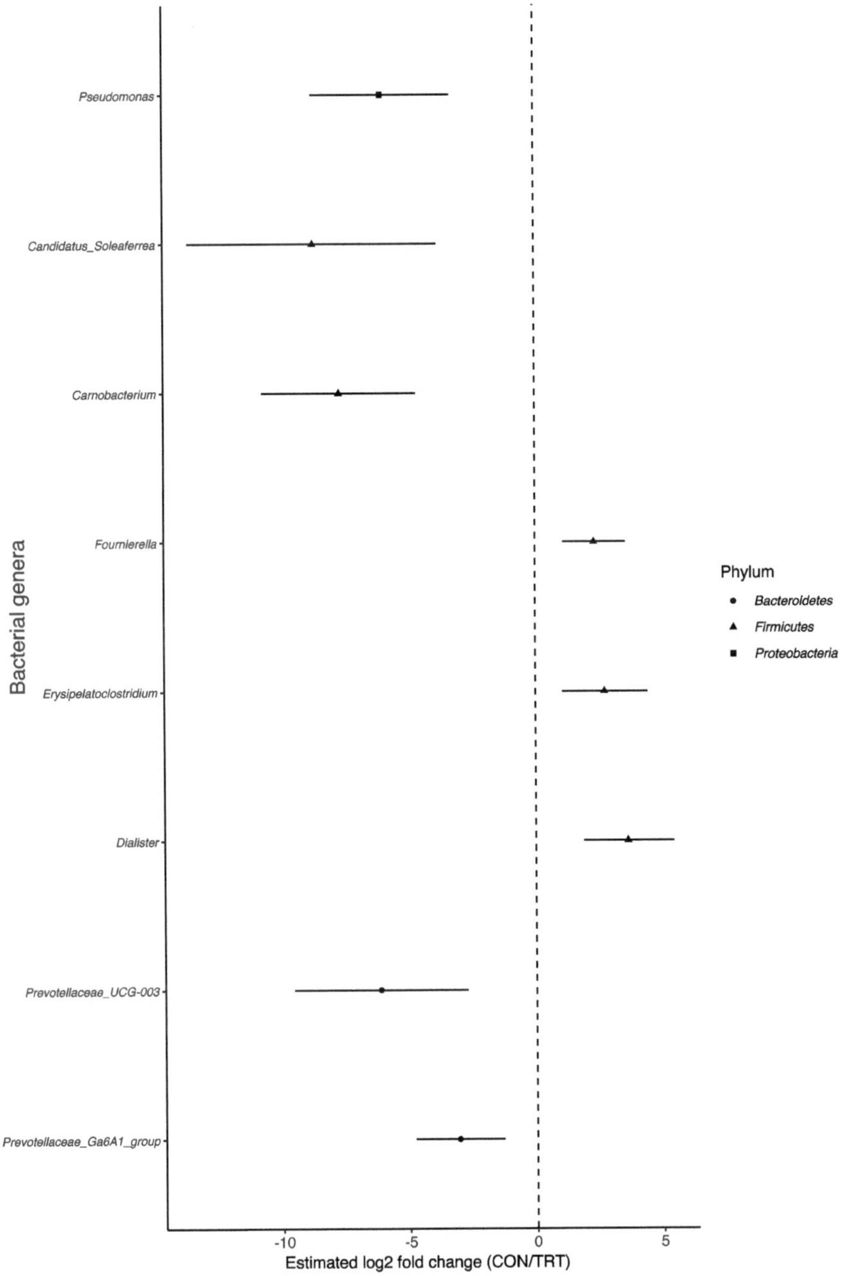

Figure 5. Significant log2-fold differences in bacterial families' abundance between CON and TRT groups. Each point is shaped according to its assigned phyla and represents a bacterial genus that showed a significant difference in abundance. Negative values denote a log2-fold reduction in that genus in the CON compared to the TRT. Vertical dashed line represent the null. Adjusted *p*-values of 0.01 or less were considered significant.

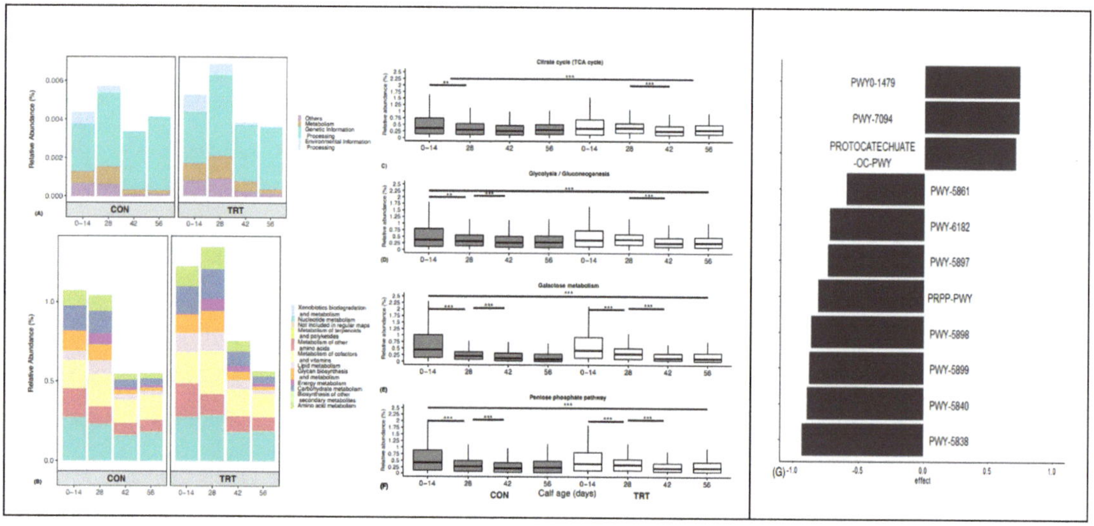

Figure 6. PICRUSt predicted summary of COG categories from 16S amplicon sequencing of faecal microbiomes. Relative abundances of level 1 (**A**) and level 2 metabolism (**B**) KEGG pathways are depicted by calf age within CON and TRT in barplots. Relative abundances of the four most abundant L3 KEGG pathways involved in carbohydrate metabolism (citrate cycle (**C**), glycolysis/gluconeogenesis (**D**), galactose metabolism (**E**), and pentose phosphate pathway (**F**)) at different ages (0–14, 28, 42, 56) for each group are presented in boxplots. Predicted functional pathways at Day 14 are shown in (**G**). ** $p < 0.01$, and *** $p < 0.001$ for the shown comparisons of L3 pathways.

4. Discussion

Early-life gastrointestinal microbial communities critically influence the growth, development, and immunity of the host [6,44]. Understanding the colonisation and succession patterns of the gut microbiota provides crucial insights into host–microbiome interactions during the post-natal phase. In the current study, we examined faecal samples of the study animals. The composition and function of the faecal microbiota are representative of the lower gut and were found to be diverse and age-dependent throughout the post-natal period, which is similar to findings reported in previous studies [6,10,45]. Changes in the microbiota structure and function are likely to be in response to the nutrient availability in the gut and could have assisted in GIT development and in the transition process from liquid to solid throughout the post-natal period until weaning [6,46].

We show here that from the first few weeks of life until weaning, there is rapid evolution and emergence of diverse species in the neonatal calf's gut microbiome. Across all faecal samples, little difference was observed between the groups in the most abundant phyla (Figure 1). Our results show that the intestinal microbiota at birth is characterised by low diversity and a relative dominance of three phyla: *Proteobacteria*, *Firmicutes*, and *Bacteroidetes*. This is in agreement with other studies in which these phyla were the most abundant in this age group [7,13,47]. The first facultative anaerobe colonisers aid the establishment of other bacterial groups, most notably *Methanobrevibacter*, *Prevotella_9*, *Succinivibrio*, and *Escherichia/Shigella*, by maintaining enteric anoxia [6,45,48]. There was numerical difference between the two groups in the most abundant phyla, notably *Proteobacteria* (a reduction of 4% in favour of the TRT group). Whilst this dominance could be considered a sign of dysbiosis [49–51], members of the *Proteobacteria* decline rapidly, and the arrival and dominance of *Firmicutes* and *Bacteroidetes* increase over time.

In the current study, calves remained healthy and did not become clinically ill during the study period. The supplementation of DFM in calves may yield the greatest advantages during periods of health challenges [52,53]. Both groups of calves in this study were pre-screened as healthy, housed in separate individual pens, and their care followed strict animal husbandry and biosecurity practices, all of which could have contributed significantly to the absence of illness. Calf's live weight at weaning (Day 56) was higher in the treatment group, in agreement with previous reports [45,48,54–56]. However, no difference in live weight was observed prior to Day 42, and differences in ADG or dry matter intake were not detected between groups. As calves grew older, those in the treatment group had a higher relative change in microbial richness and evenness indices over time compared with the control calves. These findings indicate that distinctive changes at the species level could explain differences in the live weight gains that were observed as the animal transitioned closer toward a solid feed ration [18,45,48,57]. *Carnobacterium* spp. were amongst the bacterial classes that were over-represented in the TRT group. Both *Carnobacterium* spp. and the DFM organisms are known as bacteriocin-producing bacteria that exhibit a wide spectrum of activity against known gastrointestinal pathogens such as *Listeria monocytogenes* [58–60]. As calves transition to a solid diet, the challenge to the GIT tract from exposure to the pathogens increases and puts pressure on the immune system, which in turn demands a diversion of energy and other metabolic resources. The proliferation of bacteria that are capable of producing bacteriocins increases the likelihood of inhibitory actions against GIT pathogens and spares the metabolic resource loss for growth and ongoing maturation. Further in vivo and in vitro work is required to verify these findings.

Moreover, rumen and intestinal organs' folding, crypts, papillae length, and intestinal villi density and villi length were more pronounced in the TRT calves and could have also contributed to the observed differences in weight gain at weaning. Ruminal papillae are critical in nutrient absorption [61]. The TRT calves in this study had an increased nutrient-absorptive area of their ruminal papillae compared to the control animals. The papillae growth in the DFM-treated calves in the present study could be related to increased short chain fatty acids (SCFAs) (propionate, butyrate; [52]), because SCFAs are a by-product of microbial fermentation and a healthy rumen epithelium. The observed difference in papillary growth supports the suggested study finding that the rumen is more developed in the DFM-treated calves. The effect of the DFM may have also extended to the small intestine and hind gut. In the ileum, the total height, villus height, and crypt depth were greater for the TRT calves at weaning compared with the control group. A portion of the DFM may have bypassed the developing forestomach and colonised, or otherwise influenced, the structure and perhaps function of the ileum. Taller villi, akin to papillary growth in this study, may signify an enhanced absorptive area for improved nutrient uptake within the intestinal tract.

Fournierella spp. and *Dialister* spp. were enriched in the CON calves. They have been found to interfere with the metabolism of amino acids and glucose in gastrointestinal disorders [62–64]. This, in turn, reduces the availability of nutrients that are necessary for growth and development. The predicted metagenomes (PICRUSt) of rumen and faeces microbiomes confirmed those findings and showed a significant reduction in genes associated with carbohydrate metabolism between the two groups. In agreement with previous studies [14,65], this suggests that DFM supplementation initiated a shift in the microbiome structure and function towards that of the mature rumen at an earlier point in post-natal life (approx. 6 weeks of age) compared with the control group. However, caution should be exercised when interpreting these findings. PICRUSt are dynamic predictions that depend on discerned functions within microbial communities residing in the gastrointestinal tract of humans and animals, and they are derived from comprehensive whole-genome shotgun sequencing of samples. Given the paucity of shotgun sequencing studies in ruminants, the precision of PICRUSt in gauging the functionality of the ruminant gastrointestinal microbiota may be subject to over- or under-estimation.

The results of the current study should be interpreted with care, as there is strong internal validity with limited external validity. Therefore, the findings of the current study can only be extrapolated to similar cattle populations that have also been similarly managed. In addition, the animal sample size calculation was not optimised towards microbiota analyses. Therefore, the sample size is considered small, and the study is likely to be under-powered. Larger cohorts should be evaluated in future studies. Finally, we did not incorporate biomarkers or perform microbiota analyses by other methods. Thus, it is not possible to determine the precise mechanism(s) of action or how the microbiota results correlate with the observed phenotypic and histology changes in the gastrointestinal tract. Future studies should assess these changes in a longitudinal study and work towards identifying biomarkers of interest as the animals transition onto solid feed rations. These future studies should also expand analyses such as host responses and transcriptomics to more fully understand the mechanisms by which a host's intestinal mucosa and the DFM interact.

5. Conclusions

Supplementing calves with the LAB-based DFM increased the live weight at weaning and influenced the microbiota diversity and evenness of the intestinal tract. Calves in the TRT group had greater duodenum, abomasum, and reticulum weights, as well as more pronounced rumen and intestinal development. The relative abundance of eight bacterial genera differed between the treatment groups. Future work is needed to better understand the potential association of DFM products with gut mucosa-associated microbiota.

Supplementary Materials: The following supporting information can be downloaded at https://www.mdpi.com/article/10.3390/ani14050693/s1: Figure S1: Caecum histology samples with an example of area selection for linear measurement; forestomach (reticulum, rumen, omasum) linear measurements were obtained through measurement of all intact villi to ensure that the minimum sample number needed for significance was measured (See Figure S2). To obtain an adequate number of measurements, histological recuts were acquired and measured; Figure S2: Linear measurements of forestomach example; Table S1: Descriptive statistics of live weight (kg), live weight gain (kg), average daily gain (g/d), average daily feed intake (kg/d), total feed intake (kg), and feed efficiency by experimental group (Control—CON—and Treatment—TRT). Table S2: Descriptive statistics of live weight, feed intake and body measurements of animals randomly selected from each experimental group (Control—CON—and Treatment—TRT) for postmortem and histopathology investigation; Table S3: Descriptive statistics of gastrointestinal tracts' organ weights (g) of animals randomly selected from each experimental group (Control—CON—and Treatment—TRT) for postmortem and histopathology investigation; Table S4: Descriptive statistics of gastrointestinal tracts' organ lengths (mm) of animals randomly selected from each experimental group (Control—CON—and Treatment—TRT) for postmortem and histopathology investigation; Table S5: Descriptive statistics of organ tissue's fold density and length of animals randomly selected from each experimental group (Control—CON—and Treatment—TRT) for postmortem and histopathology investigation; Table S6: Organ histopathology microscopic villi length (mm) and width (mm) and histopathological measurements of animals randomly selected from Control (CON) and Treatment (TRT) for postmortem and histopathology investigation; Table S7: Organ histopathology microscopic villi surface area (μm) of animals randomly selected from each experimental group (Control—CON—and Treatment—TRT) for postmortem and histopathology investigation; Data S1: Significant log2-fold differences in bacterial abundance between calves' age groups are presented. Bacterial abundances on Day 14 (A) and Day 28 (B) were compared to Day 0, and those at Day 42 (C) and Day 56 (D) were compared to Day 28. Each point is shaped according to its assigned phyla and represents a bacterial genus that showed a significant difference in abundance between the age groups compared. For example, in A, a negative value denotes a log2-fold reduction in that genus on Day 14 compared to Day 0. Adjusted p-values of 0.01 or less were considered significant.

Author Contributions: Conceptualisation, J.I.A.; methodology, J.I.A., R.A. and H.R.; formal analysis, J.I.A. and H.R.; investigation, J.I.A., T.O. and H.R.; interpretation of results, J.I.A., R.A., H.R. and T.O.; resources, M.S.; data curation, R.A.; histology, J.I.A. and H.R.; writing—original draft preparation, J.I.A. and H.R.; writing—review and editing, all co-authors except M.S.; visualisation, J.I.A.; supervision, J.I.A., R.A.J. and H.R.; funding acquisition, J.I.A. and M.S. All authors have read and agreed to the published version of the manuscript.

Funding: This research was funded by Terragen biotech Pty Ltd.

Institutional Review Board Statement: The animal study protocol was approved by by the University of Queensland's Animal Ethics committee (animal ethics approval No. SVS/128/18 date of approval 30 April 2018).

Informed Consent Statement: Not applicable.

Data Availability Statement: Data available on request from first author.

Acknowledgments: The authors acknowledge the inputs of Amyleigh Hewitt, Orlando Ramirerz, and all farm staff for their inputs and help in this study.

Conflicts of Interest: M.S. is employed by Terragen biotech Pty, Ltd., who funded this study. All remaining co-authors declare no conflicts of interest. The funders had no role in the design of the study; in the collection, analysis, or interpretation of data; in the writing of the manuscript; or in the decision to publish the results.

References

1. Du, W.; Wang, X.; Hu, M.; Hou, J.; Du, Y.; Si, W.; Yang, L.; Xu, L.; Xu, Q. Modulating gastrointestinal microbiota to alleviate diarrhea in calves. *Front. Microbiol.* **2023**, *14*, 1181545. [CrossRef] [PubMed]
2. Zhao, Q.; Elson, C.O. Adaptive immune education by gut microbiota antigens. *Immunology* **2018**, *154*, 28–37. [CrossRef] [PubMed]
3. Fan, P.; Bian, B.; Teng, L.; Nelson, C.D.; Driver, J.; Elzo, M.A.; Jeong, K.C. Host genetic effects upon the early gut microbiota in a bovine model with graduated spectrum of genetic variation. *ISME J.* **2020**, *14*, 302–317. [CrossRef] [PubMed]
4. Torow, N.; Hornef, M.W. The neonatal window of opportunity: Setting the stage for life-long host-microbial interaction and immune homeostasis. *J. Immunol.* **2017**, *198*, 557–563. [CrossRef] [PubMed]
5. Gensollen, T.; Iyer, S.S.; Kasper, D.L.; Blumberg, R.S. How colonization by microbiota in early life shapes the immune system. *Science* **2016**, *352*, 539–544. [CrossRef]
6. Amin, N.; Schwarzkopf, S.; Troscher-Mussotter, J.; Camarinha-Silva, A.; Danicke, S.; Huber, K.; Frahm, J.; Seifert, J. Host metabolome and faecal microbiome shows potential interactions impacted by age and weaning times in calves. *Anim. Microbiome* **2023**, *5*, 12. [CrossRef]
7. Alipour, M.J.; Jalanka, J.; Pessa-Morikawa, T.; Kokkonen, T.; Satokari, R.; Hynonen, U.; Iivanainen, A.; Niku, M. The composition of the perinatal intestinal microbiota in cattle. *Sci. Rep.* **2018**, *8*, 10437–10451. [CrossRef]
8. Dias, J.; Marcondes, M.I.; Motta de Souza, S.; Cardoso Da Mata E Silva, B.; Fontes Noronha, M.; Tassinari Resende, R.; Machado, F.S.; Cuquetto Mantovani, H.; Dill-Mcfarland, K.A.; Suen, G. Bacterial community dynamics across the gastrointestinal tracts of dairy calves during preweaning development. *Appl. Environ. Microbiol.* **2018**, *84*, 2675–2689. [CrossRef]
9. Svetlana Ferreira, L.; de Sousa Bicalho, M.; Bicalho, R.C. The *Bos taurus* maternal microbiome: Role in determining the progeny early-life upper respiratory tract microbiome and health. *PLoS ONE* **2019**, *14*, e0208014. [CrossRef]
10. Badman, J.; Daly, K.; Kelly, J.; Moran, A.W.; Cameron, J.; Watson, I.; Newbold, J.; Shirazi-Beechey, S.P. The effect of milk replacer composition on the intestinal microbiota of pre-ruminant dairy calves. *Front. Vet. Sci.* **2019**, *6*, 371. [CrossRef]
11. Robertson, R.C.; Manges, A.R.; Finlay, B.B.; Prendergast, A.J. The Human microbiome and child growth-first 1000 days and beyond. *Trends Microbiol.* **2019**, *27*, 131–147. [CrossRef]
12. Guilloteau, P.; Zabielski, R.; David, J.C.; Blum, J.W.; Morisset, J.A.; Biernat, M.; Wolinski, J.; Laubitz, D.; Hamon, Y. Sodium-butyrate as a growth promoter in milk replacer formula for young calves. *J. Dairy Sci.* **2009**, *92*, 1038–1049. [CrossRef]
13. Malmuthuge, N.; Griebel, P.J.; Guan, L.L. The gut microbiome and its potential role in the development and function of newborn calf gastrointestinal tract. *Front. Vet. Sci.* **2015**, *2*, 36. [CrossRef]
14. Meale, S.J.; Li, S.C.; Azevedo, P.; Derakhshani, H.; DeVries, T.J.; Plaizier, J.C.; Steele, M.A.; Khafipour, E. Weaning age influences the severity of gastrointestinal microbiome shifts in dairy calves. *Sci. Rep.* **2017**, *7*, 198. [CrossRef] [PubMed]
15. Bladwin VI, R.L.; McLeod, K.R.; Klotz, J.L.; Heitmann, R.N. Rumen development, intestinal growth and hepatic metabolism in the pre- and postweaning ruminant. *J. Dairy Sci.* **2004**, *87*, E55–E65. [CrossRef]
16. Cezar, A.M.; Donde, S.C.; Tomaluski, C.R.; da Silva, A.P.; Toledo, A.F.; Coelho, M.G.; Virginio Junior, G.F.; Bittar, C.M.M. Age and post-prandial variations on selected metabolites in dairy calves fed different liquid diets. *Animals* **2022**, *12*, 3063. [CrossRef] [PubMed]
17. Abecia, L.; Martin-Garcia, A.I.; Martinez, G.; Newbold, C.J.; Yanez-Ruiz, D.R. Nutritional intervention in early life to manipulate rumen microbial colonization and methane output by kid goats postweaning. *J. Anim. Sci.* **2013**, *91*, 4832–4840. [CrossRef] [PubMed]
18. Meale, S.J.; Popova, M.; Saro, C.; Martin, C.; Bernard, A.; Lagree, M.; Yanez-Ruiz, D.R.; Boudra, H.; Duval, S.; Morgavi, D.P. Early life dietary intervention in dairy calves results in a long-term reduction in methane emissions. *Sci. Rep.* **2021**, *11*, 3003. [CrossRef]
19. Jiao, J.; Wu, J.; Zhou, C.; Tang, S.; Wang, M.; Tan, Z. Composition of ileal bacterial community in grazing goats varies across non-rumination, transition and rumination stages of life. *Front. Microbiol.* **2016**, *7*, 1364. [CrossRef]

20. Walter, J. Ecological role of lactobacilli in the gastrointestinal tract: Implications for fundamental and biomedical research. *Appl. Environ. Microbiol.* **2008**, *74*, 4985–4996. [CrossRef]
21. Du, R.; Jiao, S.; Dai, Y.; An, J.; Lv, J.; Yan, X.; Wang, J.; Han, B. Probiotic *Bacillus amyloliquefaciens* C-1 improves growth performance, stimulates gh/igf-1, and regulates the gut microbiota of growth-retarded beef calves. *Front. Microbiol.* **2018**, *9*, 2006. [CrossRef]
22. Kelly, A.; McCabe, M.; Kenny, D.; Waters, S. Effect of a butyrate-fortified milk replacer on gastrointestinal microbiota and products of fermentation in artificially reared dairy calves at weaning. *Sci. Rep.* **2018**, *8*, 14901. [CrossRef]
23. Schofield, B.J.; Lachner, N.; Le, O.T.; McNeill, D.M.; Dart, P.; Ouwerkerk, D.; Hugenholtz, P.; Klieve, A.V. Beneficial changes in rumen bacterial community profile in sheep and dairy calves as a result of feeding the probiotic *Bacillus amyloliquefaciens* H57. *J. Appl. Microbiol.* **2018**, *124*, 855–866. [CrossRef] [PubMed]
24. Minato, H.; Otsuka, M.; Shirasaka, S.; Itabashi, H.; Mitsumori, M. Colonization of microorganisms in the rumen of young calves. *J. Gen. Appl. Microbiol.* **1992**, *38*, 447–456. [CrossRef]
25. Barreto, M.O.; Soust, M.; Moore, R.J.; Olchowy, T.W.J.; Alawneh, J.I. Systematic review and meta-analysis of probiotic use on inflammatory biomarkers and disease prevention in cattle. *Prev. Vet. Med.* **2021**, *194*, 105433. [CrossRef] [PubMed]
26. Virginio Junior, G.F.; Bittar, C.M.M. Microbial colonization of the gastrointestinal tract of dairy calves—A review of its importance and relationship to health and performance. *Anim. Health Res. Rev.* **2021**, *22*, 97–108. [CrossRef]
27. Mansilla, F.I.; Ficoseco, C.A.; Miranda, M.H.; Puglisi, E.; Nader-Macias, M.E.F.; Vignolo, G.M.; Fontana, C.A. Administration of probiotic lactic acid bacteria to modulate fecal microbiome in feedlot cattle. *Sci. Rep.* **2022**, *12*, 12957. [CrossRef]
28. Hewitt, A.; Olchowy, T.; James, A.S.; Fraser, B.; Ranjbar, S.; Soust, M.; Alawneh, J.I. Linear body measurements and productivity of subtropical Holstein-Friesian dairy calves. *Aust. Vet. J.* **2020**, *98*, 280–289. [CrossRef]
29. Bielmann, V.; Gillan, J.; Perkins, N.R.; Skidmore, A.L.; Godden, S.; Leslie, K.E. An evaluation of Brix refractometry instruments for measurement of colostrum quality in dairy cattle. *J. Dairy Sci.* **2010**, *93*, 3713–3721. [CrossRef]
30. Weaver, D.M.; Tyler, J.W.; VanMetre, D.C.; Hostetler, D.E.; Barrington, G.M. Passive transfer of colostral immunoglobulins in calves. *J. Vet. Int. Med.* **2000**, *14*, 569–577. [CrossRef]
31. Chigerwe, M.; Tyler, J.W.; Schultz, L.G.; Middleton, J.R.; Steevens, B.J.; Spain, J.N. Effect of colostrum administration by use of oroesophageal intubation on serum IgG concentrations in Holstein bull calves. *Am. J. Vet. Res.* **2008**, *69*, 1158–1163. [CrossRef]
32. Engelbrektson, A.; Kunin, V.; Wrighton, K.C.; Zvenigorodsky, N.; Chen, F.; Ochman, H.; Hugenholtz, P. Experimental factors affecting PCR-based estimates of microbial species richness and evenness. *ISME J.* **2010**, *4*, 642–647. [CrossRef] [PubMed]
33. Callahan, B.J.; McMurdie, P.J.; Rosen, M.J.; Han, A.W.; Johnson, A.J.; Holmes, S.P. DADA2: High-resolution sample inference from Illumina amplicon data. *Nat. Methods* **2016**, *13*, 581–583. [CrossRef] [PubMed]
34. Price, M.N.; Dehal, P.S.; Arkin, A.P. FastTree 2--approximately maximum-likelihood trees for large alignments. *PLoS ONE* **2010**, *5*, e9490. [CrossRef] [PubMed]
35. Douglas, G.M.; Maffei, V.J.; Zaneveld, J.R.; Yurgel, S.N.; Brown, J.R.; Taylor, C.M.; Huttenhower, C.; Langille, M.G.I. PICRUSt2 for prediction of metagenome functions. *Nat. Biotechnol.* **2020**, *38*, 685–688. [CrossRef] [PubMed]
36. Fernandes, A.D.; Reid, J.N.; Macklaim, J.M.; McMurrough, T.A.; Edgell, D.R.; Gloor, G.B. Unifying the analysis of high-throughput sequencing datasets: Characterizing RNA-seq, 16S rRNA gene sequencing and selective growth experiments by compositional data analysis. *Microbiome* **2014**, *2*, 15. [CrossRef] [PubMed]
37. Fernandes, A.D.; Macklaim, J.M.; Linn, T.G.; Reid, G.; Gloor, G.B. ANOVA-Like differential gene expression analysis of single-organism and meta-RNA-Seq. *PLoS ONE* **2013**, *8*, e67019. [CrossRef] [PubMed]
38. R Development Core Team. *R: A Language and Environment for Statistical Computing*; R Foundation for Statistical Computing: Vienna, Austria, 2012.
39. Kanehisa, M.; Goto, S. KEGG: Kyoto encyclopedia of genes and genomes. *Nucleic Acids Res.* **2000**, *28*, 27–30. [CrossRef]
40. Du, J.; Yuan, Z.; Ma, Z.; Song, J.; Xie, X.; Chen, Y. KEGG-PATH: Kyoto encyclopedia of genes and genomes-based pathway analysis using a path analysis model. *Mol. Biosyst.* **2014**, *10*, 2441–2447. [CrossRef]
41. Dohoo, I.R.; Martin, W.; Stryhn, W. *Veterinary Epidemiologic Research*; AVC Inc.: Charlottetown, QC, Canada, 2009.
42. Stevenson, M.; Nunes, T.; Sanchez, J.; Thornton, R. epiR: Functions for Analysing Epidemiological Data. R Package, Version 0.9-3; 2008. Epi Center, Massey University, New Zealand. Available online: http://epicentre.massey.ac.nz (accessed on 1 April 2018).
43. Bates, D. lme4: Linear Mixed-Effects Models Using S4 Classes. R Package, Version 0.99875-9. 2007. Available online: https://cran.r-project.org/web/packages/lme4/lme4.pdf (accessed on 15 February 2024).
44. Kenez, A.; Koch, C.; Korst, M.; Kesser, J.; Eder, K.; Sauerwein, H.; Huber, K. Different milk feeding intensities during the first 4 weeks of rearing dairy calves: Part 3: Plasma metabolomics analysis reveals long-term metabolic imprinting in Holstein heifers. *J. Dairy Sci.* **2018**, *101*, 8446–8460. [CrossRef]
45. Oikonomou, G.; Teixeira, A.G.V.; Foditsch, C.; Bicalho, M.L.; Machado, V.S.; Bicalho, R.C. Fecal microbial diversity in pre-weaned dairy calves as described by pyrosequencing of metagenomic 16S rDNA. Associations of *Faecalibacterium* species with health and growth. *PLoS ONE* **2013**, *8*, e63157. [CrossRef]
46. Schwarzkopf, S.; Kinoshita, A.; Kluess, J.; Kersten, S.; Meyer, U.; Huber, K.; Dänicke, S.; Frahm, J. Weaning holstein calves at 17 weeks of age enables smooth transition from liquid to solid feed. *Animals* **2019**, *9*, 1132. [CrossRef]
47. Malmuthuge, N.; Griebel, P.J.; le Guan, L. Taxonomic identification of commensal bacteria associated with the mucosa and digesta throughout the gastrointestinal tracts of preweaned calves. *Appl. Environ. Microbiol.* **2014**, *80*, 2021–2028. [CrossRef] [PubMed]

48. Kim, E.T.; Lee, S.J.; Kim, T.Y.; Lee, H.G.; Atikur, R.M.; Gu, B.H.; Kim, D.H.; Park, B.Y.; Son, J.K.; Kim, M.H. Dynamic changes in fecal microbial communities of neonatal dairy calves by aging and diarrhea. *Animals* **2021**, *11*, 1113. [CrossRef] [PubMed]
49. Hennessy, M.; Indugu, N.; Vecchiarelli, B.; Redding, L.; Bender, J.; Pappalardo, C.; Leibstein, M.; Toth, J.; Stefanovski, D.; Katepalli, A.; et al. Short communication: Comparison of the fecal bacterial communities in diarrheic and nondiarrheic dairy calves from multiple farms in southeastern Pennsylvania. *J. Dairy Sci.* **2021**, *104*, 7225–7232. [CrossRef] [PubMed]
50. Maharshak, N.; Packey, C.D.; Ellermann, M.; Manick, S.; Siddle, J.P.; Huh, E.Y.; Plevy, S.; Sartor, R.B.; Carroll, I.M. Altered enteric microbiota ecology in interleukin 10-deficient mice during development and progression of intestinal inflammation. *Gut Microbes* **2013**, *4*, 316–324. [CrossRef] [PubMed]
51. Singh, P.; Teal, T.K.; Marsh, T.L.; Tiedje, J.M.; Mosci, R.; Jernigan, K.; Zell, A.; Newton, D.W.; Salimnia, H.; Lephart, P.; et al. Intestinal microbial communities associated with acute enteric infections and disease recovery. *Microbiome* **2015**, *3*, 45. [CrossRef] [PubMed]
52. Cappellozza, B.I.; Copani, G.; Boll, E.J.; Queiroz, O. Supplementation of direct-fed microbial *Enterococcus faecium* 669 affects performance of preweaning dairy calves. *JDS Commun.* **2023**, *4*, 284–287. [CrossRef]
53. Roodposhti, P.M.; Dabiri, N. Effects of probiotic and prebiotic on average daily gain, fecal shedding of *Escherichia Coli*, and immune system status in newborn female calves. *Asian-Australas. J. Anim. Sci.* **2012**, *25*, 1255–1261. [CrossRef] [PubMed]
54. Wang, L.; Sun, H.; Gao, H.; Xia, Y.; Zan, L.; Zhao, C. A meta-analysis on the effects of probiotics on the performance of pre-weaning dairy calves. *J. Anim. Sci. Biotechnol.* **2023**, *14*, 3. [CrossRef]
55. Dick, K.J.; Duff, G.C.; Limesand, S.W.; Cuneo, S.P.; Knudson, D.K.; McMurphy, C.P.; Hall, L.W.; Bernal-Rigoli, J.C.; Marchello, J.A. Effects of a direct-fed microbial on digestive-tract morphology of Holstein bull calves and performance and carcass characteristics of Holstein steers. *Prof. Anim. Sci.* **2013**, *29*, 107–115. [CrossRef]
56. Amin, N.; Seifert, J. Dynamic progression of the calf's microbiome and its influence on host health. *Comput. Struct. Biotechnol. J.* **2021**, *19*, 989–1001. [CrossRef] [PubMed]
57. Dill-McFarland, K.A.; Breaker, J.D.; Suen, G. Microbial succession in the gastrointestinal tract of dairy cows from 2 weeks to first lactation. *Sci. Rep.* **2017**, *7*, 40864. [CrossRef] [PubMed]
58. Pilchova, T.; Pilet, M.F.; Cappelier, J.M.; Pazlarova, J.; Tresse, O. Protective effect of *Carnobacterium* spp. against *Listeria monocytogenes* during host cell invasion using in vitro HT29 model. *Front. Cell Infect. Microbiol.* **2016**, *6*, 88. [CrossRef] [PubMed]
59. Józefiak, D.; Sip, A.; Kaczmarek, S.; Rutkowski, A. The effects of *Carnobacterium divergens* AS7 bacteriocin on gastrointestinal microflora in vitro and on nutrient retention in broiler chickens. *J. Anim. Feed. Sci.* **2010**, *19*, 460–467. [CrossRef]
60. Pilet, M.F.; Dousset, X.; Barré, R.; Novel, G.; Desmazeaud, M.; Piard, J.C. Evidence for two bacteriocins produced by *Carnobacterium pisciola* and *Carnobacterium divergens* isolated from fish and active against *Listeria monocytogenes*. *J. Food Prot.* **1995**, *58*, 256–262. [CrossRef]
61. Lesmeister, K.E.; Heinrichs, A.J. Effects of corn processing on growth characteristics, rumen development, and rumen parameters in neonatal dairy calves. *J. Dairy Sci.* **2004**, *87*, 3439–3450. [CrossRef]
62. Yan, Z.; Zhang, K.; Zhang, K.; Wang, G.; Wang, L.; Zhang, J.; Qiu, Z.; Guo, Z.; Song, X.; Li, J. Integrated 16S rDNA Gene sequencing and untargeted metabolomics analyses to investigate the gut microbial composition and plasma metabolic phenotype in calves with dampness-heat diarrhea. *Front. Vet. Sci.* **2022**, *9*, 703051. [CrossRef]
63. Wright, E.K.; Kamm, M.A.; Teo, S.M.; Inouye, M.; Wagner, J.; Kirkwood, C.D. Recent advances in characterizing the gastrointestinal microbiome in Crohn's disease: A systematic review. *Inflamm. Bowel Dis.* **2015**, *21*, 1219–1228. [CrossRef] [PubMed]
64. An, G.; Zhang, Y.; Fan, L.; Chen, J.; Wei, M.; Li, C.; Chen, X.; Zhang, L.; Yang, D.; Wang, J. Integrative analysis of vaginal microorganisms and serum metabolomics in rats with estrous cycle disorder induced by long-term heat exposure based on 16s rDNA gene sequencing and LC/MS-based metabolomics. *Front. Cell Infect. Microbiol.* **2021**, *11*, 595716. [CrossRef] [PubMed]
65. Meale, S.J.; Li, S.; Azevedo, P.; Derakhshani, H.; Plaizier, J.C.; Khafipour, E.; Steele, M.A. Development of ruminal and fecal microbiomes are affected by weaning but not weaning strategy in dairy calves. *Front. Microbiol.* **2016**, *7*, 582. [CrossRef] [PubMed]

animals

MDPI

Article

Comparative Analysis of Maternal Colostrum and Colostrum Replacer Effects on Immunity, Growth, and Health of Japanese Black Calves

Marimu Urakawa [1,†], Mahmoud Baakhtari [1,2,†], Amany Ramah [1,3], Shoichiro Imatake [1], Parnian Ahmadi [2,4], Yuichiro Deguchi [5], Mizuho Uematsu [5], Yoshiki Nakama [5], Kazunari Imabeppu [5], Yusuke Nomura [5] and Masahiro Yasuda [1,4,*]

1 Laboratory of Veterinary Anatomy, Faculty of Agriculture, University of Miyazaki, Miyazaki 889-2192, Japan; mahmoud.baakhtari@yahoo.com (M.B.)
2 Faculty of Veterinary Science, Balkh University, Mazar-i-Sharif 1703, Afghanistan
3 Department of Forensic Medicine and Toxicology, Faculty of Veterinary Medicine, Benha 13518, Egypt
4 Graduate School of Medicine and Veterinary Medicine, University of Miyazaki, Miyazaki 889-2192, Japan
5 Miyazaki Agricultural Mutual Aid Association, Miyazaki 880-0877, Japan
* Correspondence: yasudaja@cc.miyazaki-u.ac.jp; Tel.: +81-985-58-7264
† These authors contributed equally to this work.

Simple Summary: The immunity, growth, and health of Japanese black calves was evaluated via feeding maternal colostrum versus colostrum replacer. Calves that received maternal colostrum showed significant increases in T cells and B cells in the peripheral blood. In addition, the expression levels of interleukin-1β–, interleukin-2–, and interferon-γ–encoding mRNAs were significantly higher in the maternal colostrum group. A lower incidence of disease and higher carcass weight in calves fed maternal colostrum were observed. These results suggest that, compared to maternal colostrum, the use of colostrum replacers may result in delayed immune system activation and slower growth in calves.

Abstract: Maternal colostrum (MC) is an important source of nutrients and immune factors for newborn calves. However, when colostrum is unavailable or of poor quality, a colostrum replacer (CR) may be a suitable alternative to MC. As stock-raising farmers must make informed decisions about colostrum feeding management, this study was conducted to determine the effect of feeding MC versus CR on the promotion of immunological status, growth, and health in pre-weaned Japanese black (JB) calves. Sixteen newborn JB calves were fed MC after birth, and 16 JB calves were fed CR. For the MC group, the numbers of γδ T cells, $CD4^+$ cells, $CD8^+$ cells, $CD4^+CD8^+$ cells, B cells, and MHC class II^+ cells were significantly higher compared with the CR group. Furthermore, the expression levels of interleukin (IL)-1β-, IL-2-, and interferon-γ (IFN-γ)-encoding mRNAs were significantly higher in the MC group compared with the CR group. A lower incidence of disease in 1-month-old calves and higher carcass weight in the MC group were observed compared with the CR group. These results suggest that CR activates the immune system delayed in calves compared with MC. MC increases populations of various immunocompetent cells, which can reduce infection rates and improve body weight gain.

Keywords: colostrum; growth; health; immunity; Japanese black calf

1. Introduction

During the gestation period in cows, the transplacental transmission of immunoglobulins is impeded by the placental structure. Consequently, the acquisition of passive immunity in neonatal calves depends on maternal colostrum (MC) [1]. Calf health, future productive life, and farm profitability are directly related to the management of colostrum and the successful transfer of passive immunity [2]. Colostrum contains high concentrations of nutrients, immunoglobulins, and cytokines and is widely acknowledged for its role in bolstering disease resistance [3,4]. However, the quality and quantity of colostrum are contingent upon on the health status of the producing cow [5].

In instances in which colostrum is either unavailable or of poor quality, an appropriate alternative for MC would be a colostrum replacer (CR) [6]. The prevalent adoption of early weaning techniques has led to the widespread use of CR among livestock farmers. CR is designed to replace MC and encapsulates bovine immunoglobulins derived typically from colostrum or plasma [7]. However, the utilization of CR entails additional costs associated with milk replacement and requires supplementary efforts for the suckling of calves [8]. Although CR has advantages in terms of consistent composition and quality, its impact on the well-being of calves is characterized by variability [9,10].

The use of CR emerged from concerns regarding the potential presence of pathogenic agents in colostrum [7]. Previous studies documented a reduction in the incidence of Johne's disease through the administration of CR [11]. Similarly, reports indicate a nearly 50% reduction in the chance of infection with *Mycobacterium avium* ssp. *paratuberculosis* through the feeding of CR [11]. However, reports conflict regarding the success of passive transfer when utilizing CR. Failure in the transfer of passive immunity is one of the most important risk factors and predictors of morbidity and mortality among dairy calves [12]. Poulsen et al. (2010) conducted a comparative analysis of passive transfer in bovines fed MC and CR and found no significant differences [13]. Conversely, Swan et al. (2007) reported the failure of passive transfer in calves fed CR due to the insufficient absorption of immunoglobulins, thereby potentially impairing health and survival [8]. To prevent or mitigate the failure of passive immunity transfer in neonatal Japanese black (JB) calves, information regarding the impact of MC and CR on immunological parameters is essential and would enable producers to make more-informed decisions regarding colostrum feeding management. Therefore, the purpose of this study was to determine the effect of feeding a commercially available CR versus MC on the immunological profile, growth, and health of pre-weaned JB calves.

2. Materials and Methods

All procedures used in this study were conducted according to protocols approved by the Animal Care and Use Committee of the University of Miyazaki (approval no. 2019-001-04).

2.1. Animals and Experimental Design

Newborn JB calves, sourced from a commercial JB dairy company in Miyazaki, Japan, which operates four farms (A, B, C, and D), were methodically allocated into two groups. Sixteen calves from farms A and B received MC. Simultaneously, 16 calves from farms C and D were administered CR. The dams were typically grouped in a large room of 3 to 4 cows and housed in separate rooms at the end of pregnancy and post-calving. Cows in maternity rooms were monitored continuously by veterinary staff to determine the calving time. Delivery rooms dedicated to pregnant cows were specifically designed on each farm. The enrolled calves were delivered naturally without any abnormalities and consistently monitored thereafter. Calves were allowed to stay in the maternity stall for the initial days. Calves were weaned at 60 days post-birth and then housed in groups of 5 to 6 animals in a designated room. All four farms adhered to a unified management system for JB calves from calving to market age. The history of respiratory diseases and treatment times during the first month of life were recorded in both groups by veterinary staff, and the carcass weight at 30 months of age was estimated after slaughtering at slaughterhouses.

2.2. Colostrum Feeding

MC was fed within the initial 10 h post-birth (maximum 3 L of colostrum) until weaning started. CR was fed within the initial 12 h post-birth (maximum 4.2 L containing 150 g of colostrum-derived IgG, which is a requisite amount of IgG for achieving successful passive transfer in newborn calves [14]) until weaning started. The CR (Elanco Co., Tokyo, Japan) was stored under refrigerated conditions before use and administered in three 750 g doses (between 2 h, between 6 h, and between 12 h) post-birth. Prior to administering CR, a container of bottled water was placed in a bucket filled with hot tap water for warming. Temperature was verified using a thermometer 10 min before feeding to ensure that the water reached the specified target of 41 °C. The mixing bucket, containing water, received 750 g of CR powder in three doses. Subsequently, the prepared mixture was transferred into feeder containers. Research technicians were responsible for the tasks associated with CR preparation. The CR was manufactured using carefully selected high-quality cow colostrum as a raw material. The inherent variability of colostrum is standardized to a certain quality to ensure a high level of safety, and the positive outcomes attributed to the utilization of the CR were evidenced in previous study [15].

2.3. Feeding Management

A uniform feeding regimen, consisting of starter grain, hay, and water, was introduced to calves 2 days post-birth, maintaining consistency throughout the perinatal period on all four farms. Dams and calves were treated using the standard feeding system for beef cattle [16]. During the fattening period, all calves were moved to a single large fattening area and fed a uniform amount of food. Throughout the pregnancy period, mother cows received 6 kg of hay and 1 kg of concentrated food, 113% total digestible nutrients (TDN) and 96% crude protein (CP). In the final 30 days before calving, the regimen was adjusted to 6 kg of hay and 2.5 kg of concentrated food (112% TDN and 94% CP). During the postpartum period following calving, dams were fed 6 kg of hay and 1 kg of concentrated food (TDN 141% and CP 133%) (Table S1).

2.4. Sampling and Purification of Peripheral Blood Mononuclear Cells (PBMCs)

Blood samples were collected from newborn calves at the ages of 1, 2, and 3 months and transported from the farms to the University of Miyazaki in sterile tubes. Ficoll-Paque PLUS (GE Healthcare UK, Little Chalfont, Buckinghamshire, UK) and NH_4Cl lysis buffer [17] were used to purify PBMCs and remove red blood cells, respectively. Finally, phosphate-buffered saline (PBS) (Fujifilm Wako Pure Chemical Co., Tokyo, Japan) was used to wash the remaining cells.

2.5. Lymphocyte Populations

PBMCs were suspended in PBS with 0.5% bovine serum albumin (Nacalai Tesque, Inc., Kyoto, Japan) and 0.05% sodium azide (Fujifilm Wako Pure Chemical Co., Tokyo, Japan) (BSA-PBS). The trypan blue exclusion test was used to detect and count viable cells (at densities ranging from 1×10^5 to 1×10^6 cells). Viable cells were then treated at 4 °C for 1 h with fluorescently labeled monoclonal antibodies (mAbs). After washing three times with BSA-PBS, the cells incubated with mAbs were resuspended in BSA-PBS. A FACS CantoTM II system (Becton Dickinson, Franklin Lakes, NJ, USA) was used to determine the relative immunofluorescence intensities of labeled cell populations. Anti-CD4 ILA11A, anti-CD8 CC63, anti-γδ TCR GB21A, anti-MHC class II TH14B (200-fold dilution, Monoclonal Antibody Center at Washington State University, Pullman, WA, USA), and anti-IgM BIG73A (100-fold dilution, Monoclonal Antibody Center at Washington State University) antibodies were used (Table S2). A FITC labeling kit-NH2 (Dojindo Laboratories, Kumamoto, Japan), HiLyteTM Fluor 555 (F555) labeling kit-NH2 (Dojindo Laboratories), and HiLyteTM Fluor 647 (F647) labeling kit-NH2 (Dojindo Laboratories) were utilized to mark fluorescently labeled mAbs, as directed by the manufacturer. Positive cell density (cells/μL) was defined as the number of white blood cells (WBCs) counted using pocH-100iV Diff (Sysmex, Hyogo, Japan) × the percentage of lymphocytes counted by Giemsa staining of blood smears × the subset population determined by FACS.

2.6. Calculation of the Phagocytosis Index

At the age of 1 month, buffy coat was collected from blood samples. NH_4Cl lysis solution was used to remove red blood cells. The remaining cells were rinsed with PBS, and RPMI 1640 medium (Fujifilm Wako Pure Chemical Co.) with 10% fetal calf serum (Sigma-Aldrich, St. Louis, MO, USA) and antibiotics (Fujifilm Wako Pure Chemical Co.) was used to suspend the cells (4.0×10^6 cells/mL). The specific conditions (37 °C for 1 h in a 5% CO_2 humidified atmosphere) were adjusted to incubate the mixture of cells with 2.5% suspension of latex beads suspension labeled with FITC (1 μm diameter, L1030, Sigma-Aldrich). Ice-cold 1 mM EDTA-PBS was used to remove cell-free beads from the mixture. The remaining cells were incubated with F647-labeled anti–MHC class II (TH14B) mAbs at 200-fold dilution, and F555-labeled anti-granulocyte (CH138A, Monoclonal Antibody Center at Washington State University) mAbs at 100-fold dilution (Table S2), followed by analysis on a FACS CantoTM II system. The percentage of $FITC^+$ $TH14B^+$ cells relative to all $TH14B^+$ cells and the percentage of $FITC^+$ $CH138A^+$ cells relative to all $CH138A^+$ cells were used to determine the phagocytic index.

2.7. Proliferation of Lymphocytes

PBMCs (2×10^5 cells/well) in RPMI 1640 medium from 1-month-old calves were stimulated at 37 °C for 72 h in a 5% CO_2 humidified atmosphere with concanavalin A (Con A, Sigma-Aldrich) at a final concentration of 3.13 μg/mL and phytohemagglutinin (PHA; Sigma-Aldrich) at a final concentration of 15.6 μg/mL. A total of 20 μL of 3-(4,5-dimethylthiazol-2-yl)-2,5-diphenyl tetrazolium bromide (MTT, 5 mg/mL, Fujifilm Wako Pure Chemical Co.) was added to each well 4 h before the end of culture. At the end of culture, formazan crystals were collected and dissolved with dimethyl sulfoxide. A microplate reader (Benchmark Plus, measurement wavelength 570 nm, reference wavelength 610 nm, Bio-Rad, Hercules, CA, USA) was used to measure the optical density (OD), and the OD of the experimental group divided by the OD of the control group was defined as the stimulation index (SI) as a measure of lymphocyte proliferation.

2.8. RNA Extraction and Expression of mRNAs

An RNeasy Mini kit (Qiagen, Valencia, CA, USA) and one-step TB Green PrimeScript PLUS RT-PCR kit (Takara Bio., Tokyo, Japan) were used according to the manufacturers' instructions to extract RNA from PBMCs from 1-month-old calves and perform real-time RT-PCR, respectively. Oligo7 software (Molecular Biology Insights, Colorado Springs, CO, USA) was used to design the primer pairs (Table 1). A QuantStudio™ Real-Time PCR system (Applied Biosystems, Carlsbad, CA, USA) was used, and the conditions for real-time PCR were as follows: reverse transcription for 5 min at 42 °C, initial PCR activation for 10 s at 95 °C, and 40 cycles of 5 s at 95 °C, 30 s at 57 °C, and 30 s at 70 °C; a dissociation curve was then generated according to the results. Glyceraldehyde phosphate dehydrogenase (GAPDH) was used to normalize the mRNA expression levels of other target genes, and the expression levels of GAPDH were not significantly different among the samples. The comparative Ct method ($2^{-\Delta\Delta Ct}$ method/Livak method) and QuantStudio™ software v1.x series (Thermo Fisher Scientific, Waltham, MA, USA) were used to measure target mRNAs and analyze the data, respectively [18].

2.9. Statistical Analysis

Statistical analysis of differences between the MC and CR groups was performed using the Mann–Whitney *U*-test. Results are expressed as the mean ± SD. p-values < 0.05 were considered statistically significant. Statistical analysis was performed using the statistical software package SPSS for Windows (version 20.0, SPSS Inc., Chicago, IL, USA).

Table 1. Sequences of primers used for real-time PCR.

Gene	Primer	Sequences	Accession Number	Length (Base Pairs)
GAPDH	F R	GTTCAACGGCACAGTCAAGGCAGAG ACCACATACTCAGCACCAGCATCAC	NM_001034034	123
IL-1 β	F R	GCCTACGCACATGTCTTCCA TGCGTCACACAGAAACTCGTC	NM_174093	111
IL-2	F R	TGCTGGATTTACAGTTGCTT TCAATTCTGTAGCGTTAACCT	XM_024976996	111
IL-4	F R	ATCAAAACGCTGAACATCCTC TCCTGTAGATACGCCTAAGCTC	NM_173921	142
IL-6	F R	AGCTCTCATTAAGCGCATGG ATCGCCTGATTGAACCCAG	NM_173923	168
IL-10	F R	GGCCTGACATCAAGGAGCAC CTCTTGTTTTCGCAGGGCAGA	NM_174088	103
IFN-γ	F R	TGATTCAAATTCCGGTGGAT TCTTCCGCTTTCTGAGGTT	NM_174086	108

3. Results

3.1. Lymphocyte Subset Analysis

The numbers of $CD4^+$ (Figure 1A), $CD8^+$ (Figure 1B), and $CD4^+CD8^+$ (Figure 1C) cells in the MC group were significantly higher ($p < 0.05$) compared with the CR group at 1 month of age; however, there were no significant differences between the groups at 2 and 3 months of age. Additionally, the number of γδ T cells in the MC group was significantly higher ($p < 0.05$) than that in the CR group at 1 and 2 months of age (Figure 1D). Furthermore, the numbers of B cells (Figure 2A) and MHC class II^+ cells (Figure 2B) in the MC group were significantly higher ($p < 0.01$ and $p < 0.05$, respectively) compared with the CR group at 3 months of age, but there were no significant differences at 1 and 2 months of age. The number of WBCs was slightly higher in the MC group for all 3 months compared to the CR group; however, the difference between groups was not significant (Figure S1).

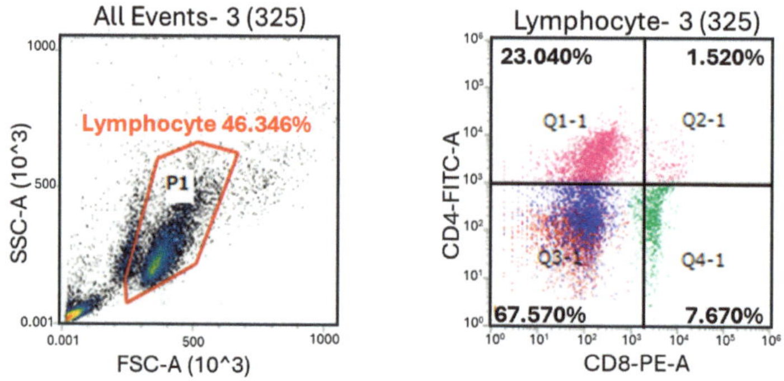

Figure 1. *Cont.*

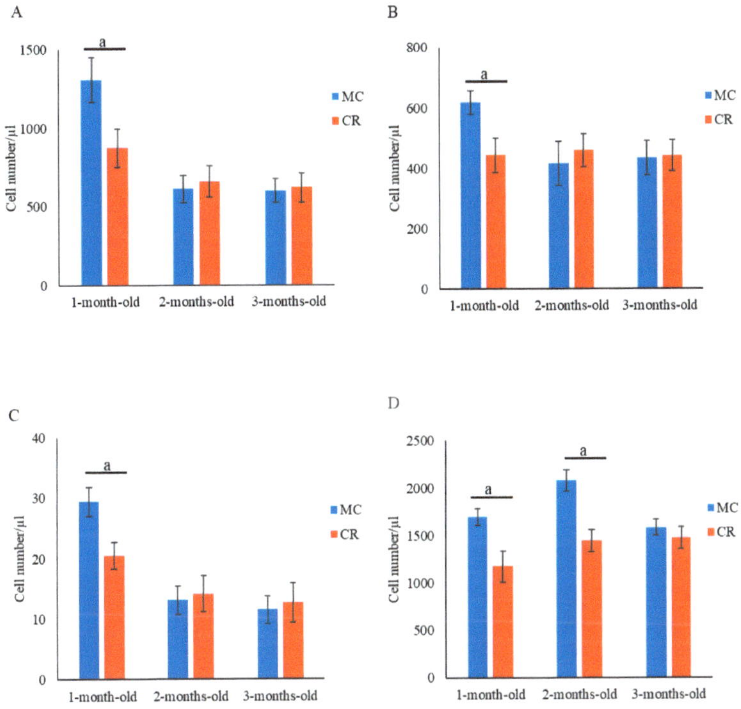

Figure 1. Relative population of CD4$^+$ and CD8$^+$ cells in the blood of newborn JB calves. The lymphocyte gate (P1) using forward scatter (FSC) and side scatter (SSC) for flow cytometry is shown in the upper left. Two-parameter dot pots used for gating the single- (Q1-1 and Q4-1) and double-positive (Q2-1) cell populations are shown in the upper right. The effect of MC and CR feeding on T cell populations in JB calves and populations of CD4$^+$ cells (**A**), CD8$^+$ cells (**B**), CD4$^+$CD8$^+$ cells (**C**), and γδ T cells (**D**) were recorded in 1-, 2-, and 3-month-old calves. Data are presented as the mean ± SD. Letters (a: $p < 0.05$ MC vs. CR) indicate significant differences.

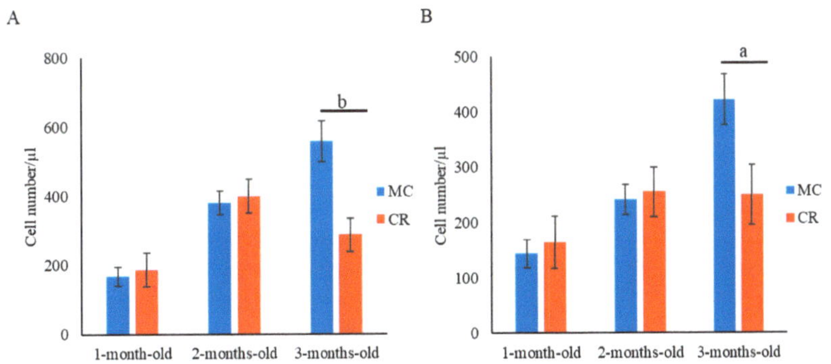

Figure 2. Effect of MC and CR feeding on B cell populations in JB calves. Populations of B cells (**A**) and MHC class II$^+$ cells (**B**) were recorded in 1-, 2-, and 3-month-old calves. Data are presented as the mean ± SD. Letters (a: $p < 0.05$ and b: $p < 0.01$ MC vs. CR) indicate significant differences.

3.2. Phagocytosis and Lymphocyte Proliferation

The phagocytic index of granulocytes and MHC class II$^+$ monocytes did not differ significantly between the MC and CR groups (Figure 3A). Similarly, there was no significant change in the lymphocyte proliferative response to Con A and PHA between groups, but the responses to Con A and PHA were slightly higher in the MC group than the CR group (Figure 3B).

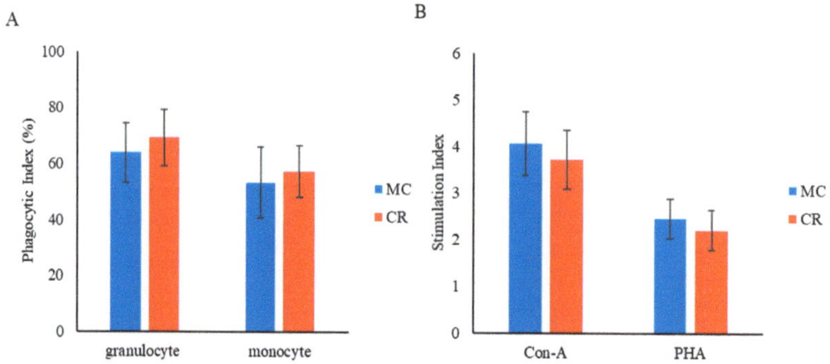

Figure 3. Effect of MC and CR feeding on the phagocytic index of granulocytes and monocytes (**A**) and effect of MC and CR feeding on lymphocyte proliferation in response to Con A and PHA treatment in JB calves (**B**). The phagocytic and stimulation indexes of mitogens were recorded in 1-month-old calves. Data are presented as the mean ± SD.

3.3. Expression of Cytokine-Encoding mRNAs

The expression of cytokine-encoding mRNAs in the MC and CR groups was analyzed at 1 month of age. The levels of IL-1β, IL-2, and IFN-γ mRNAs were significantly higher ($p < 0.01$, $p < 0.01$, and $p < 0.05$, respectively) in the MC group than the CR group (Figure 4).

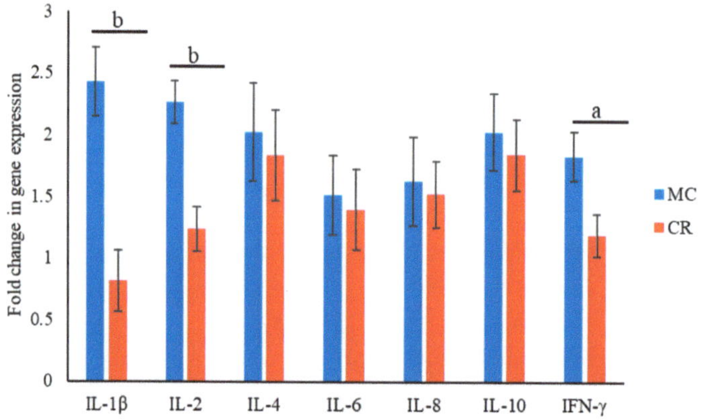

Figure 4. Effect of MC and CR feeding on the expression of mRNAs encoding IL-1β, IL-2, IL-4, IL-6, IL-10, and IFN-γ. The fold-change in the gene expression was measured in 1-month-old JB calves. Expression levels were measured using real-time PCR. Data are presented as the mean ± SD. Letters (a: $p < 0.05$ and b: $p < 0.01$ MC vs. CR) indicate significant differences.

3.4. Respiratory Disease Treatment and Carcass Weight Measurement

The frequency of treatment of calves for respiratory diseases decreased in the MC group. In 1-month-old calves, the frequency of treatments (Figure 5A) for respiratory diseases was significantly lower ($p < 0.01$) in the MC group than the CR group. Moreover, the carcass weight of calves in the MC group was significantly higher ($p < 0.05$) than that of calves in the CR group (Figure 5B).

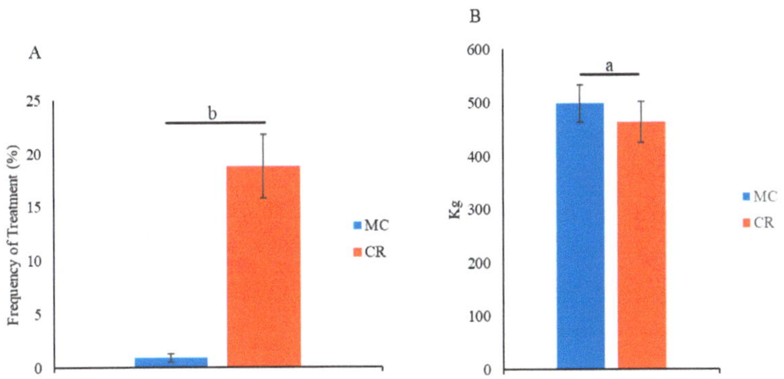

Figure 5. Effect of MC and CR feeding on treatment for respiratory diseases and carcass weight in JB calves. The frequency of treatments was recorded in 1-month-old JB calves by the Japanese Mutual Aid Association (NOSAI) veterinary staff (**A**), and carcass weight was recorded at the age of 30 months after slaughtering (**B**). Data are presented as the mean ± SD. Letters (a: $p < 0.05$ and b: $p < 0.01$ MC vs. CR) indicate significant differences.

4. Discussion

Colostrum immunity-related components such as lymphocytes are readily transferred to newborn calves, in which they circulate and stimulate neonatal immune functions [19–21]. In the present study, the numbers of $CD4^+$, $CD8^+$, and $CD4^+CD8^+$ cells in calves during the first month of life were significantly higher in the MC group. $CD4^+$ cells are the primary cell fraction for acquired immunity, and memory $CD8^+$ T cells are the principal component of immunity in the defense against viral infection [22]. Likewise, $CD4^+CD8^+$ cells reportedly play a role in several autoimmune diseases, virus infections, and cancer [23–25]. Regarding the pro-inflammatory and anti-inflammatory functions of $CD4^+CD8^+$ cells, Diedrich et al. [26] suggested in response to *Mycobacterium tuberculosis* infection that $CD4^+CD8^+$ cells play an important role in immune reactions. The observed increases in lymphocyte subset populations in newborn calves suggest that components present in MC enhance calf immune responses.

A high proportion of $\gamma\delta$ T cells among the circulating lymphocyte population is a characteristic of healthy immune conditions in young ruminants [27]. In our study, the number of $\gamma\delta$ T cells in the MC group was significantly higher in 1- and 2-month-old calves. Lundberg et al. [28] and Murakami et al. [29] reported that $\gamma\delta$ T cells in peripheral blood play an important role in anti-viral responses. $\gamma\delta$ T cells also appear to play an important role in bridging the innate and adaptive immune responses [30]. The low number of $\gamma\delta$ T cells observed in the CR group in the present study might be unfavorable for the immune system. Fewer $\gamma\delta$ T cells may be associated with susceptibility to infection in JB calves [31]. Therefore, the increase in the number of $\gamma\delta$ T cells resulting from MC consumption may enhance immunity in calves after birth and increase their disease resistance.

Blood mononuclear leukocyte populations in newborn calves are characterized by a higher proportion of $\gamma\delta$ T cells [32,33] and lower proportion of B cells [34]. In our study, the numbers of B cells and MHC class II^+ cells in MC group calves were lower at 1 and 2 months of age compared with 3 months. At the age of 3 months, the numbers of these cells were significantly higher compared with the CR group. The percentage of B cells increased over time, most notably between 32 and 60 days of age. Similar age-dependent increases in the proportion of circulating B cells in young calves have been reported [34], which is typical of the maturing immune system in calves. Thus, the increases in B cell and MHC class II^+ cell populations in 3-month-old calves suggest that MC contributes to the development of calf immunity even after passive immunity has elapsed.

Phagocytosis plays an important role in early host defense, and some bioactive factors in colostrum, such as immunoglobulins [35], complement components [36], lactoferrin [37], cytokines [33], and other as yet unknown factors [38], are thought to function as phagocytosis-activating agents. In this study, the phagocytic indexes of granulocytes and monocytes did not differ significantly between the MC and CR groups. It was previously reported that human colostrum contains a phagocytosis-promoting factor [39], and the absence of any significant changes in phagocytosis between the MC and CR groups could be related to the short lifespan of neutrophils. Neutrophils, with a shorter lifespan compared to lymphocytes [40], may exert less of an effect in colostrum on phagocytic efficacy by the time samples are collected at 1 month of age. In addition, in response to stimulation of lymphocytes by Con A and PHA, no significant changes were observed between the MC and CR groups. A previous study reported that increasing the nutrition regime to a higher level in young calves has no effect on improving the general responsiveness of PBMC populations in response to mitogenic stimulation [41]. Therefore, the present results suggest that both MC and CR may provide sufficient protein and energy to maintain or promote the general responsiveness of lymphocyte proliferation.

Immunoglobulins are one of the main components of total protein in blood and play a pivotal role in fostering both passive and adaptive immune responses in neonatal animals. MC is an important source of diverse immunoglobulins and facilitates their transfer to newborn animals through feeding. Therefore, to evaluate the efficacy of colostrum, it is essential to measure gene expression levels and the protein concentration. Recent studies in calves and piglets have demonstrated a significant increase in protein concentrations, particularly the concentration of immunoglobulins, following the administration of MC to neonatal animals [42,43]. In the context of immune enhancement by colostrum, we extended the investigation to the expression levels of cytokine-encoding mRNAs in the present study. MC significantly stimulated the systemic expression of transcripts encoding IFN-γ, IL-1β, and IL-2 compared with CR. IFN-γ and IL-1β are secreted by Th1 cells [44] and innate immune cells such as macrophages [45], respectively. IL-1β contributes to the immune response against pathogens by inducing fever, activating lymphocytes, and promoting the infiltration of leukocytes into sites of infection [46]. Likewise, IFN-γ is recognized as an important factor for preventing bacterial invasion of mucous membranes [47]. Moreover, IL-2 is necessary for the development of T cell memory and the development and maturation of regulatory T cells in the thymus [48,49]. Therefore, it is possible that calves in the CR group with a lower exposure to IL-2 during the first month of life might exhibit the reduced development of T cell memory and differentiation. During the first month of life, in which the development of Th1 cells and regulation of innate immune cell responses occur, MC plays a fundamental role that may assist in reducing the risk of infectious illness.

Quigley et al. (2001) and Jones et al. (2004) observed higher feed efficiency in calves fed MC. A tendency for MC-fed calves to gain more BW in the first week of life has also been reported [9,50]. In our study, calves fed MC had significantly greater carcass weight than calves fed CR. Therefore, the present study confirms the growth-promoting benefits of feeding MC to JB calves. In addition, when disease was present, the CR group exhibited a higher number of treatment days and frequency of treatments compared with the MC group at 1 month of age. It was previously reported that 93% of CR-fed calves exhibited a failure of passive transfer, and the negative effect of failure of passive transfer on morbidity and mortality is well known in neonatal calves [51,52]. Similarly, another study reported significantly lower morbidity and mortality in calves fed MC (46.9%) compared with calves fed plasma-derived (71.4%) or colostrum-derived (67.3%) CR [53]. The greatest risk of calf morbidity and mortality is observed during the first month of life [54–56]; therefore, one of our study objectives focused on reducing mortality and morbidity in 1-month-old JB calves. It is thus important to pay attention to the frequency of treatments in the first month of a study. In the present study, local and systemic protection provided by immunoglobulins and immunomodulatory molecules contained in MC might have reduced the incidence of infection and therefore the need for antibiotic therapy in calves fed MC. The decreased incidence rates of respiratory diseases in MC-fed JB calves suggest that MC increases disease resistance by enhancing both natural and acquired immunity in the first month after birth.

5. Conclusions

JB calves fed CR exhibited the delayed activation of the immune system and higher chances of failure in the transfer of passive immunity. In addition, the feeding of CR promoted decreases in populations of various immunocompetent cells not only during the period of colostrum feeding but also after passive immunity disappeared. Feeding JB calves CR may increase the chances of bacterial infection in early life while simultaneously diminished calf performance post-colostrum, as evidenced by reduced disease resistance, increased infection rates, and decreased carcass weight.

Supplementary Materials: The following supporting information can be downloaded at https://www.mdpi.com/article/10.3390/ani14020346/s1. Figure S1: White blood cell (WBC) populations in 1-, 2-, and 3-month-old calves in the MC and CR groups; Table S1: Daily feeding during pregnancy and after pregnancy; Table S2: Monoclonal antibodies used for flow cytometry and phagocytosis assays.

Author Contributions: M.U. (Marimu Urakawa) and M.B. contributed to data collection, analysis of the data, and writing the first draft of the manuscript. A.R., S.I. and P.A. contributed to performing the experiments. Y.D., M.U. (Mizuho Uematsu), Y.N. (Yoshiki Nakama), K.I. and Y.N. (Yusuke Nomura) assisted in sample collection and designing the experiments. M.Y. aided in the study design, the writing of the first draft of the manuscript, and the critical revision of the manuscript. All authors have read and agreed to the published version of the manuscript.

Funding: This study was supported by funding from the University of Miyazaki.

Institutional Review Board Statement: The animal study protocol was approved by the Animal Care and Use Committee of the University of Miyazaki (approval no. 2019-001-04).

Informed Consent Statement: Not applicable.

Data Availability Statement: The data presented in this study are available on request from the corresponding author.

Conflicts of Interest: The authors declare no conflict of interest.

References

1. Weaver, D.M.; Tyler, J.W.; VanMetre, D.C.; Hostetler, D.E.; Barrington, G.M. Passive Transfer of Colostral Immunoglobulins in Calves. *J. Vet. Intern. Med.* **2000**, *14*, 569–577. [CrossRef] [PubMed]
2. Faber, S.N.; Faber, N.E.; Mccauley, T.C. Effects of Colostrum Ingestion on Lactational Performance. *Prof. Anim. Sci.* **2005**, *21*, 420–425. [CrossRef]
3. Donovan, D.C.; Reber, A.J.; Gabbard, J.D.; Aceves-Avila, M.; Galland, K.L.; Holbert, K.A.; Ely, L.O.; Hurley, D.J. Effect of Maternal Cells Transferred with Colostrum on Cellular Responses to Pathogen Antigens in Neonatal Calves. *Am. J. Vet. Res.* **2007**, *68*, 778–782. [CrossRef] [PubMed]
4. Mallard, B.A.; Dekkers, J.C.; Ireland, M.J.; Leslie, K.E.; Sharif, S.; Vankampen, C.L.; Wagter, L.; Wilkie, B.N. Alteration in Immune Responsiveness during the Peripartum Period and Its Ramification on Dairy Cow and Calf Health. *J. Dairy Sci.* **1998**, *81*, 585–595. [CrossRef]
5. McGuirk, S.M.; Collins, M. Managing the Production, Storage, and Delivery of Colostrum. *Vet. Clin. N. Am.—Food Anim. Pract.* **2004**, *20*, 593–603. [CrossRef]
6. Godden, S.M.; Lombard, J.E.; Woolums, A.R. Colostrum Management for Dairy Calves. *Vet. Clin. N. Am.—Food Anim. Pract.* **2019**, *35*, 535–556. [CrossRef]
7. Godden, S.M.; Haines, D.M.; Hagman, D. Improving Passive Transfer of Immunoglobulins in Calves. I: Dose Effect of Feeding a Commercial Colostrum Replacer. *J. Dairy Sci.* **2009**, *92*, 1750–1757. [CrossRef]
8. Swan, H.; Godden, S.; Bey, R.; Wells, S.; Fetrow, J.; Chester-Jonest, H. Passive Transfer of Immunoglobulin G and Preweaning Health in Holstein Calves Fed a Commercial Colostrum Replacer. *J. Dairy Sci.* **2007**, *90*, 3857–3866. [CrossRef]
9. Jones, C.M.; James, R.E.; Quigley, J.D.; McGilliard, M.L. Influence of Pooled Colostrum or Colostrum Replacement on IgG and Evaluation of Animal Plasma in Milk Replacer. *J. Dairy Sci.* **2004**, *87*, 1806–1814. [CrossRef]
10. Foster, D.M.; Smith, G.W.; Sanner, T.R.; Busso, G.V. Serum IgG and Total [Protein] in Dairy Calves Fed Two Colostrum Replacement Products. *J. Am. Vet. Med. Assoc.* **2006**, *27606*, 1282–1285. [CrossRef]

11. Pithua, P.; Godden, S.M.; Wells, S.J.; Oakes, M.J. Efficacy of Feeding Plasma-Derived Commercial Colostrum Replacer for the Prevention of Transmission of *Mycobacterium avium* Subsp *paratuberculosis* in Holstein Calves. *J. Am. Vet. Med. Assoc.* **2009**, *234*, 1167–1176. [CrossRef] [PubMed]
12. Windeyer, M.C.; Leslie, K.E.; Godden, S.M.; Hodgins, D.C.; Lissemore, K.D.; LeBlanc, S.J. Factors Associated with Morbidity, Mortality, and Growth of Dairy Heifer Calves up to 3 Months of Age. *Prev. Vet. Med.* **2014**, *113*, 231–240. [CrossRef] [PubMed]
13. Poulsen, K.P.; Foley, A.L.; Collins, M.T.; McGuirk, S.M. Comparison of Passive Transfer of Immunity in Neonatal Dairy Calves Fed Colostrum or Bovine Serum-Based Colostrum Replacement and Colostrum Supplement Products. *J. Am. Vet. Med. Assoc.* **2010**, *237*, 949–954. [CrossRef] [PubMed]
14. Chigerwe, M.; Tyler, J.W.; Schultz, L.G.; Middleton, J.R.; Steevens, B.J.; Spain, J.N. Effect of Colostrum Administration by Use of Oroesophageal Intubation on Serum IgG Concentrations in Holstein Bull Calves. *Am. J. Vet. Res.* **2008**, *69*, 1158–1163. [CrossRef] [PubMed]
15. Kawashima, C.; Oie, H.; Mashimo, R.; Goto, A.; Kusaba, N.; Yamagishi, N. The Effects of Maternal Supplementation of Rumen-Protected Lysine during the Close-up Dry Period on Newborn Metabolism and Growth in Holstein Calves. *J. Anim. Physiol. Anim. Nutr.* **2023**, *107*, 28–36. [CrossRef] [PubMed]
16. NRC. *Nutrient Requirements of Beef Cattle*, 8th ed.; National Academies Press: Washington, DC, USA, 2015; ISBN 978-0-309-31702-3.
17. Baakhtari, M.; Imaizumi, N.; Kida, T.; Yanagita, T.; Ramah, A.; Ahmadi, P.; Takebe, N.; Iwamoto, Y.; Korosue, K.; Tsuzuki, N.; et al. Effects of Branched-Chain Amino Acids on Immune Status of Young Racing Horses. *J. Vet. Med. Sci.* **2022**, *84*, 558–565. [CrossRef]
18. Livak, K.J.; Schmittgen, T.D. Analysis of Relative Gene Expression Data Using Real-Time Quantitative PCR and the $2^{-\Delta\Delta CT}$ Method. *Methods* **2001**, *25*, 402–408. [CrossRef]
19. Hagiwara, K.; Yamanaka, H.; Higuchi, H.; Nagahata, H.; Kirisawa, R.; Iwai, H. Oral Administration of IL-1β Enhanced the Proliferation of Lymphocytes and the O_2 Production of Neutrophil in Newborn Calf. *Vet. Immunol. Immunopathol.* **2001**, *81*, 59–69. [CrossRef]
20. Yamanaka, H.; Hagiwara, K.; Kirisawa, R.; Iwai, H. Transient Detection of Proinflammatory Cytokines in Sera of Colostrum-Fed Newborn Calves. *J. Vet. Med. Sci.* **2003**, *65*, 813–816. [CrossRef]
21. Yamanaka, H.; Hagiwara, K.; Kirisawa, R.; Iwai, H. Proinflammatory Cytokines in Bovine Colostrum Potentiate the Mitogenic Response of Peripheral Blood Mononuclear Cells from Newborn Calves through IL-2 and CD25 Expression. *Microbiol. Immunol.* **2003**, *47*, 461–468. [CrossRef]
22. Laidlaw, B.J.; Craft, J.E.; Kaech, S.M. The Multifaceted Role of CD4+ T Cells in CD8+ T Cell Memory. *Nat. Rev. Immunol.* **2016**, *16*, 102–111. [CrossRef] [PubMed]
23. Nascimbeni, M.; Shin, E.C.; Chiriboga, L.; Kleiner, D.E.; Rehermann, B. Peripheral $CD4^+CD8^+$ T Cells Are Differentiated Effector Memory Cells with Antiviral Functions. *Blood* **2004**, *104*, 478–486. [CrossRef] [PubMed]
24. Quandt, D.; Rothe, K.; Scholz, R.; Baerwald, C.W.; Wagner, U. Peripheral CD4CD8 Double Positive T Cells with a Distinct Helper Cytokine Profile Are Increased in Rheumatoid Arthritis. *PLoS ONE* **2014**, *9*, e93293. [CrossRef] [PubMed]
25. Wu, Y.; Cai, B.; Feng, W.; Yang, B.; Huang, Z.; Zuo, C.; Wang, L. Double Positive $CD4^+CD8^+$ T Cells: Key Suppressive Role in the Production of Autoantibodies in Systemic Lupus Erythematosus. *Indian J. Med. Res.* **2014**, *140*, 513–519. [PubMed]
26. Diedrich, C.R.; Gideon, H.P.; Rutledge, T.; Baranowski, T.M.; Maiello, P.; Myers, A.J.; Lin, P.L. CD4CD8 Double Positive T Cell Responses during Mycobacterium Tuberculosis Infection in Cynomolgus Macaques. *J. Med. Primatol.* **2019**, *48*, 82–89. [CrossRef] [PubMed]
27. Hein, W.R.; Mackay, C.R. Prominence of Γδ T Cells in the Ruminant Immune System. *Immunol. Today* **1991**, *12*, 30–34. [CrossRef] [PubMed]
28. Lundberg, P.; Splitter, G.A. $\Gamma\delta^+$ T Lymphocyte Cytotoxicity against Envelope-Expressing Target Cells Is Unique to the Alymphocytic State of Bovine Leukemia Virus Infection in the Natural Host. *J. Virol.* **2000**, *74*, 8299–8306. [CrossRef]
29. Murakami, K.; Sentsui, H.; Inoshima, Y.; Inumaru, S. Increase in Γδ T Cells in the Blood of Cattle Persistently Infected with Bovine Leukemia Virus Following Administration of Recombinant Bovine IFN-γ. *Vet. Immunol. Immunopathol.* **2004**, *101*, 61–71. [CrossRef]
30. Guzman, E.; Price, S.; Poulsom, H.; Hope, J. Bovine Γδ T Cells: Cells with Multiple Functions and Important Roles in Immunity. *Vet. Immunol. Immunopathol.* **2012**, *148*, 161–167. [CrossRef]
31. Ohtsuka, H.; Fukunaga, N.; Hara, H.; Fukuda, S.; Hayashi, T.; Hoshi, F.; Yoshino, T.; Koiwa, M.; Kawamura, S. Changes in Peripheral Leukocyte Populations of Weak Calf Syndrome of Japanese Black Calves. *J. Vet. Med. Sci.* **2003**, *65*, 793–796. [CrossRef]
32. Wyatt, C.R.; Madruga, C.; Cluff, C.; Parish, S.; Hamilton, M.J.; Goff, W.; Davis, W.C. Differential Distribution of ΓδT-Cell Receptor Lymphocyte Subpopulations in Blood and Spleen of Young and Adult Cattle. *Vet. Immunol. Immunopathol.* **1994**, *40*, 187–199. [CrossRef] [PubMed]
33. Wilson, R.A.; Zolnai, A.; Rudas, P.; Frenyo, L.V. T-Cell Subsets in Blood and Lymphoid Tissues Obtained from Fetal Calves, Maturing Calves, and Adult Bovine. *Vet. Immunol. Immunopathol.* **1996**, *53*, 49–60. [CrossRef]
34. Nonnecke, B.J.; Horst, R.L.; Waters, W.R.; Dubeski, P.; Harp, J.A. Modulation of Fat-Soluble Vitamin Concentrations and Blood Mononuclear Leukocyte Populations in Milk Replacer-Fed Calves by Dietary Vitamin A and β-Carotene. *J. Dairy Sci.* **1999**, *82*, 2632–2641. [CrossRef] [PubMed]
35. Maidment, D.C. Changes in Immunoglobulin Levels during Storage of Fermented Bovine Colostrum. *Br. Vet. J.* **1982**, *138*, 18–22. [CrossRef]

36. Eckblad, W.P.; Hendrix, K.M.; Olson, D.P. Total Complement Hemolytic Activity of Colostral Whey and Sera from Dairy Cows. *Cornell Vet.* **1981**, *71*, 54–58.
37. Tsuji, S.; Hirata, Y.; Matsuoka, K. Two Apparent Molecular Forms of Bovine Lactoferrin. *J. Dairy Sci.* **1989**, *72*, 1130–1136. [CrossRef]
38. Kishikawa, Y.; Watanabe, T.; Watanabe, T.; Kubo, S. Purification and Characterization of Cell Growth Factor in Bovine Colostrum. *J. Vet. Med. Sci.* **1996**, *58*, 47–53. [CrossRef]
39. Straussberg, R.; Sirota, L.; Hart, J.; Amir, Y.S.; Djaldetti, M.; Bessler, H. Phagocytosis-Promoting Factor in Human Colostrum. *Biol. Neonate* **1995**, *681*, 15–18. [CrossRef]
40. Hidalgo, A.; Chilvers, E.R.; Summers, C.; Koenderman, L. The Neutrophil Life Cycle. *Trends Immunol.* **2019**, *40*, 584–597. [CrossRef] [PubMed]
41. Nonnecke, B.J.; Foote, M.R.; Smith, J.M.; Pesch, B.A.; Van Amburgh, M.E. Composition and Functional Capacity of Blood Mononuclear Leukocyte Populations from Neonatal Calves on Standard and Intensified Milk Replacer Diets. *J. Dairy Sci.* **2003**, *86*, 3592–3604. [CrossRef]
42. Maciag, S.S.; Bellaver, F.V.; Bombassaro, G.; Haach, V.; Morés, M.A.Z.; Baron, L.F.; Coldebella, A.; Bastos, A.P. On the Influence of the Source of Porcine Colostrum in the Development of Early Immune Ontogeny in Piglets. *Sci. Rep.* **2022**, *12*, 15630. [CrossRef] [PubMed]
43. Grigaleviciute, R.; Planciuniene, R.; Prikockyte, I.; Radzeviciute-Valciuke, E.; Baleviciute, A.; Zelvys, A.; Zinkeviciene, A.; Zigmantaite, V.; Kucinskas, A.; Matusevicius, P.; et al. The Influence of Feeding with Colostrum and Colostrum Replacer on Major Blood Biomarkers and Growth Performance in Dairy Calves. *Vet. Sci.* **2023**, *10*, 128. [CrossRef] [PubMed]
44. O'Garra, A. Cytokines Induce the Development of Functionally Heterogeneous T Helper Cell Subsets. *Immunity* **1998**, *8*, 275–283. [CrossRef] [PubMed]
45. Lopez-Castejon, G.; Brough, D. Understanding the Mechanism of IL-1β Secretion. *Cytokine Growth Factor Rev.* **2011**, *22*, 189–195. [CrossRef] [PubMed]
46. Dinarello, C.A. Biologic Basis for Interleukin-1 in Disease. *J. Am. Soc. Hematol.* **1996**, *87*, 2095–2147. [CrossRef]
47. Ferrick, D.A.; Schrenzel, M.D.; Mulvania, T.; Hsieh, B.; Ferlin, W.G.; Lepper, H. Differential Production of Interferon-γ and Interleukin-4 in Response to Th1- and Th2-Stimulating Pathogens by $\Gamma\delta$ T Cells in Vivo. *Nature* **1995**, *373*, 255–257. [CrossRef]
48. Sakaguchi, S.; Sakaguchi, N.; Asano, M.; Itoh, M.; Toda, M. Immunologic Self-Tolerance Maintained by Activated T Cells Expressing IL-2 Receptor Alpha-Chains (CD25). Breakdown of a Single Mechanism of Self-Tolerance Causes Various Autoimmune Diseases. *J. Immunol.* **1995**, *155*, 1151–1164. [CrossRef] [PubMed]
49. Thornton, A.M.; Shevach, E.M. $CD4^+CD25^+$ Immunoregulatory T Cells Suppress Polyclonal T Cell Activation in vitro by Inhibiting Interleukin 2 Production. *J. Exp.* **1998**, *188*, 287–296. [CrossRef]
50. Quigley, J.D.; Strohbehn, R.E.; Kost, C.J.; O'Brien, M.M. Formulation of Colostrum Supplements, Colostrum Replacers and Acquisition of Passive Immunity in Neonatal Calves. *J. Dairy Sci.* **2001**, *84*, 2059–2065. [CrossRef]
51. Robison, J.D.; Stott, G.H.; DeNise, S.K. Effects of Passive Immunity on Growth and Survival in the Dairy Heifer. *J. Dairy Sci.* **1988**, *71*, 1283–1287. [CrossRef]
52. Donovan, G.A.; Dohoo, I.R.; Montgomery, D.M.; Bennett, F.L. Associations between Passive Immunity and Morbidity and Mortality in Dairy Heifers in Florida, USA. *Prev. Vet. Med.* **1998**, *34*, 31–46. [CrossRef] [PubMed]
53. Priestley, D.; Bittar, J.H.; Ibarbia, L.; Risco, C.A.; Galvão, K.N. Effect of Feeding Maternal Colostrum or Plasma-Derived or Colostrum-Derived Colostrum Replacer on Passive Transfer of Immunity, Health, and Performance of Preweaning Heifer Calves. *J. Dairy Sci.* **2013**, *96*, 3247–3256. [CrossRef] [PubMed]
54. Misaka, M.; Uematsu, M.; Hashimoto, K.; Kitahara, G.; Osawa, T.; Sasaki, Y. Impact of Dystocia and Cow/Calf Characteristics on Mortality from 0 to 120 Days of Age in Japanese Black Calves in Commercial Cow-Calf Operations. *Prev. Vet. Med.* **2022**, *207*, 105716. [CrossRef] [PubMed]
55. Uetake, K. Newborn Calf Welfare: A Review Focusing on Mortality Rates. *Anim. Sci. J.* **2013**, *84*, 101–105. [CrossRef]
56. Hordofa, D.; Abunna, F.; Megersa, B.; Abebe, R. Incidence of Morbidity and Mortality in Calves from Birth to Six Months of Age and Associated Risk Factors on Dairy Farms in Hawassa City, Southern Ethiopia. *Heliyon* **2021**, *7*, e08546. [CrossRef]

animals

MDPI

Article

Effects of Supplementing Milk Replacer with Sodium Butyrate on Dairy Calves

Anna Luiza Belli [1], Sandra G. Coelho [1], Joana P. Campolina [1], Luiz F. M. Neves [1], Hilton C. Diniz Neto [1], Camila S. Silva [2], Fernanda S. Machado [2], Luiz Gustavo R. Pereira [2], Thierry R. Tomich [2], Wanessa A. Carvalho [2], Suely de Fátima Costa [3] and Mariana M. Campos [2,*]

1 Departamento de Zootecnia, Escola de Veterinária, Universidade Federal de Minas Gerais, Belo Horizonte 30161-970, MG, Brazil; annabelliac@gmail.com (A.L.B.); sandragesteiracoelho@gmail.com (S.G.C.); joana.campolina@yahoo.com.br (J.P.C.); luizneveszoo@yahoo.com.br (L.F.M.N.); hiltondinizneto@gmail.com (H.C.D.N.)

2 Empresa Brasileira de Pesquisa Agropecuária—Embrapa Gado de Leite, Juiz de Fora 36038-330, MG, Brazil; camilaszootecnia@gmail.com (C.S.S.); fernanda.machado@embrapa.br (F.S.M.); luiz.gustavo@embrapa.br (L.G.R.P.); thierry.tomich@embrapa.br (T.R.T.); wanessa.carvalho@embrapa.br (W.A.C.)

3 Departmento de Medicina Veterinária Preventiva, Universidade Federal de Lavras, Lavras 37203-202, MG, Brazil; sfcosta@ufla.br

* Correspondence: mariana.campos@embrapa.br; Tel.: +55-32-9909-9737

Simple Summary: Raising dairy calves poses a dual challenge for the dairy industry—economically and physiologically. The industry seeks innovative solutions that can not only reduce dependence on antibiotic treatments but also enhance the overall wellbeing, gastrointestinal tract development, and performance of young calves. Among the potential solutions, butyrate supplementation has emerged as a promising tool. To validate this hypothesis, the present study meticulously assesses the impact of butyrate supplementation on key parameters such as feed intake, gastrointestinal tract development, health, and performance of pre-weaned dairy calves. Furthermore, this investigation delves into the lasting effects of butyrate supplementation post weaning.

Abstract: Diarrhea and respiratory diseases pose significant challenges in the rearing of pre-weaned calves, motivating the investigation of tools to improve gastrointestinal tract development, health, and overall performance in young calves. Consequently, the primary objective of this study was to assess the effectiveness of an additive incorporated into milk replacer to promote the development and health of the animals. Forty-six dairy calves were randomly assigned into two treatments: control (CON, $n = 23$; with 15 females and 8 males), and sodium butyrate (SB, $n = 23$; with 15 females and 8 males). The calves in the SB treatment group were supplemented with 4 g/d of unprotected sodium butyrate (Adimix, Adisseo, China), added to the milk replacer from 4 to 60 days of age. Water and starter were fed ad libitum. The study evaluated several parameters, including feed intake, nutrient digestibility, ruminal pH, ammonia and volatile fatty acids, blood metabolites (glucose, insulin-like growth factor type 1, urea, β–hydroxybutyrate), hemogram, health scores, performance, and feed efficiency. Bull calves were euthanized at 60 days of age for organ comparison, while heifer calves were assessed for carryover effects up to 90 days of age. Data were analyzed independently using linear mixed models using the nlme package in R, and the Artools package for non-parametric categorical outcomes. Although the feed intake and performance variables exhibited differences within weeks, no divergence was observed between treatment groups. Notably, a positive treatment-by-week interaction was identified for starter feed intake ($p = 0.02$) and total dry matter intake ($p = 0.04$) during pre-weaning for CON animals. Ruminal parameters, blood metabolites, and hemogram values such as glucose, urea, insulin-like growth factor type 1, mean corpuscular value, lymphocytes, and neutrophils displayed differences within weeks during the pre-weaning stage, but similar results within groups. No differences between supplemented and non-supplemented calves were found across nutrient digestibility, organ development, and histology. Regarding health scores, differences were noted within weeks for fecal and respiratory scores during the pre-weaning stage, and only the respiratory score during the post-weaning stage. Consequently, butyrate supplementation did not elicit improvements or negative effects in the body development or health status of dairy calves.

Keywords: additive; butyrate supplementation; dairy calf; gastrointestinal tract development; growth promoter

1. Introduction

Antimicrobial feed additives have been included in the diets of livestock since the 1950s, becoming a popular tool to improve performance, animal health, and farm profitability ever since. Nonetheless, the appearance of antimicrobial resistance over time became a dangerously common occurrence for both animals and humans [1,2], leading to the European Union's ban on growth-promoting additives in 2006 [3]. As a result, researchers have attempted to find alternatives for enhancing animal performance while improving the herd's health status and the food security of animal-derived products.

Diarrhea and respiratory diseases are the major causes of economic losses at the rearing stage on a dairy farm, sometimes accounting for more than 10% of the causes of mortality for young calves [4,5]. Short-chain fatty acids (SCFAs) naturally present in the mammalian gastrointestinal tract (GIT) have been pointed out as a promising option to reduce disease incidence in the early stages of a calf's life [6]. Butyrate, one of the SCFAs found in milk and synthesized from the microbial fermentation of carbohydrates in the rumen and large intestines [7,8], plays a pivotal role in transepithelial fluid transport, the inhibition of mucosa inflammation, oxidative status, epithelial barrier defense, visceral sensibility, and intestinal motility [8,9] and can be used as an energetic source for rumen epithelial cells and colonocytes, which are vital for nutrient absorption [8,10].

Antibiotics administered during clinical disease episodes severely jeopardize the ability of intestinal cells to utilize butyrate because of disturbances to the intestinal microbiota [11]. This is especially critical for newborn calves, who have a high susceptibility to GIT disturbances and diseases and whose GIT is still underdeveloped. Recent research has indicated that this effect could be minimized by feeding exogenous sources of butyrate [12,13], which include sodium or calcium salts of butyric acid, usually encapsulated when added to solid feeds and non-encapsulated when added to a liquid diet [14]. These sources are designed to bypass the rumen to prevent butyrate's degradation in the forestomach, thereby helping develop and mature the small and large intestines [15,16]. Several studies have demonstrated that the inclusion of sodium butyrate in a liquid diet improved the cell cycle of dairy calves, reducing apoptosis, stimulating cell proliferation and body antioxidant function, and strengthening the intestinal barrier and mucosal thickness [14,17,18].

Since encapsulated or non-encapsulated butyrate given to pre-weaned calves have improved GIT development and decreased the incidence of diarrhea [17,19–21], the objectives of this study were to evaluate the effects of supplementing pre-weaning calves with unprotected sodium butyrate through milk replacer (MR) on feed intake, ruminal pH, nitrogen ammonia (NH_3-N) and volatile fatty acids (VFA) concentrations in rumen fluid, nutrient digestibility, blood concentrations of glucose, IGF-1, urea, and β–hydroxybutyrate (BHB), hemogram, health scores, performance, feed efficiency (FE), organ and GIT development during the pre-weaning phase, and the carryover effects during the post-weaning phase. The hypothesis was that supplementation with sodium butyrate in MR for dairy calves would improve their feed intake, nutrient digestibility, GIT development, health parameters, and performance.

2. Material and Methods

The experiment was conducted at the Embrapa Dairy Cattle, Minas Gerais, Brazil, from March to August 2018. All procedures were approved by the Ethics Committee of Embrapa Dairy Cattle, Brazil (protocol number 9078250118). Researchers were entrusted with the daily responsibilities of animal feeding, measurements, and health checks, ensuring the wellbeing and comfort of all calves throughout the study.

2.1. Animals, Housing, and Treatments

Thirty female and sixteen male, Holstein and Holstein × Gyr crossbred, dairy calves were enrolled in a randomized complete block design. After birth, calves were weighed and immediately moved to individual sand-bedded pens (1.25 × 1.75 m) tethered with 1.2 m chains, and allocated in an open-sided barn. All animals received 10% of their body weight in good-quality colostrum (Brix > 23%) and had their umbilical cord immersed in an iodine solution (10%) within two hours after birth. After the first colostrum feeding and until the third day of life, calves received 5 L/d of transition milk split into two meals (0800 and 1600 h), fed via a commercial milk feeder (Milkbar, Waipu, New Zealand). Blood was collected on the third day of life via jugular venipuncture into a tube without anticoagulant (Labor Import, Osasco, Brazil), and centrifuged at 3000× *g* for 10 min at room temperature (22–25 °C) to evaluate passive immune transfer using a Brix refractometer (Aichose refractometer, Xindacheng, Jiaozhou City, China). Only calves with a Brix value above 8.3% were enrolled in the study.

On the 4th of age, animals were blocked by sex and genetic composition and assigned to one of two treatments: control (CON, nothing added to the feed or to MR; n = 15 heifer and 8 bull calves); and unprotected sodium butyrate (SB; 4 g/d, Adimix, Adisseo, Nanjing, China) added to the MR; n = 15 heifer and 8 bull calves). The dose of butyrate was determined according to the levels of inclusion reported by other authors [10,19]. In every meal, the additive was diluted in the first 0.5 L of MR to ensure total ingestion of the product, resulting in supplementation of 2 g in the morning and 2 g in the afternoon.

All calves received 5 L per day of MR (Kalvolak, Nutrifeed, The Netherlands; Table 1) divided into two equal daily meals (0800 and 1600 h). The MR was mixed to provide 15% of total solids, providing 145.5 g of crude protein, and 105.8 g of fat, and offered to calves using a milk calf feeder bucket (Milkbar, New Zealand). Calf starter (Soymax Rumen pre-initial Flocculated, Total Alimentos, Três Corações, Brazil, Table 1) and water were individually offered ad libitum from the first day of life until weaning, at 60 days of age.

Table 1. Nutrient composition (% DM ± SD) of milk replacer (MR), starter, and corn silage.

Item	MR [1]	Starter [2]	Corn Silage
DM (%)	96.0 ± 0.4	86.7 ± 0.7	36.1 ± 3.1
CP (% of DM)	19.4 ± 0.5	17.1 ± 0.5	7.9 ± 0.7
Ether extract (% of DM)	14.1 ± 0.6	3.9 ± 1.2	4.3 ± 0.5
Organic matter (% of DM)	9.7 ± 0.2	7.2 ± 1.5	6.0 ± 1.1
NDF (% of DM)	-	22.1 ± 2.9	46.1 ± 4.1
ADF (% of DM)	-	10.6 ± 0.9	28.9 ± 3.5
Metabolizable energy (Mcal/kg DM) [3]	3.6 ± 0.1	2.8 ± 0.0	2.4 ± 0.1

[1] Powder integral milk, wheat isolated protein, acidifying additive, whey, coconut oil, palm oil, vitamin A, Vitamin D3, Vitamin E, Vitamin C. [2] Basic composition: oats (rolled grains), calcitic limestone, sodium chloride, corn gluten meal, defatted corn germ, wheat bran, soybean meal, rice hulls, kaolin, molasses, flocculated corn, ground corn, corn grain, alfalfa hay, monensin, citrus pulp, dried sugarcane yeast, whole toasted soybean, sodium selenite, copper sulfate, manganese sulfate, cobalt sulfate, iron sulfate, zinc sulfate, calcium iodate, vitamin A, vitamin B1, vitamin B12, vitamin B2, vitamin B6, vitamin C, vitamin D3, vitamin E, vitamin K, niacin, pantolenic acid, folic acid, biotin, propionic acid, caramel aroma, milk aroma, and probiotic additive. [3] Metabolizable energy was calculated from NRC (2001).

2.2. Feed Intake, Rumen Samples, Digestibility, Performance, and Nutritional Composition

Feed intake (MR, starter, water and, after weaning, corn silage) was measured daily using the difference between offers and refusals. Feed efficiency (FE) was calculated using the ratio between average daily gain (ADG) and dry matter intake (DMI) [22]. Samples of MR, starter, and corn silage (offered after weaning) were collected thrice a week and weekly pooled into individual subsamples and kept at −20 °C for further analysis.

Samples of ruminal fluid were collected on days 14, 28, 42, and 60 for all calves and on days 74 and 90 for heifers only, using an esophageal tube, four hours after the morning MR feeding. Ruminal fluid was strained through four layers of gauze to separate solid and liquid fractions. Rumen pH was immediately measured using a pH meter (Phmetro T-1000, Tekna, Araucária, Brazil), and two samples (10 mL) of ruminal fluid were obtained. One sample was acidified with 1 mL of 20% metaphosphoric acid, and the other with 2 mL of 50% sulfuric acid. The samples were stored at −20 °C for further VFA and NH_3-N analyses. Samples used to test VFA concentrations were centrifuged at 1800× *g* for 10 min at room temperature (22–25 °C) and analyzed via high-performance liquid chromatography (Waters Alliance e2695 Chromatograph, Waters Technologies do Brazil LTDA, São Paulo, Brazil). Concentrations of NH_3-N were quantified using a colorimetric distillation method [23]. Absorbance was measured at 630 nm (Thermo Fisher Scientific, Madison, WI, USA) after Kjeldahl distillation with magnesium oxide and calcium chloride according to Method 920.03 [24].

Digestibility was conducted only with the male calves. A rubber mat (WingFlex, Kraiburg TPE GmbH & Co., Ltd., Waldkraiburg, Germany) was placed atop the sand bed beneath each animal to facilitate the complete collection of feces over a period of five days. On the last day of the digestibility trial, animals were transferred to metabolic cages (1.50 m × 0.80 m; Intergado Ltd.a., Contagem, Brazil) for total urine collection and the last day of fecal collection. After the end of the collection (24 h), the urines' total volume, weight, and density were measured. Two samples (50 mL) were collected from the total urine and kept at −20 °C for further analysis of urea. Samples of feces and the starter provided during the digestibility test were collected daily. After the end of the trial, a composite sample corresponding to the 5 days of digestibility was frozen at −20 °C. The apparent digestibility values of the nutrients (%) were determined using the amount consumed and the amount of each nutritional component recovered in the feces.

The pool of feed and fecal samples was ground through a 1 mm screen in a mill (model 3, Arthur H. Thomas Co., Philadelphia, PA, USA) before analysis. The dry matter content of the ground starter, corn silage, MR, and feces was oven-dried at 55 °C for 72 h (Method 934.01). Crude protein (CP) (Method 988.05), ether extract (EE) (Method 920.39), ash (Method 942.05), neutral detergent fiber (NDF), and acid detergent fiber (ADF) were determined using the methods [25].

For performance measurements, body weight (BW) was initially measured at birth and on day 3. After that, BW was measured once a week before the morning meal, using a portable scale (ICS 300, Coimma Comércio Indústria de Madeiras e Metalúrgica São Cristóvão Ltd.a., Dracena, Brazil). Wither height, hip height, hip width, and heart girth were measured at birth and every 7 d until the end of the experimental period. These measurements were taken on a flat surface using a portable hipometer (Walmur Instrumentos Veterinários Ltd.a., Porto Alegre, Brazil) and a metric tape.

2.3. Health Scores

Rectal temperature was evaluated daily at 0600h using a digital thermometer (Ombo Eletronics, iColor, modelo G-Tech, Shenzhen, China). Fecal [4] and respiratory scoring [26] was performed daily. A calf was considered to have diarrhea if the fecal score was 2 or 3. Diarrhea was considered severe when the fecal score was 3. Calves were treated with antibiotics only if the respiratory score was above 4 or if they had a fever for two consecutive days.

2.4. Blood Sampling and Weaning

Blood was drawn via jugular venipuncture using 10 mL tubes with anticoagulant and sodium fluoride at birth, before colostrum ingestion, and every 7 d thereafter, three hours after morning meals, for all animals and for the heifers until 90 d of age (Labor Import, Osasco, Brazil) for analysis of BHB, urea, and glucose concentrations. Also, every 14 days, 10 mL tubes containing heparin (Labor Import, Osasco, Brazil) were used to collect samples for IGF-1 concentrations. The samples were centrifuged at 3000× g for 10 min, at room temperature (22–25 °C), and two aliquots from each sample were individually stored frozen at −20 °C.

Serum concentrations of the BHB and urea were determined with an auto-analyzer (Cobas Mira Plus, Roche Diagnostic Systems, Risch-Rotkreuz, Switzerland) using commercial kits (Ranbut-D-3-Hidroxibutyrate, Randox Laboratories Ltd., Antrim, UK; Urea UV, Kovalent do Brazil Ltd.a., Bom Retiro São Gonçalo, Brazil). The concentration of plasma glucose was measured via a Spectrophotometer EON microplate (Biotek Instruments Inc., Winooski, VT, USA) using an enzymatic colorimetric method (Kovalent do Brazil Ltd.a., Rio de Janeiro, Brazil). Plasma concentrations of IGF-1 were determined using a chemiluminescence assay (Immulite 2000 Systems 1038144, IGF-1 200, Siemens Healthcare Diagnostics Products Ltd., Llanberis, Gwynedd, UK).

At days 0, 30, and 60, blood from all animals was sampled in 2 mL tubes containing EDTA (Labor Import, Osasco, Brazil), stored at 4 °C, and immediately sent to the laboratory for hemogram analysis.

At 60 d of age, heifer calves were weaned, kept in the experiment and, after weaning, at d 61, started receiving corn silage as part of the diet (Table 1). The amount of corn silage provided was enough to assure at least 10% of orts, and the starter intake was limited to 3.0 kg heifer/d, fed divided into two meals. Water continued to be offered ad libitum.

2.5. Internal Organ Weight, Gastrointestinal Tract Development, and Histology

At day 60 of age, male calves were euthanized to evaluate the weight of organs and tissue sampling, using the procedures recommended by the Brazilian Federal Veterinary Medicine Council [27] Briefly, the chose method consisted of the administration of sedative (0.05 mg/kg BW of xylazine intramuscularly) followed by the administration of intravenous barbituric acid (20 mL of pentobarbital).

The tongue, lungs with trachea, heart, liver, spleen, kidneys, mesentery, omental and peri-kidney fat were weighed and discarded. The rumen–reticulum, omasum, abomasum, small and large intestines, and bladder were weighted, full of biological content, and emptied. All variables were evaluated in proportion to the weight of the empty body. The length of the small and large intestines was measured using a metric tape.

Samples of 9 cm^2 (3 × 3 cm) were obtained from the rumen (dorsal sac, ventral sac), omasum, abomasum, and portions of the small and large intestines (duodenum, ileum, and colon). The samples were fixed in formalin, and after fixation, they were kept in an ethanol solution (alcohol 70%) until the analysis was performed. The samples were incorporated into paraffin blocks, being sectioned using a manual microtome (Olympus, Tokyo, Japan) in sections 5 μm thick. For morphometric analysis, the slides were stained with hematoxylin–eosin. The images were captured with the aid of a microscope (Olympus CX31, Tokyo, Japan), coupled with a camera (Olympus OSIS SC30, Tokyo, Japan), using the Cell-B software version 3.4 (Olympus, Tokyo, Japan). Papillary height (μm) and papillary area (μm^2) were measured in regions of the rumen's ventral and dorsal sac, and the omasum. Villus height (μm) and villus area (μm^2) were measured in regions of the duodenum and ileum. Fossette depth (μm) and gastric gland depth (μm) were measured in the abomasum. The crypt depth (μm) was also measured in regions of the duodenum, ileum, and colon.

Mitotic index (MI) was determined in the epithelium of the rumen (ventral and dorsal sac) and omasum, using an optical microscope (Olympus OSIS SC30, Tokyo, Japan) with 400× enlargement. Approximately 2000 cells from the epithelium basal layer were counted, including those presenting mitotic figures in the nucleus. The index was calculated using the quotient between the number of nuclei in division and the total of nuclei counted [28]. The cell proliferation in regions of the abomasum, duodenum, ileum, and colon was determined by counting the number of mitotic figures in the epithelium of gastric and intestinal glands, in 10 fields, using a 400× enlargement.

2.6. Statistical Analysis

The data were analyzed using the software R version 4.3.1 [29]. The pre-weaning (1–60 d) and post-weaning (61–90 d) data were analyzed separately. A randomized complete block design with repeated measures was used to test the effect of sodium butyrate on each outcome measured. Treatment, week, and the interaction of treatment and week were used as fixed effects. The effect of calf within treatment was included in the models to account for individual variability.

Feed intake, ruminal parameters, digestibility, rectal temperature, blood parameters, body measurements, performance, and feed efficiency, were analyzed using the linear mixed model (package: nlme) [30], including calf as the random term, treatment, week, and their interaction as fixed variables.

Respiratory and fecal score were analyzed using a non-parametric aligned rank transformation methodology implemented via the software package ARTool version 0.11.1 [31]. Differences between averages of treatments were evaluated using the Fisher test, and weekly and interactions weekly × treatment were evaluated through the Tukey test.

Variables with a single measurement during the study, such as final and initial BW, organ/viscera weight and GIT development were analyzed using the linear mixed model (package: nlme) [30], including calf as the random term and treatment as a fixed variable. All outcomes underwent testing for homoscedasticity and normality using Bartlett and Shapiro tests to satisfy the necessary assumptions of this model. Significance was asserted at $p \leq 0.05$.

3. Results

3.1. Feed Intake, Rumen Parameters, Performance, and Feed Efficiency

In the current study, milk replacer intake, water intake, body measurements, feed efficiency, and average daily gain were not different ($p > 0.05$) among treatments during all the evaluation periods (Tables 2 and 3). Water intake was higher ($p = 0.01$) for CON animals compared to SB animals in the post-weaning period (Table 2). There was a significant effect of week ($p < 0.05$) on feed intake and performance variables. The milk replacer intake was smaller in weeks 2 and 3, while the average dairy gain, starter, and water intakes, as well as the total DMI showed significant increases in the subsequent weeks both during pre- and post-weaning ($p < 0.001$, Table 2). There was a positive treatment-by-week interaction for starter feed intake and total DMI during pre-weaning ($p < 0.05$, Table 2), whereas in weeks 3 to 5 ($p < 0.05$, Table 2), CON animals ate, on average, 44 g more starter than SB animals. As for the total DMI, the CON animals presented higher intake values compared to SB between weeks 4 and 7, eating, on average 5.7% more than SB calves.

Table 2. Feed intake, feed efficiency, and performance of dairy calves in the control group (CON, n = 23) and supplemented group (SB, n = 23) in the pre-weaning period, and the control (CON, n = 15) and supplemented (SB, n = 15) groups in the post-weaning period.

Item	Treatment [1]		SEM	*p*-Value [2]		
	CON	SB		T	W	T × W
Feed intake—pre-weaning						
Starter (g of DM/day)	259	223	171	0.88	<0.001	0.02
Milk replacer (g of DM/day)	729	726	76	0.67	<0.001	0.85
Water (mL)	1176	1271	667	0.48	<0.001	0.40
Total DMI (g/day)	987	945	186	0.88	<0.001	0.04
Performance—pre-weaning						
ADG (g/day)	527	509	15	0.62	<0.001	0.69
Feed efficiency	0.5	0.5	0.02	0.45	<0.001	0.85

Table 2. *Cont.*

Item	Treatment [1]		SEM	*p*-Value [2]		
	CON	SB		T	W	T × W
Feed intake—post-weaning						
Starter (g of DM/day)	1843	1468	342	0.09	<0.001	0.59
Corn silage (g of DM/day)	145	128	66	0.27	<0.001	0.86
Water (mL/day)	5695	4848	1310	0.01	<0.001	0.85
Total DMI (g/day)	1957	1312	345	0.15	<0.001	0.35
Performance—post-weaning						
ADG (g/day)	880	796	0	0.58	<0.001	0.31
Feed efficiency	0.4	0.4	0.03	0.69	0.55	0.37

[1] CON = control; SB = 4 g of sodium butyrate (Adimix, Adisseo, China) supplemented from 4 to 60 d of age;
[2] T = treatment effect; W= week effect, T × W = treatment by week interactions.

Table 3. Body measurements of control (CON, *n* = 23) and supplemented (SB, *n* = 23) groups in the pre-weaning period and the control (CON, *n* = 15) and supplemented (SB, *n* = 15) groups in the post-weaning period.

Item		Treatment [1]		SEM	*p*-Value [2]
		CON	SB		T
At Birth	Weight (kg)	33.1	32.2	0.61	0.57
	Wither height (cm)	68.3	67.7	0.38	0.56
	Rump height (cm)	71.3	70.8	0.39	0.67
	Rump width (cm)	17.0	16.9	0.16	0.72
	Heart girth (cm)	74.6	72.9	0.46	0.16
Weaning	Weight (kg)	64.8	62.7	1.07	0.47
	Wither height (cm)	80.6	80.2	0.41	0.71
	Rump height (cm)	83.7	83.9	0.40	0.89
	Rump width (cm)	21.6	21.5	0.13	0.77
	Heart girth (cm)	91.6	91.5	0.45	0.99
Final	Weight (kg)	82.7	82.9	2.32	0.33
	Wither height (cm)	86.0	85.5	0.98	0.16
	Rump height (cm)	89.0	87.9	0.98	0.89
	Rump width (cm)	23.8	23.1	0.45	0.17
	Heart girth (cm)	102.0	98.0	1.55	0.14

[1] CON = control; SB = 4 g of sodium butyrate (Adimix, Adisseo, China) supplemented from 4 to 60 d of age;
[2] T = treatment effect.

For ruminal parameters, there were no differences among treatments during the pre- and post-weaning periods ($p > 0.05$, Table 4). The ruminal NH_3-N and VFA were influenced by the week during the pre-weaning period, mirroring the observed pattern in intake ($p < 0.001$, Table 4). The cecum concentrations of VFA, evaluated only for the bull calves at 60 days of age, did not show any differences among treatments ($p > 0.05$, Table 4).

3.2. Digestibility, Comparative Slaughter, and Histology

The digestibility and nitrogen balance performed with the bull calves at the end of the pre-weaning period were not different ($p > 0.05$, Table 5) among treatments. As for comparative slaughter, there were no differences regarding organ weight, intestine length, or histological parameters among the CON and SB calves ($p > 0.05$, Tables 6 and 7).

Table 4. Rumen parameters of the control (CON, n = 23) and supplemented (SB, n = 23) groups in the pre-weaning period, and cecum parameters of 60-day bull calves in the control (CON, n = 8) and supplemented (SB, n = 8) groups.

Item	Treatment [1]		SEM	*p*-Value [2]		
	CON	SB		T	W	T × W
Pre-weaning						
pH	6.1	6.2	0.47	0.53	0.16	0.72
Ammonia-N (mg/dL)	11.4	12.8	0.07	0.31	<0.001	0.35
Volatile fatty acids (mmol/L)						
Acetic (C2)	27.8	25.5	7.78	1.0	<0.001	0.57
Propionic (C3)	15.0	16.4	7.18	0.72	<0.001	0.90
Butyric (C4)	3.8	3.6	1.55	0.52	<0.001	0.35
C2:C3	1.9	1.9	0.54	0.74	<0.001	0.65
Post-weaning						
pH	6.2	6.1	0.53	0.44	0.09	0.82
Ammonia-N (mg/dL)	11.0	9.0	9.17	0.20	0.75	0.56
Volatile fatty acids (mmol/L)						
Acetic (C2)	37.7	38.6	9.50	0.33	0.64	0.28
Propionic (C3)	27.7	27.9	7.04	0.52	0.39	0.48
Butyric (C4)	5.6	6.3	1.61	0.91	0.39	0.76
C2:C3	1.4	1.4	0.20	0.60	0.32	0.58
Cecum						
Acetic (C2)	21.3	20.5	0.33	0.90	-	-
Propionic (C3)	13.1	12.3	0.41	0.84	-	-
Butyric (C4)	2.2	3.0	0.07	0.14	-	-
C2:C3	1.7	1.7	0.07	0.63	-	-

[1] CON = control; SB = 4 g of sodium butyrate (Adimix, Adisseo, China) supplemented from 4 to 60 d of age; [2] T = treatment effect; W= week effect, T × W = treatment by week interactions. The "-" means that this outcome was not tested W or T × W.

Table 5. Nutrient apparent digestibility, from 54 to 59 days of age, of bull dairy calves in the control (CON, n = 8) and supplemented (SB, n = 8) groups.

Item	Treatment [1]		SEM	*p*-Value [2]
	CON	SB		T
Dry matter (g/day)	877	889	4.02	0.45
Organic matter (g/day)	919	911	3.01	0.57
Crude protein (mg/day)	921	932	1.98	0.60
Ether extract (mg/day)	957	965	1.58	0.53
Ingested nitrogen (g/kg of MW [3]/day)	2.1	2.0	0.03	0.79
Fecal nitrogen/(g/kg of MW [3]/day)	0.2	0.2	0.01	0.66
Urine nitrogen/(g/kg of MW [3]/day)	0.4	0.4	0.02	0.87
Retained nitrogen (g/kg of MW [3]/day)	1.6	1.5	0.01	0.79

[1] CON = control; SB = 4 g of sodium butyrate (Adimix, Adisseo, China) supplemented from 4 to 60 d of age; [2] T = treatment effect. [3] MW = metabolic weight.

Table 6. Results of comparative euthanasia and measurement of organs of bull dairy calves in the control (CON, n = 8) and supplemented (SB, n = 8) groups.

Item	Treatment [1]		SEM	*p*-Value [2]
	CON	SB		T
Live body weight (kg)	65.4	67.7	0.77	0.52
Empty carcass (kg)	53.3	53.8	0.57	0.84
Tong (kg)	0.47	0.55	0.002	0.77
Lungs and trachea (kg)	2.49	2.28	0.08	0.51

Table 6. *Cont.*

Item	Treatment [1]		SEM	*p*-Value [2]
	CON	SB		T
Heart (kg)	0.77	0.77	0.003	0.97
Spleen (kg)	0.72	0.74	0.02	0.86
Liver (kg)	2.57	2.6	0.09	0.87
Pancreas (kg)	0.05	0.06	0.003	0.57
Kidneys (kg)	0.55	0.65	0.029	0.31
Bladder (kg)	0.06	0.05	0.002	0.54
Reticulum–rumen (kg)	1.08	1.21	0.35	0.39
Omasum (kg)	0.22	0.20	0.06	0.68
Abomasum (kg)	0.35	0.35	0.09	0.90
Small intestine (kg)	1.84	1.73	0.13	0.31
Small intestine length (m)	21.97	22.58	1.34	0.66
Large intestine (kg)	0.63	0.55	0.08	0.17
Large intestine length (m)	3.48	3.23	0.47	0.22
Omental fat (kg)	0.38	0.36	0.003	0.12
Peri-kidney fat (kg)	0.42	0.55	0.009	0.11
Mesentery + omental fat (kg)	0.46	0.48	0.007	0.82

[1] CON = control; SB = 4 g of sodium butyrate (Adimix, Adisseo, China) supplemented from 4 to 60 d of age.
[2] T = treatment effect.

Table 7. Results of the histology analysis of the gastrointestinal organs of bull calves in the control (CON, $n = 8$) and supplemented (SB, $n = 8$) groups.

Item		Treatment [1]		SEM	*p*-Value [2]
		CON	SB		T
Rumen—ventral sac	Cell proliferation	11.21	9.75	0.67	0.63
	Total cells	2011.12	2010.75	0.66	0.63
	Mitotic index	0.005	0.005	0.0003	0.67
	Papillary height (mm)	1.24	1.49	0.039	0.18
	Papillary area (mm)	9.30	6.50	0.83	0.45
Rumen—dorsal sac	Papillary height (mm)	2.68	2.35	0.006	0.27
	Papillary area (mm)	4.60	3.70	0.01	0.15
Omasum	Cell proliferation	18.50	14.50	0.87	0.33
	Total cells	2019.0	2014.5	0.87	0.32
	Mitotic index	0.007	0.009	0.0004	0.35
	Papillary area (mm)	0.13	0.09	0.01	0.25
	Papillary height (mm)	0.13	0.22	0.002	0.34
Abomasum	Cell proliferation	8.56	10.12	0.36	0.35
	Fossette depth (mm)	0.27	0.22	0.006	0.08
	Glandular depth (mm)	0.15	0.12	0.003	0.20
Duodenum	Cell proliferation	23.3	22.8	0.77	0.87
	Villus height (mm)	0.37	0.41	0.01	0.43
	Villus area (mm)	0.51	0.65	0.012	0.51
	Crypt depth (mm)	0.31	0.28	0.05	0.27
Ileum	Cell proliferation	40.5	40.4	2.54	0.99
	Villus height (mm)	0.28	0.28	0.007	0.82
	Villus area (mm)	0.32	0.35	0.001	0.65
	Crypt depth (mm)	0.28	0.27	0.005	0.64
Colon	Cell proliferation	13.0	10.10	0.71	0.39
	Crypt depth (mm)	0.37	0.35	0.008	0.48

[1] CON = control; SB = 4 g of sodium butyrate (Adimix, Adisseo, China) supplemented from 4 to 60 d of age;
[2] T = treatment effect.

3.3. Blood Parameters and Health Scores

No differences were found regarding the evaluated blood parameters ($p > 0.05$, Tables 8 and 9). A week effect was observed on urea, glucose, and IGF-1 concentration, and mean corpuscular volume, absolute lymphocytes, and segmented neutrophils during the pre-weaning phase ($p < 0.05$, Tables 8 and 9), and on glucose in the post-weaning period ($p < 0.05$, Table 8). The IGF-1 values increased over time. In contrast, glucose levels followed a divergent pattern, decreasing, and this trend persisted post weaning. Urea exhibited higher values in the third and fourth weeks but decreased thereafter. The mean corpuscular volume and neutrophils exhibited a decreasing trend over time. In contrast, lymphocyte levels showed an increase throughout the weeks.

Table 8. Pre- and post-weaning blood concentrations of β–hydroxybutyrate (BHB), urea, glucose, and insulin-like growth factor type I (IGF-1) for the control (CON, $n = 23$) and supplemented (SB, $n = 23$) groups in the pre-weaning period and the control (CON, $n = 15$) and supplemented (SB, $n = 15$) groups in the post-weaning period.

Item	Treatment [1]		SEM	p-Value [2]		
	CON	SB		T	W	T × W
Pre-weaning						
BHB (mmol/L)	0.2	0.1	0.03	0.65	0.17	0.92
Urea (mg/dL)	11.4	12.2	4.0	0.13	<0.001	0.37
Glucose (mg/dL)	100.4	104.9	12.4	0.21	<0.001	0.27
IGF-1 (ng/mL)	94	85.4	34.6	0.89	<0.001	0.36
Post-weaning						
BHB (mmol/L)	0.3	0.3	0.02	0.68	0.13	0.59
Urea (mg/dL)	22.5	32.5	5.6	0.23	0.29	0.46
Glucose (mg/dL)	88.5	90.7	7.5	0.21	<0.001	0.19
IGF-1 (ng/mL)	160.7	179.7	21.6	0.16	0.18	0.12

[1] CON = control; SB = 4 g of sodium butyrate (Adimix, Adisseo, China) supplemented from 4 to 60 d of age;
[2] T = treatment effect; W= week effect, T × W = treatment by week interactions.

Table 9. Hemogram of samples collected on day 0, 30, and 60, of dairy calves in the control (CON, $n = 23$) and supplemented (SB, $n = 23$) groups in the pre-weaning period.

Item	Treatment [1]		SEM	p-Value [2]		
	CON	SB		T	W	T × W
Erythrocytes (×10^6/μL)	7.8	7.8	1.1	0.93	0.30	0.81
Hemoglobin (q/dL)	10.8	10.9	1.5	0.93	0.69	0.69
Mean corpuscular volume (fL)	44.6	45	4.1	0.61	0.002	0.62
Packet cell volume (%)	34.7	35	4.8	0.83	0.75	0.73
Mean corpuscular hemoglobin concentration (%)	31.3	31.2	1.1	0.55	0.96	0.57
Platelets (×10^3/μL)	389	376.1	107.4	0.32	0.24	0.12
Total leucocytes	10,357.2	10,441.8	2330.0	0.97	0.26	0.65
Absolute monocytes (/μL)	438.7	519.4	376.0	0.67	0.19	0.05
Absolute lymphocytes (/μL)	4872.6	5262.3	1450.0	0.23	<0.001	0.62
Segmented neutrophils (/μL)	4920.8	4534.1	1920.0	0.19	<0.001	0.48
Band neutrophils (/μL)	29.2	42.4	5.9	0.17	0.22	0.54
Eosinophils (/μL)	86.3	72.4	25.7	0.64	0.75	0.03
Plasmatic protein (g/dL)	6.0	6.0	0.7	0.36	0.19	0.30

[1] CON = control; SB = 4 g of sodium butyrate (Adimix, Adisseo, China) supplemented from 4 to 60 d of age.
[2] T = treatment effect; W = week effect, T × W = treatment by week interactions.

While assessing the impact of sodium butyrate on clinical responses, no treatment effect was observed on the health scores of calves. However, a significant week effect was identified for fecal scores ($p < 0.001$) during the pre-weaning period, particularly in weeks

2 and 3. Additionally, respiratory scores showed significant variations during both the pre- and post-weaning periods, specifically in weeks 3, 4, and the 9th week (one week after weaning) ($p < 0.001$, Table 10). No diarrhea treatments were administered for SB animals, while for the CON group, two animals had to be treated with antibiotics, non-steroidal anti-inflammatory, and oral hydration fluids, with an average of 2.0 ± 1.4 days in treatment.

Table 10. Health scores during the pre- and post-weaning period of dairy calves in the control (CON, n = 23) and supplemented (SB, n = 23) groups in the pre-weaning period and the control (CON, n = 15) and supplemented (SB, n = 15) groups in the post-weaning period.

Item	Treatment [1]		SEM	p-Value [2]		
	CON	SB		T	W	T × W
Pre-weaning						
Fecal score	0.49	0.52	0.02	0.85	<0.001	0.96
Respiratory score	0.68	0.66	0.04	0.64	<0.001	0.67
Days with respiratory score > 4	1.65	1.26	0.18	0.36	-	-
Days with fever	1.39	1.39	0.2	0.81	-	-
Days with diarrhea	8.78	8.96	0.75	0.99	-	-
Days with severe diarrhea	3.22	2.61	0.27	0.35	-	-
Post-weaning						
Fecal score	0.04	0.05	0.007	0.36	0.61	0.89
Respiratory score	1.10	1.01	0.04	0.47	<0.001	0.69
Days with respiratory score > 4	0	0	-	-	-	-
Days with fever	0.52	0.49	0.13	0.18	-	-

[1] CON = control; SB = 4 g of sodium butyrate (Adimix, Adisseo, China) supplemented from 4 to 60 d of age. [2] T = treatment effect; W = week effect, T × W = treatment by week interactions. The "-" means that this outcome was not tested W or T × W.

4. Discussion

The use of additives in dairy management has increased widely, especially of products that could be a potential alternative to antimicrobial growth promoters and that can assist animal development [2]. Previous work suggested that the inclusion of exogenous butyrate could be an alternative to help GIT development in calves [6,12] and improve animals' health [13]. However, the type of exogenous butyrate, such as protected sources, could preserve it from ruminal microbial action. However, even when feeding unprotected sources of butyrate in the MR, feeding it with a milk calf feeder bucket facilitates ruminal bypass and increases butyrate activity in the GIT [14]. Additionally, besides the dose and the source type of butyrate used, the duration of administration can exert a major impact on the practical results of butyrate supplementation [9,32]. Even so, the route, dosage, and source of butyrate still vary a lot among researchers. In the present study, an unprotected source of butyrate was chosen to be fed in the liquid diet. The chosen dosage was like the average reported by other authors [10,17,32–35]. However, the period of its supplementation through MR was longer than most of what has been published so far, with additional evaluation of outcomes and carryover effects. Thus, this study aimed to evaluate the supplementation in MR of unprotected sodium butyrate for pre-weaned dairy calves and its impact on performance, gut development, and health.

4.1. Feed Intake, Rumen Parameters, Performance, and Feed Efficiency

Contrary to what was expected, sodium butyrate supplementation did not exert a positive effect on the feed intake, weight gain, performance, and digestibility of the supplemented calves. These findings corroborate others that did not demonstrate increases in performance [36,37], DMI [19,37,38], or digestibility [34,36,39] through butyrate supplementation. Reinforcing our findings, previous works found no differences in ADG and biometrical evaluations such as WH, HG, and RH between supplemented and control animals when evaluating the same product and dosage used in this present work [35]. As for nutrient digestibility, it is also important to mention that previous findings showed an increase in the digestibility of some components like fat, ash, and calcium for calves fed with a soy-based MR [40]. The MR used in the present trial had a high percentage of protein derived from milk and wheat protein. Therefore, the diet protein source could also have impacted the results. Thus, not all supplementation conditions seem to favor the use of butyrate as an additive for calves, and other experiments using it as a growth promoter supplement for this animal category are not consistent with body weight gain and performance results [16,41,42]. Thus, although the digestibility was similar between treatment groups, the positive treatment-by-week interaction for starter intake, and consequently the total DMI for the CON group, suggest a possible detrimental effect of butyrate supplementation. Therefore, given the different routes of supplementation, different doses, and differences in additive protection or not, it is still difficult to affirm the possibility of using sodium butyrate as an additive to increase performance, intake, and digestibility in dairy calves.

Most of the butyric acid produced in the rumen as a result of the feed fermentation by microorganisms, especially fibrous carbohydrates, is metabolized in the rumen epithelium cells and transformed into BHB [43]. This metabolization provides energy for these cells in acetyl-CoA and decreases the load of butyric acid to the liver and peripheral tissues [44]. The effects of time on the concentrations of urea, butyric acid, and C2:C3 ratio in rumen content can be directly explained by the increase in DMI as calves grow, increasing the availability of substrates for ruminal fermentation. Since no differences were found for ruminal parameters and development, no differences in blood parameters were expected.

4.2. Digestibility, Comparative Slaughter, and Histology

Dietary butyrate is mostly metabolized by the ruminal and gut epithelium, providing energy to the local cells and increasing mitotic indices and cellular growth [43]. Since, in the present experiment, butyrate was administered through the liquid diet, therefore bypassing the rumen, changes in GIT development were expected. However, the histological results found herein indicate that the supplementation strategy used was ineffective at promoting the beneficial effects of butyrate on the cellular cycle at the intestinal level of the calves. Previous work reported that calves fed rumen-protected butyrate in MR before weaning had increased ruminal papilla length and width [19]. Other research reported that the supplementation of sodium butyrate, even produced through the MR, may increase the concentration of butyrate in the rumen [45], probably due to an impaired closure of the esophageal leak [46,47]. This effect on the ruminal papillae was not observed in our study using unprotected sodium butyrate. In the present study, nipple buckets were used to feed MR to the calves, which may have resulted in less milk being directed to the rumen. Previous studies comparing protected and unprotected sodium butyrate in the starter found that the unprotected source increased the papilla surface in the rumen and the protected source increased mucosa thickness in the abomasum and tended to increase the whole organ weight [13]. A similar study, using the same commercial sodium butyrate and the same dose as in the present study, found that animals that were euthanized earlier, at 30 days of age, had higher ruminal and duodenum development [35]. Thus, the addition of supplements could potentially enhance health during the initial weeks of life, particularly in light of the occurrence of neonatal diseases.

4.3. Health Scores

Lastly, since one of the hypotheses of the present work was that sodium butyrate would influence health parameters, health scores were evaluated daily to check if the additive would have a positive impact on these outcomes. Nevertheless, no difference was found for the health results, contradicting the hypothesis that the sodium butyrate could help overcome neonatal disease challenges, as well as helping to regulate the inflammation process and status. Similar to the present study, earlier studies did not find consistent effects from supplementing butyrate in the MR on the health status of evaluated calves [41].

Using the same commercial additive as the current research, earlier findings reported lower diarrhea morbidity and recurrence; however, there is no information on the type of milk given in the liquid diet, diet management, facilities, and time of the year the trial was conducted [35]. Thus, these differences in the health status of supplemented animals could be due to management challenges. Since diarrhea is a clinical sign of several diseases frequently caused by bacterial, protozoans, and viral pathogens, we can expect to find controversial results in the literature. Unfortunately, little is known about sodium butyrate's effects on the ruminant immune response. New experiments should be performed, especially focusing on the first 30 days of life, to verify the supplementation effects on the immunomodulation, performances, and physiology of calves to overcome neonatal disease.

5. Conclusions

The absence of significant changes in response to the 4g dose of sodium butyrate (0.5% of MR DM) in this study highlights the nuanced nature of animal physiology. This may suggest that the intervention did not elicit the anticipated physiological responses or that the physiological systems demonstrated stability and adaptability. The present findings contribute to a comprehensive understanding, guiding future research, and acknowledging the complexity of biological systems. These insights underscore the need for refined approaches, optimal dosages, and a holistic interpretation of experimental outcomes in the study of animal physiology.

Author Contributions: Conceptualization, S.G.C. and M.M.C.; methodology, A.L.B., J.P.C. and L.F.M.N.; software, J.P.C. and H.C.D.N.; validation, A.L.B., S.G.C., J.P.C., L.F.M.N. and H.C.D.N.; formal analysis, J.P.C. and H.C.D.N.; investigation, A.L.B., S.G.C., J.P.C., L.F.M.N., F.S.M., L.G.R.P., T.R.T., W.A.C. and S.d.F.C.; resources, S.G.C., F.S.M., L.G.R.P., T.R.T., W.A.C. and M.M.C.; data curation, A.L.B. and J.P.C.; writing—original draft preparation, A.L.B., S.G.C. and J.P.C.; writing—review and editing, S.G.C., J.P.C., H.C.D.N., C.S.S., F.S.M., L.G.R.P., T.R.T., W.A.C. and M.M.C.; supervision, S.G.C., J.P.C., C.S.S. and M.M.C.; project administration, S.G.C., F.S.M., L.G.R.P., T.R.T. and M.M.C.; funding acquisition, S.G.C., F.S.M., L.G.R.P., T.R.T. and M.M.C. All authors have read and agreed to the published version of the manuscript.

Funding: This research was funded by Embrapa Dairy Cattle and Coordenação de Aperfeiçoamento de Pessoal de Nível Superior (CAPES, Brasília, Brazil), project number: 20500.18/0005-2. The APC was funded by Fundação de Amparo à Pesquisa do Estado de Minas Gerais (FAPEMIG, Project number: RED-00131-23).

Institutional Review Board Statement: All procedures were approved by the Ethics Committee of Embrapa Dairy Cattle, Brazil (protocol number 9078250118).

Informed Consent Statement: Not Applicable.

Data Availability Statement: Data are contained within the article.

Acknowledgments: The authors thank the Brazilian Agricultural Research Corporation (EMBRAPA), Embrapa Dairy Cattle for funding this research and providing its facilities for funding this research (project number: 20500.18/0005-2), Coordenação de Aperfeiçoamento de Pessoal de Nível Superior (CAPES, Brasília, Brazil) for granting the scholarship and Fundação de Amparo à Pesquisa do Estado de Minas Gerais (FAPEMIG) for funding the APC (project number: RED-00131-23).

Conflicts of Interest: The authors declare no conflict of interest.

References

1. Kühn, I.; Iversen, A.; Finn, M.; Greko, C.; Burman, L.G.; Blanch, A.R.; Vilanova, X.; Manero, A.; Taylor, H.; Caplin, J.; et al. Occurrence and Relatedness of Vancomycin-Resistant Enterococci in Animals, Humans, and the Environment in Different European Regions. *Appl. Environ. Microbiol.* **2005**, *71*, 5383–5390. [CrossRef]
2. Sneeringer, S.; MacDonald, J.M.; Key, N.; McBride, W.D.; Mathews, K. *Economics of Antibiotic Use in U.S. Livestock Production*; USDA ERS: Washington, DC, USA, 2015.
3. Litherland, N.B. Dairy Calf Liquid Feeding Strategies to Cope with New Antibiotic Use Regulations. In Proceedings of the Four State Dairy Nutrition & Management Conference, Dubuque, IA, USA, 9–10 June 2010; pp. 96–101.

4. McGuirk, S.M. Disease Management of Dairy Calves and Heifers. *Vet. Clin. N. Am. Food Anim. Pract.* **2008**, *24*, 139–153. [CrossRef] [PubMed]
5. Urie, N.J.; Lombard, J.E.; Shivley, C.B.; Kopral, C.A.; Adams, A.E.; Earleywine, T.J.; Olson, J.D.; Garry, F.B. Preweaned Heifer Management on US Dairy Operations: Part I. Descriptive Characteristics of Preweaned Heifer Raising Practices. *J. Dairy Sci.* **2018**, *101*, 9168–9184. [CrossRef] [PubMed]
6. O'Hara, E.; Kelly, A.; McCabe, M.S.; Kenny, D.A.; Guan, L.L.; Waters, S.M. Effect of a Butyrate-Fortified Milk Replacer on Gastrointestinal Microbiota and Products of Fermentation in Artificially Reared Dairy Calves at Weaning. *Sci. Rep.* **2018**, *8*, 14901. [CrossRef]
7. Bourriaud, C.; Robins, R.J.; Martin, L.; Kozlowski, F.; Tenailleau, E.; Cherbut, C.; Michel, C. Lactate Is Mainly Fermented to Butyrate by Human Intestinal Microfloras but Inter-Individual Variation Is Evident. *J. Appl. Microbiol.* **2005**, *99*, 201–212. [CrossRef]
8. Hamer, H.M.; Jonkers, D.; Venema, K.; Vanhoutvin, S.; Troost, F.J.; Brummer, R.-J. Review Article: The Role of Butyrate on Colonic Function. *Aliment. Pharmacol. Ther.* **2007**, *27*, 104–119. [CrossRef] [PubMed]
9. Canani, R.B.; Di Costanzo, M.; Leone, L.; Pedata, M.; Meli, R.; Calignano, A. Potential Beneficial Effects of Butyrate in Intestinal and Extraintestinal Diseases. *World J. Gastroenterol.* **2011**, *17*, 1519–1528. [CrossRef]
10. Niwińska, B.; Hanczakowska, E.; Arciszewski, M.B.; Klebaniuk, R. Review: Exogenous Butyrate: Implications for the Functional Development of Ruminal Epithelium and Calf Performance. *Animal* **2017**, *11*, 1522–1530. [CrossRef]
11. Manrique Vergara, D.; Sánchez, M.E.G. Ácidos Grasos de Cadena Corta (Ácido Butírico) y Patologías Intestinales. *Nutr. Hosp.* **2017**, *34*, 58–61. [CrossRef]
12. Koch, C.; Gerbert, C.; Frieten, D.; Dusel, G.; Eder, K.; Zitnan, R.; Hammon, H.M. Effects of Ad Libitum Milk Replacer Feeding and Butyrate Supplementation on the Epithelial Growth and Development of the Gastrointestinal Tract in Holstein Calves. *J. Dairy Sci.* **2019**, *102*, 8513–8526. [CrossRef]
13. Wu, D.L.; Meng, Q.S.; Wang, Y.D.; Wang, M.Y.; Xu, E.H.; Xiao, L.; Xu, M. Dietary Supplementation of Free or Two Fat-Coated Sodium Butyrate with Varying Release Times on Gastrointestinal Development and Tight Junctions in Preweaning Holstein Calves. *Anim. Feed Sci. Technol.* **2022**, *285*, 115224. [CrossRef]
14. Amin, A.B.; Trabi, E.B.; Zhu, C.; Mao, S. Role of Butyrate as Part of Milk Replacer and Starter Diet on Intestinal Development in Pre-Weaned Calves. A Systematic Review. *Anim. Feed Sci. Technol.* **2022**, *292*, 115423. [CrossRef]
15. Stahl, T.C.; Hatungimana, E.; Klanderman, K.D.; Moreland, S.C.; Erickson, P.S. Sodium Butyrate and Monensin Supplementation to Postweaning Heifer Diets: Effects on Growth Performance, Nutrient Digestibility, and Health. *J. Dairy Sci.* **2020**, *103*, 10207–10218. [CrossRef]
16. Roh, S.; Kimura, N.; Sakamoto, K.; Nishihara, K.; Suzuki, K.; Katoh, K. Effects of Butyrate Supplementation in Antibiotic-Free Milk Replacer and Starter on Growth Performance in Suckling Calves. *Anim. Sci. J.* **2018**, *89*, 1486–1491. [CrossRef]
17. Górka, P.; Kowalski, Z.M.; Pietrzak, P.; Kotunia, A.; Kiljanczyk, R.; Flaga, J.; Holst, J.J.; Guilloteau, P.; Zabielski, R. Effect of Sodium Butyrate Supplementation in Milk Replacer and Starter Diet on Rumen Development in Calves. *J. Physiol. Pharmacol.* **2009**, *60*, 47–53. [PubMed]
18. Liu, W.; La, A.L.T.Z.; Evans, A.; Gao, S.; Yu, Z.; Bu, D.; Ma, L. Supplementation with Sodium Butyrate Improves Growth and Antioxidant Function in Dairy Calves before Weaning. *J. Anim. Sci. Biotechnol.* **2021**, *12*, 2. [CrossRef] [PubMed]
19. Górka, P.; Kowalski, Z.M.; Pietrzak, P.; Kotunia, A.; Jagusiak, W.; Holst, J.J.; Guilloteau, P.; Zabielski, R. Effect of Method of Delivery of Sodium Butyrate on Rumen Development in Newborn Calves. *J. Dairy Sci.* **2011**, *94*, 5578–5588. [CrossRef] [PubMed]
20. Górka, P.; Pietrzak, P.; Kotunia, A.; Zabielski, R.; Kowalski, Z.M. Effect of Method of Delivery of Sodium Butyrate on Maturation of the Small Intestine in Newborn Calves. *J. Dairy Sci.* **2014**, *97*, 1026–1035. [CrossRef]
21. McCurdy, D.E.; Wilkins, K.R.; Hiltz, R.L.; Moreland, S.; Klanderman, K.; Laarman, A.H. Effects of Supplemental Butyrate and Weaning on Rumen Fermentation in Holstein Calves. *J. Dairy Sci.* **2019**, *102*, 8874–8882. [CrossRef]
22. Khan, M.A.; Lee, H.J.; Lee, W.S.; Kim, H.S.; Kim, S.B.; Ki, K.S.; Park, S.J.; Ha, J.K.; Choi, Y.J. Starch Source Evaluation in Calf Starter: I. Feed Consumption, Body Weight Gain, Structural Growth, and Blood Metabolites in Holstein Calves. *J. Dairy Sci.* **2007**, *90*, 5259–5268. [CrossRef]
23. Chaney, A.L.; Marbach, E.P. Modified Reagents for Determination of Urea and Ammonia. *Clin. Chem.* **1962**, *8*, 130–132. [CrossRef] [PubMed]
24. AOAC International. *Official Methods of Analysis*, 19th ed.; AOAC International: Gaithersburg, MD, USA, 2012.
25. Van Soest, P.J.; Robertson, J.B.; Lewis, B.A. Methods for Dietary Fiber, Neutral Detergent Fiber, and Nonstarch Polysaccharides in Relation to Animal Nutrition. *J. Dairy Sci.* **1991**, *74*, 3583–3597. [CrossRef] [PubMed]
26. McGuirk, S.M.; Peek, S.F. Timely Diagnosis of Dairy Calf Respiratory Disease Using a Standardized Scoring System. *Anim. Health Res. Rev.* **2014**, *15*, 145–147. [CrossRef]
27. CFMV; Conselho Federal de Medicina Veterinária. *Guia Brasileiro de Boas Práticas Para Eutanásia Em Animais—Conceitos e Procedimentos*, 1st ed.; CFMV: Brasília, Brazil, 2013.
28. Costa, S.F.; Pereira, M.N.; Melo, L.Q.; Resende Júnior, J.C.; Chaves, M.L. Alterações Morfológicas Induzidas Por Butirato, Propionato e Lactato Sobre a Mucosa Ruminal e a Epiderme de Bezerros: I Aspectos Histológicos. *Arq. Bras. Med. Veterinária e Zootec.* **2008**, *60*, 1–9. [CrossRef]

29. R Core Team. *R: A Language and Environment for Statistical Computing*; R Foundation for Statistical Computing: Vienna, Austria, 2019.
30. Suk, H.W.; West, S.G.; Fine, K.L.; Grimm, K.J. Nonlinear Growth Curve Modeling Using Penalized Spline Models: A Gentle Introduction. *Psychol. Methods* **2019**, *24*, 269–290. [CrossRef]
31. Wobbrock, J.O.; Findlater, L.; Gergle, D.; Higgins, J.J. The Aligned Rank Transform for Nonparametric Factorial Analyses Using Only Anova Procedures. In Proceedings of the SIGCHI Conference on Human Factors in Computing Systems, Vancouver, BC Canada, 7–12 May 2011; ACM: New York, NY, USA, 2011; pp. 143–146.
32. Górka, P.; Kowalski, Z.M.; Zabielski, R.; Guilloteau, P. Invited Review: Use of Butyrate to Promote Gastrointestinal Tract Development in Calves. *J. Dairy Sci.* **2018**, *101*, 4785–4800. [CrossRef] [PubMed]
33. Bedford, A.; Gong, J. Implications of Butyrate and Its Derivatives for Gut Health and Animal Production. *Anim. Nutr.* **2018**, *4*, 151–159. [CrossRef] [PubMed]
34. Rice, E.M.; Aragona, K.M.; Moreland, S.C.; Erickson, P.S. Supplementation of Sodium Butyrate to Postweaned Heifer Diets: Effects on Growth Performance, Nutrient Digestibility, and Health. *J. Dairy Sci.* **2019**, *102*, 3121–3130. [CrossRef]
35. Nicola, M.S.; Kalb, A.L.; Barbosa, A.A.; Velasquez, B.E.S.; Rincon, J.A.A.; Feijó, J.O.; Dellagostin, E.N.; Martins, A.W.S.; Blödorn, E.B.; Domingues, W.B.; et al. Butyrate Supplementation in the Liquid Diet of Dairy Calves Leads to a Rapid Recovery from Diarrhea and Reduces Its Occurrence and Relapses in the Preweaning Period. *J. Dairy Sci.* **2023**, *106*, 7908–7923. [CrossRef]
36. Burakowska, K.; Penner, G.B.; Korytkowski, Ł.; Flaga, J.; Kowalski, Z.M.; Górka, P. Canola Meal or Soybean Meal as Protein Source and the Effect of Microencapsulated Sodium Butyrate Supplementation in Calf Starter Mixture. I. Performance, Digestibility, and Selected Blood Variables. *J. Dairy Sci.* **2021**, *104*, 6646–6662. [CrossRef]
37. Ghaffari, M.H.; Hammon, H.M.; Frieten, D.; Gerbert, C.; Dusel, G.; Koch, C. Effects of Milk Replacer Meal Size on Feed Intake, Growth Performance, and Blood Metabolites and Hormones of Calves Fed Milk Replacer with or without Butyrate Ad Libitum: A Cluster-Analytic Approach. *J. Dairy Sci.* **2021**, *104*, 4650–4664. [CrossRef] [PubMed]
38. Guilloteau, P.; Zabielski, R.; David, J.C.; Blum, J.W.; Morisset, J.A.; Biernat, M.; Woliński, J.; Laubitz, D.; Hamon, Y. Sodium-Butyrate as a Growth Promoter in Milk Replacer Formula for Young Calves. *J. Dairy Sci.* **2009**, *92*, 1038–1049. [CrossRef] [PubMed]
39. Rodríguez, G.B.; Carmona, D.A.C.; Elghandour, M.M.Y.; Salem, A.Z.M.; Soto, H.R.; Sánchez, R.R.; Adegbeye, M.J. Sustainable Use of Sodium Butyrate as a Source of Bioactive Additive: Impact on Calf Growth Performance, Rumen Fermentation Characteristics, and Microbial Count. *Biomass Convers. Biorefinery* **2022**. [CrossRef]
40. Guilloteau, P.; Savary, G.; Jaguelin-Peyrault, Y.; Romé, V.; Le Normand, L.; Zabielski, R. Dietary Sodium Butyrate Supplementation Increases Digestibility and Pancreatic Secretion in Young Milk-Fed Calves. *J. Dairy Sci.* **2010**, *93*, 5842–5850. [CrossRef] [PubMed]
41. Gerbert, C.; Frieten, D.; Koch, C.; Dusel, G.; Eder, K.; Stefaniak, T.; Bajzert, J.; Jawor, P.; Tuchscherer, A.; Hammon, H.M. Effects of Ad Libitum Milk Replacer Feeding and Butyrate Supplementation on Behavior, Immune Status, and Health of Holstein Calves in the Postnatal Period. *J. Dairy Sci.* **2018**, *101*, 7348–7360. [CrossRef]
42. Hiltz, R.L.; Laarman, A.H. Effect of Butyrate on Passive Transfer of Immunity in Dairy Calves. *J. Dairy Sci.* **2019**, *102*, 4190–4197. [CrossRef]
43. Bergman, E.N. Energy Contributions of Volatile Fatty Acids from the Gastrointestinal Tract in Various Species. *Physiol. Rev.* **1990**, *70*, 567–590. [CrossRef]
44. Kristensen, N.B.; Harmon, D.L. Splanchnic Metabolism of Volatile Fatty Acids Absorbed from the Washed Reticulorumen of Steers1. *J. Anim. Sci.* **2004**, *82*, 2033–2042. [CrossRef]
45. Nazari, M.; Karkoodi, K.; Alizadeh, A. Performance and Physiological Responses of Milk-Fed Calves to Coated Calcium Butyrate Supplementation. *S. Afr. J. Anim. Sci.* **2012**, *42*, 296–303. [CrossRef]
46. Berends, H.; van Reenen, C.G.; Stockhofe-Zurwieden, N.; Gerrits, W.J.J. Effects of Early Rumen Development and Solid Feed Composition on Growth Performance and Abomasal Health in Veal Calves. *J. Dairy Sci.* **2012**, *95*, 3190–3199. [CrossRef]
47. Ellingsen, K.; Mejdell, C.M.; Ottesen, N.; Larsen, S.; Grøndahl, A.M. The Effect of Large Milk Meals on Digestive Physiology and Behaviour in Dairy Calves. *Physiol. Behav.* **2016**, *154*, 169–174. [CrossRef] [PubMed]

animals

MDPI

Article

Effects of Dietary Fiber Level and Forage Particle Size on Growth, Nutrient Digestion, Ruminal Fermentation, and Behavior of Weaned Holstein Calves under Heat Stress

Mohammad-Hossein Izadbakhsh [1], Farzad Hashemzadeh [1], Masoud Alikhani [1], Gholam-Reza Ghorbani [1], Mohammad Khorvash [1], Mostafa Heidari [1], Morteza Hosseini Ghaffari [2] and Farhad Ahmadi [3,4,*]

1 Department of Animal Sciences, College of Agriculture, Isfahan University of Technology, Isfahan 84156-83111, Iran; f.hashemzadeh@cc.iut.ac.ir (F.H.)
2 Institute of Animal Science, University of Bonn, 53115 Bonn, Germany
3 School of Agriculture, Food and Ecosystem Sciences, Faculty of Science, The University of Melbourne, Parkville, VIC 3010, Australia
4 Department of Eco-Friendly Livestock Science, Institute of Green Bio Science and Technology, Seoul National University, Pyeongchang 25354, Republic of Korea
* Correspondence: farhad.ahmadi@unimelb.edu.au

Simple Summary: Global temperatures are on the rise, and this poses a significant threat to animal farming. However, there are certain nutritional strategies that can help mitigate the harmful effects of heat stress, particularly for the growing heifer calves who play a crucial role in the future profitability and sustainability of a dairy farm. In this study, growing calves experiencing heat stress consumed more feed and digested nutrients more efficiently when offered diets with moderate content of dietary fiber and shorter alfalfa hay particle size. Feeding a diet with a moderate level of fiber resulted in better growth, lower rectal temperature, and longer lying behavior, an indication of better animal comfort. Our findings highlight the importance of dietary fiber level and forage particle size in shaping calf performance and behavior under challenging environmental conditions.

Abstract: This experiment was designed to investigate the effects of feeding diets with different fiber content and forage particle size on the performance, health, nutrient digestion, rumen fermentation, and behavioral and sorting activity of Holstein dairy calves kept under elevated environmental temperature. Sixty weaned Holstein female calves (age = 96.7 ± 7.62 days old; body weight = 82.4 ± 10.4 kg) were randomly assigned to one of 4 treatments arranged in a 2-by-2 factorial design in a 70-day experiment. Dietary forage content (moderate, 22.5%; or high, 40.0% on DM basis) and alfalfa hay particle size (short, 4.39 mm; or long, 7.22 mm as geometric mean) were the experimental factors, resulting in the following combinations: (1) high-fiber (HF) diets with forage-to-concentrate ratio of 40:60 and long particle-sized alfalfa hay (LPS; HF-LPS); (2) HF diets with short particle-sized alfalfa hay (SPS; HF-SPS); (3) moderate-fiber (MF) diets with forage-to-concentrate ratio of 22.5:77.5 with LPS (MF-LPS); and (4) MF diets with SPS (MF-SPS). The temperature–humidity index averaged 73.0 ± 1.86, indicating that weaned calves experienced a moderate extent of heat stress. Fiber level and AH particle size interacted and affected dry matter intake, with the greatest intake (4.83 kg/d) observed in MF-SPS-fed calves. Final body weight was greater in calves receiving MF vs. HF diets (164 vs. 152 kg; $p < 0.01$). Respiration rate decreased when SPS vs. LPS AH was included in HF but not MF diet. Lower rectal temperature was recorded in calves fed MF vs. HF diet. Digestibility of dry matter and crude protein was greater in calves fed MF than HF diets, resulting in lower ruminal pH (6.12 vs. 6.30; $p = 0.03$). Fiber digestibility was greater in calves fed SPS compared with those fed LPS alfalfa hay. Feeding HF compared with MF diet increased acetate but lowered propionate molar proportions. The inclusion of SPS vs. LPS alfalfa hay decreased lying time in HF diet (920 vs. 861 min; $p < 0.01$). Calves fed MF vs. HF diets spent less time eating but more time lying, which is likely indicative of better animal comfort. Dietary fiber level and forage particle size interacted and affected sorting against 19 mm particles, the extent of which was greater in HF-SPS diet. Overall, dietary fiber level had a stronger effect than forage particle size on the performance of weaned calves exposed to a moderate degree of heat stress as feeding MF vs. HF diet resulted in greater feed intake, final body weight, structural growth measures, nutrient digestion, as well as longer lying behavior. The inclusion of SPS alfalfa hay in MF diets increased feed consumption.

Keywords: calf health; environmental temperature; sorting behavior; nutrient utilization

1. Introduction

Appropriate management strategies during pre- and post-weaning phases are crucial, laying the foundations for long-term productivity, behavior, and health of heifer calves, and ultimately contribute to a successful heifer-rearing program [1,2]. The rise in global temperature hampers the profitability and sustainability of animal production systems, a problem that is expected to worsen in the future [3,4]. Due to decreased metabolic heat production and more efficient heat dissipation mechanisms, growing calves are more resistant to thermal stress than mature cows [5]. However, there is evidence that heat stress may affect the calf physiology, productivity, rumen function, and welfare [6–8]. Newer findings suggest the long-term consequences of exposure to high environmental temperatures on the health and future productivity of young animals, including calves and growing heifers [9–11]. Calves kept in hot climates may experience reduced gastrointestinal motility and a series of physiological and metabolic reactions diverting energy towards thermoregulation, which is associated with decreased feed intake and growth efficiency [10]. The nutrient digestibility was greater in heifers exposed to heat stress, likely because of declined feed intake and increased retention time in the whole gastrointestinal tract [12]. However, other studies have not reported such responses [13,14].

There are certain dietary approaches that promote feed intake and growth rate in growing animals, which may help mitigate the harmful effects of heat stress. For example, the proper balance between forage and concentrate proportion in the diet of growing calves improved growth efficiency and promoted rumen development. The forage component of the diet promotes the muscular development of the rumen [15], while concentrate intake fuels growth [16] and stimulates rumen papillae formation [17]. However, the dietary fiber content must be balanced to ensure efficient rumen functioning and to minimize metabolic dysfunctions [5,10]. Feeding excessive amounts of readily fermentable carbohydrates might raise the risk of ruminal acidosis and parakeratosis, which can have negative implications on calf health, feed utilization efficiency, and gastrointestinal function [18,19]. Identifying the proper forage-to-concentrate proportion in the diet of growing calves ensures adequate rumen development and functionality, which is associated with improved performance.

Conflicting evidence exists in the literature about the proper forage-to-concentrate ratio in the diet of weaned calves. Several studies have reported that decreasing the forage-to-concentrate ratio in the calf diet increased feed intake [20–22]. For example, Karami et al. [23] have reported that weaned calves fed diets with a forage-to-concentrate ratio of 20:80 vs. 50:50 or 35:65 ratios from 70 to 120 days of age consumed more feed, grew faster, and were more efficient in converting feed nutrients toward growth. However, Namboothiri et al. [24] have reported that decreasing the forage-to-concentrate ratio in the diet of growing calves from 6 to 12 months of age had no effect on intake but improved growth performance.

The physical form of forage is another nutritional factor that may influence the rumen function and performance of young calves [25]. Providing a diet with appropriate forage particle size is necessary to promote chewing and salivation, which increases rumen pH and supports rumen health and proper function [26,27]. According to Montoro et al. [25], feeding coarsely chopped (3 to 4 cm) grass hay compared with finely chopped (2 mm) grass hay increased feed intake and nutrient digestibility during week 8 (the week after weaning) but decreased the time calves spent in non-nutritive oral behaviors. Nemati et al. [26] have reported that feeding alfalfa hay (AH) with medium particle size (geometric particle size = 3 mm) improved feed intake and growth of weaned calves (aged 51 to 70 days) compared with a diet containing AH with a geometric particle size of 1 mm. Increased consumption of diets containing AH with medium vs. fine particle size was attributed to increased capacity and muscle development of the reticulorumen [28–30]. In contrast, minor changes in rumen development and fermentation indicators were observed in dairy calves when straw particle length increased from 3.04 to 12.7 mm (as a geometric mean) [31]. The inconsistencies could possibly be due to the physical form of the starter feed, the amount of feed supplied, and the growth stage of the calves. Previous studies have not paid much attention to the potential effects of forage particle size on post-weaning calf nutrition, especially under heat-stress conditions. Therefore, further research is needed to determine the effects of forage particle size on intake and growth, and this research aims to fill this knowledge gap and highlight the importance of forage particle size as a factor when formulating diets for post-weaned calves.

To promote optimal nutrition in weaned calves, it is important to formulate diets to provide sufficient levels of physically effective fiber, as concentrate intake increases sharply after weaning. We hypothesized that feeding a diet with moderate fiber content and long AH particles would improve growth performance and increase nutrient digestion compared with a diet with high fiber content and long AH particles. Therefore, our objective was to investigate the interaction between dietary fiber content (medium or high) and AH particle size (long or short) on performance, nutrient digestion, rumen fermentation, and the behavior of weaned calves under elevated environmental temperatures.

2. Materials and Methods

Prior to the beginning of the study, all procedures involving animals were reviewed and approved by the Animal Care and Use Committee of the Isfahan University of Technology, and the animal-involved methods were conducted in accordance with the regulations of the Iranian Council of Animal Care (1995). The experiment was carried out at a commercial dairy farm (Namfar) located in Isfahan, Iran. The trial was conducted from June to September 2021. Humidity and air temperature in the stall area were recorded daily using a hygrometer/thermometer (model: HTC-1; Zhejiang Junkaishun Industrial and Trade Co., Ltd., Jinhua, China). Temperature–humidity index (THI) was calculated using the following equation [32]: THI = 0.8 × maximum T + (minimum relative humidity/100) × (maximum T − 14.4) + 46.4, where T is the temperature (°C).

2.1. Animals, Management, and Dietary Treatments

2.1.1. Pre-Trial Handling of Calves

After birth, Holstein calves were separated from their mothers, weighed, and kept in separate indoor pens with bedding made of wheat straw. Colostrum was fed using a plastic bottle with a nipple attachment after being heat-treated for 60 min at 60 °C. The calves were kept in indoor pens for five days, and on days 2 and 3, their second and third milkings after calving (at 6:00 a.m. and 3:00 p.m.), they were given transition milk. Then, whole pasteurized milk was fed for the next two days. On their sixth day of life, calves were placed in separate outdoor hutches (1.3 m × 2.5 m) and kept there until day 70. The hutches were bedded with wood shaving and wheat straw. Whole milk was provided using an aluminum bucket at an approximate temperature of 38 °C with the following composition: 11.7 ± 0.18% total solids, 2.92 ± 0.09% protein, and 3.33 ± 0.12% fat (mean ± standard deviation). Milk was offered three times a day at 6:00 a.m., 3:00 p.m., and 8:00 p.m. from day 6 to 30 (9 L/d), twice a day at 6:00 a.m. and 3:00 p.m. from day 31 to 44 (8 L/d) and 45 to 65 (5 L/d), and then once a day (only morning, 6:00 a.m.) from day 66 to 70 (3 L/d). Weaning occurred on day 70. Animals had unlimited access to starter feed and fresh water. The starter feed included 5% forage (50% AH and 50% wheat straw) and 95% concentrate mix (5% ground barley, 55.17% ground corn, 30% soybean meal, 3% whole soybean, 1% fat powder, 1% mineral supplement, 1% vitamin supplement, 1.5%

calcium carbonate, 1.5% sodium bicarbonate, 0.3% dicalcium phosphate, 0.5% salt, and 0.03% monensin).

2.1.2. Trial Handling of Calves

Sixty weaned Holstein female calves (mean age = 96.7 ± 7.62 days old) were weighed (mean body weight = 82.4 ± 10.4 kg) and, after balancing for age and body weight, were moved to twelve group pens with five calves housed in each pen (3.0 m × 6.0 m) that were in turn each bedded with sawdust. Manure was collected and fresh bedding was added to each pen daily. Each pen was randomly allocated into 1 of 4 treatments, as follows: (1) high-fiber (HF) diets with forage-to-concentrate ratio of 40:60 with long particle-sized AH (LPS; HF-LPS); (2) HF diets with short particle-sized AH (SPS; HF-SPS); (3) moderate-fiber (MF) diets with forage-to-concentrate ratio of 22.5:77.5 with LPS AH (MF-LPS); and (4) MF diets with SPS AH (MF-SPS). Second-cut AH was harvested at 10% flowering, wilted, and baled (small rectangular bales). Subsequently, AH was chopped with a forage harvester (Golchin Trasher Hay Co., Isfahan, Iran) to obtain the long and short particle sizes. The chemical composition and particle size distribution of corn silage and AH are presented in Table 1 and Table 2, respectively. The ingredients and chemical compositions of the experimental diets are listed in Table 3 and Table 4, respectively. Diets were formulated with CPM-Dairy version 3.0 software (Cornell University, University of Pennsylvania, and the Miner Institute). Animals were fed once daily at 8:00 a.m. Each day, concentrate and forage portions were thoroughly mixed and an appropriate amount in the form of a total mixed ration was manually weighed. Fresh water and diet were freely available to all calves. According to the orts in the previous day, the total amount of feed supplied to each pen was adjusted to provide 5–10% refusals.

2.2. Sampling and Laboratory Analyses

Approximately 2 kg of each forage, including corn silage and AH, and weekly samples of feed and refusal of each pen were collected on two consecutive days of each week and stored at −20 °C until further analysis. The samples were mixed thoroughly, dried at 55 °C for 48 h in a forced-air oven, and then ground to pass through a 1 mm screen. Chemical compositional analysis of ground samples was made using the AOAC [33] standard methods: ether extract (AOAC International 920.39), N (AOAC International 990.03), and ash (AOAC International 942.05). Analyses of neutral detergent fiber (NDF) and acid detergent fiber (ADF) were performed using ANKOM^{220} Fiber Analyzer (ANKOM Technology, Macedon, NY, USA). The NDF procedure was inclusive of heat-stable α-amylase and residual ash. Non-fiber carbohydrates (NFC) were calculated as NFC = 100 − (NDF + crude protein + ether extract + ash). Forage, feed, and refusal samples were sieved to determine the particle size distribution. The Penn State Particle Separator, which consists of 19, 8, and 1.18 mm diameter sieves and a pan, was used to monitor the particle size distribution. Geometric mean particle length (GMPL) was calculated in accordance with the ASAE [34].

Daily feed intake for each group was measured by weighing the amount of feed supplied and refused. Dry matter intake (DMI; kg/d) was then calculated on a pen basis. Individual body weight measurements were made before morning feeding at 0, 10, 25, 40, 55, and 70 days of the study using an electronic scale. Body weights recorded for each calf were used to calculate individual average daily gain (ADG). Feed conversion ratio was calculated as average daily feed intake/ADG of the pen. Skeletal growth measures, including withers height, hip height, hip width, belly girth, heart girth, and body length, were measured before morning feeding at 0, 10, 25, 40, 55, and 70 days of the study, as described by Khan et al. [35]. Calves were evaluated weekly for body condition score (BCS) using a 5-point scale [36].

Respiratory rate (breaths/min) and rectal temperature and were recorded thrice each week. The frequency of movement of the abdominal muscles in the flank during respiration was quantified visually to determine the respiratory rate and was expressed as breaths/min.

Rectal temperature was determined between 2:00 and 3:00 p.m. with the insertion of a thermometer (PIC Vedodigit II, Pic Solution Co., Como, Italy) directly into the rectum for 1 min. Calf health was monitored daily during the experiments. The University of Wisconsin Calf Health Scoring chart [37] was utilized to visually assess and score the health status of animals. For the ear and eye score categories, a score of 0 indicates a normal, healthy animal.

To determine nutrient digestibility, fecal samples were taken directly from the calf rectum 3 h after morning feeding on 5 consecutive days in weeks 5 and 10. Fecal samples were kept frozen (−20 °C) until analysis. The samples were later thawed overnight and dried at 55 °C for 72 h in a forced-air oven. These samples were ground to pass through a 1 mm screen and analyzed for nutrients, as described earlier. Acid-insoluble ash was used as an internal marker to measure the apparent total-tract digestibility of nutrients [38].

During the trial, rumen fluid was collected twice—on day 35 and day 70—4 h after the morning feeding. The collection was undertaken by using a stomach tube attached to a vacuum pump. To prepare the fluid for analysis of VFA via gas chromatography, 4 mL of the rumen fluid, filtered through four layers of cheesecloth, was mixed with 1 mL of 25% metaphosphoric acid. VFA analysis was undertaken using CP-9002 gas chromatography (Chrompack, Middelburg, The Netherlands). The chromatography used a 50 m (0.32 mm i.d.) silica-fused column (CP-Wax Chrompack Capillary Column by Varian, Palo Alto, CA, USA) with crotonic acid (1:7, v/v) as an internal standard, as described by Hashemzadeh-Cigari et al. [39].

Behavioral activities, including eating, rumination, lying, standing drinking, and nonnutritive oral behavior, were recorded at d 34 and 69 of the study for 24 h starting from 8:00 a.m., as detailed previously [40]. Four trained researchers who were not aware of the treatments were employed to record the behavioral activities. Sorting index was computed as a fraction of actual intake to expected intake of particles retained on each sieve [41].

2.3. Statistical Analysis

The MIXED PROC in SAS (2003, version 9.4, SAS Institute Inc., Cary, NC, USA) was used for data analysis. The outcome assessed varied depending on whether the observational unit was the pen or the individual calf. The model included the random effect of individual animal as the observational unit for calf body weight, ADG, structural growth measures, body condition score, health status parameters, nutrient digestion, rumen fermentation parameters, and feeding behaviors. The model included the random effect of the pen as the experimental unit for nutrient intake and feed conversion ratio, based on the experimental design and statistical models for pen studies described by St-Pierre [42], Tempelman [43], and Bello et al. [44]. Time served as a repeated measure for parameters including nutrient intake, performance, structural growth, nutrient digestion, rumen fermentation, feeding behaviors and sorting index, and health parameters. The model included fixed effects of fiber content, AH particle size, sampling time, and their interactions. An assessment was conducted on three variance–covariance structures (autoregressive type 1, Toeplitz, and compound symmetry), and, upon considering the Bayesian information criterion, the autoregressive type 1 covariance structure was identified as the most fitting choice. Initial calf measurements, including body weight and skeletal growth measures, were considered covariates for the analysis of body weight and skeletal growth data. The normality of all residuals was assessed using the Shapiro–Wilk statistic via the UNIVARIATE PROC of SAS [45]. Additionally, the homogeneity of variance was examined using Levene's test and visually evaluated through quantile–quantile plots. Where the data did not adhere to the assumptions of normality, a log transformation (base 10) was applied. The Tukey–Kramer adjustment was used for means comparison. Significance was established at a threshold of $p \leq 0.05$, while trends were identified within the range of $0.05 < p \leq 0.10$.

Table 1. Chemical composition (±SD) of corn silage and alfalfa hay.

Chemical Composition, % of DM	Corn Silage	Alfalfa Hay
DM, %	25.4 ± 0.16	97.7 ± 0.22
Organic matter	93.4 ± 0.37	88.6 ± 0.32
Crude protein	7.20 ± 0.04	13.8 ± 0.42
Crude protein	2.41 ± 0.31	1.82 ± 0.28
NDF	53.4 ± 0.91	46.2 ± 1.10
NDF	32.3 ± 0.54	37.8 ± 0.22

Table 2. Physical characteristics of corn silage and alfalfa hay.

Items	Corn Silage	Alfalfa Hay Particle Size		SEM	p Value [2]
		Long	Short		
% DM retained on sieve [1]					
19 mm	22.9 ± 4.42	20.9	3.70	1.42	<0.01
8 mm	61.1 ± 2.22	27.8	33.4	1.66	0.02
1.18 mm	14.9 ± 0.93	38.6	45.3	1.58	<0.01
<1.18 mm (pan)	1.10 ± 1.77	12.7	17.6	2.32	0.14
$peNDF_8$	44.8 ± 1.38	22.5	17.1	1.16	<0.01
$peNDF_{1.18}$	52.8 ± 0.98	40.2	38.0	1.07	0.14
GMPL [3], mm	12.0 ± 0.95	7.22	4.39	0.41	<0.01

[1] Particle size distribution was made using the Penn State Particle Separator [46]. PeNDF denotes physically effective NDF. $PeNDF_8$ and $peNDF_{1.18}$ were determined as multiplication of dietary NDF content by pef_8 and $pef_{1.18}$, respectively. Pef_8 and $pef_{1.18}$ are physical effectiveness factors as proportions of particles retained on 2 [47] and 3 sieves [46], respectively. [2] p value showing the level of significance between short and long alfalfa hay. [3] GMPL = geometric mean particle length. SEM = standard error of the means.

Table 3. Ingredients of the experimental diets.

Ingredients, % of DM	Dietary Fiber Content	
	Moderate Fiber	High Fiber
Alfalfa hay	18.0	32.0
Corn silage	4.50	8.00
Corn grain, ground	16.5	12.38
Barley grain, ground	26.6	19.88
Soybean meal	14.25	14.25
Wheat bran	15.63	8.00
Fat powder [1]	0.25	1.63
Vitamin supplement [2]	0.75	0.75
Mineral supplement [3]	0.75	0.75
Salt	0.25	0.25
Calcium carbonate	1.25	0.88
Bicarbonate	1.00	1.00
Magnesium oxide	0.25	0.25

[1] Fat powder contained 99.5% total fat, 40% C16:0, 50% C18:0, and 10% C18:1. [2] Contained (per kg): vitamin A = 1,300,000 IU, vitamin D_3 = 360,000 IU, vitamin E = 12,000 IU. [3] Contained (per kg): zinc = 16 g, manganese = 10 g, copper = 4 g, iodine = 0.15 g, cobalt = 0.12 g, iron = 0.8 g, and selenium = 0.08 mg.

Table 4. Chemical composition and physical characteristics of experimental diets differing in fiber level (F) and alfalfa hay particle size (PS) *.

Items	Moderate Fiber		High Fiber		SEM	p Values		
	LPS	SPS	LPS	SPS		F	PS	F × PS
Chemical composition								
DM, % as fed	53.6	52.7	53.4	53.3	0.96	0.83	0.33	0.28

Table 4. *Cont.*

Items	Moderate Fiber		High Fiber		SEM	*p* Values		
	LPS	SPS	LPS	SPS		F	PS	F × PS
Organic matter, % of DM	90.9	90.6	90.4	90.9	0.69	0.47	0.88	0.71
Crude protein, % of DM	16.4	16.3	15.9	16.2	0.46	0.58	0.82	0.76
Ether extract, % of DM	3.11	3.20	4.82	5.01	0.36	<0.01	<0.01	<0.01
NDF, % of DM	28.9	28.5	33.6	32.4	0.24	<0.01	<0.01	<0.01
ADF, % of DM	13.9	13.2	18.7	17.5	0.21	<0.01	<0.01	<0.01
Forage NDF, % of DM	10.7	10.7	19.0	19.0	–	–	–	–
NFC [1], % of DM	42.4	43.0	35.8	36.4	–	–	–	–
ME, Mcal/kg DM	2.17	2.17	2.16	2.16	–	–	–	–
% DM retained on sieve [2]								
19 mm	6.4 [b]	2.4 [c]	11.3 [a]	3.6 [c]	0.48	<0.01	<0.01	<0.01
8 mm	15.3 [d]	17.6 [c]	20.3 [b]	25.2 [a]	0.40	<0.01	<0.01	<0.01
1.18 mm	60.6	62.9	51.6	55.0	0.79	<0.01	<0.01	0.29
<1.18 mm (pan)	17.7	17.1	16.8	16.2	0.61	0.20	0.37	0.82
$peNDF_8$	6.35	5.75	10.61	9.47	0.20	<0.01	<0.01	0.17
$peNDF_{1.18}$	23.9	23.8	27.8	27.9	0.19	<0.01	0.42	0.97
GMPL [3] (mm)	3.91	3.71	4.66	4.23	0.07	<0.01	<0.01	0.10

* The dietary treatments were as follows: (1) high-fiber (HF) diets with forage-to-concentrate ratio of 40:60 with long particle-sized alfalfa hay (LPS; HF-LPS); (2) HF diets with short particle-sized alfalfa hay (SPS; HF-SPS); (3) moderate-fiber (MF) diets with forage-to-concentrate ratio of 22.5:77.5 with LPS alfalfa hay (MF-LPS); and (4) MF diets with SPS alfalfa hay (MF-SPS). [1] Nonfibrous carbohydrate = 100 − (crude protein + NDF + ether extract + ash). [2] Particle size distribution was made using the Penn State Particle Separator [46]. PeNDF denotes physically effective NDF. $PeNDF_8$ and $peNDF_{1.18}$ were determined as multiplications of dietary NDF content by pef_8 and $pef_{1.18}$, respectively. Pef_8 and $pef_{1.18}$ denote physical effectiveness factors as proportions of particles retained on 2 [47] and 3 sieves [46], respectively. [3] GMPL = geometric mean particle length. SEM = standard error of the means; [a–d] Within a row, least-square means with dissimilar superscripts differ ($p \leq 0.05$).

3. Results

3.1. Meteorological Data

Throughout the experiment, the mean, maximum, and minimum THI values averaged 73.0 ± 1.86, 86.2 ± 2.79, and 62.4 ± 1.79, respectively (Figure 1).

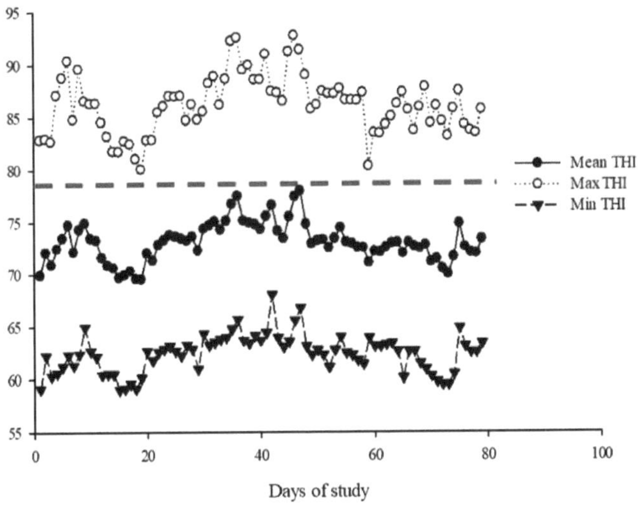

Figure 1. Temperature–humidity index (THI) during the experimental period. The cutoff point for heat stress is indicated by the dotted red line (THI ≥ 78).

3.2. Nutrient Composition and Particle Size Distribution

A large difference was observed in the particle size distribution of AH with both sizes. LPS AH had more particles retained on 19 mm screens, less particles retained on the 8.0 and 1.18 mm screens, and similar amounts of particles shorter than the 1.18 mm compared with SPS AH. GMPL and $peNDF_8$ were greater for LPS vs. SPS AH (Table 2).

Chemical composition and particle size distributions of dietary treatments are listed in Table 4. There were interactions between fiber level and AH particle size on particles retained on 19 and 8 mm sieves. Particles retained on the 19 mm screen followed the decreasing order of HF-LPS > MF-LPS > HF-SPS = MF-SPS. Particles retained on the 8 mm screen were in the order of HF-SPS > HF-LPS > MF-SPS > MF-LPS. HF vs. MF diet resulted in a lesser proportion of particles retained on the 1.18 mm sieve (61.8 vs. 53.1%, $p < 0.01$) but greater GMPL (4.45 vs. 3.81 mm; $p < 0.01$), the top two sieves (30.2 vs. 20.9%; $p < 0.01$), and $peNDF_8$ (10.0 vs. 6.0% of DM, $p < 0.01$). Diets containing SPS AH had a greater proportion of particles retained on the 1.18 mm sieve but had shorter GMPL (3.97 vs. 4.30 mm; $p < 0.01$), lower particles retained on the top two sieves, and $peNDF_8$ (7.61 vs. 8.48% of DM; $p < 0.01$) compared with diets containing LPS AH.

3.3. Performance Responses

Nutrient intakes, weight gain, feed conversion ratio, skeletal growth measures, and BCS results are listed in Table 5. Calf body weight, ADG, and DM intake at different time points (days 10, 25, 40, 55, and 70 of the experiment) are also presented in Figure 2, Figure 3 and Figure 4, respectively. Fiber levels and AH particle size interacted and affected DM, NDF, and NFC intakes, with the greatest values of DM and NFC observed in calves fed the MF-SPS diet. Forage NDF intake was greater in calves receiving HF than in MF diets (0.79 vs. 0.48 kg/d; $p < 0.01$). Calves fed MF vs. HF diet had a greater final body weight (164 vs. 152 kg; $p < 0.01$). There was a fiber level by time interaction for body weight, with calves fed the MF diets having a greater body weight than those fed HF diet from day 40 to 70 of the trial. Further, calves fed MF diets had greater ADG (1.19 vs. 1.05 kg/d; $p < 0.01$) than those fed HF diets throughout the study. Calves fed MF vs. HF diet tended to be more efficient in converting feed nutrients toward growth (3.84 vs. 3.94; $p = 0.08$).

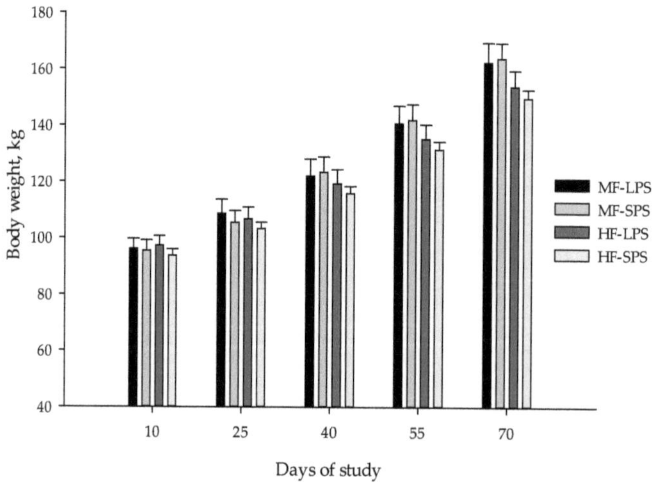

Figure 2. Calf body weight (means ± SE) at different time points. The dietary treatments were as follows: (1) high-fiber (HF) diets with forage-to-concentrate ratio of 40:60 with long particle-sized alfalfa hay (LPS; HF-LPS); (2) HF diets with short particle-sized alfalfa hay (SPS; HF-SPS); (3) moderate-fiber (MF) diets with forage-to-concentrate ratio of 22.5:77.5 with LPS alfalfa hay (MF-LPS); and (4) MF diets with SPS alfalfa hay (MF-SPS).

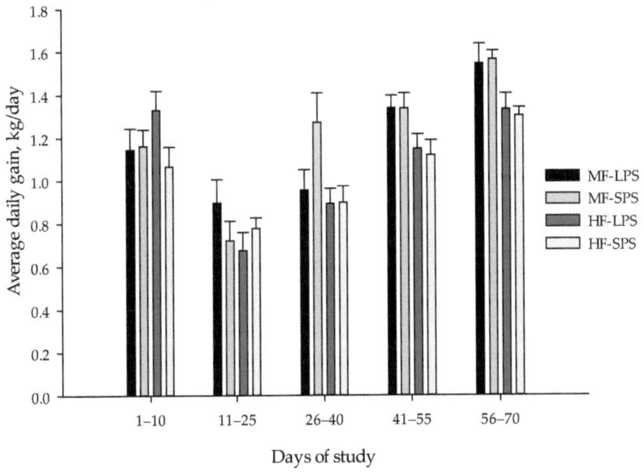

Figure 3. Calf average daily gain (means ± SE) at different time points. The dietary treatments were as follows: (1) high-fiber (HF) diets with forage-to-concentrate ratio of 40:60 with long particle-sized alfalfa hay (LPS; HF-LPS); (2) HF diets with short particle-sized alfalfa hay (SPS; HF-SPS); (3) moderate-fiber (MF) diets with forage-to-concentrate ratio of 22.5:77.5 with LPS alfalfa hay (MF-LPS); and (4) MF diets with SPS alfalfa hay (MF-SPS).

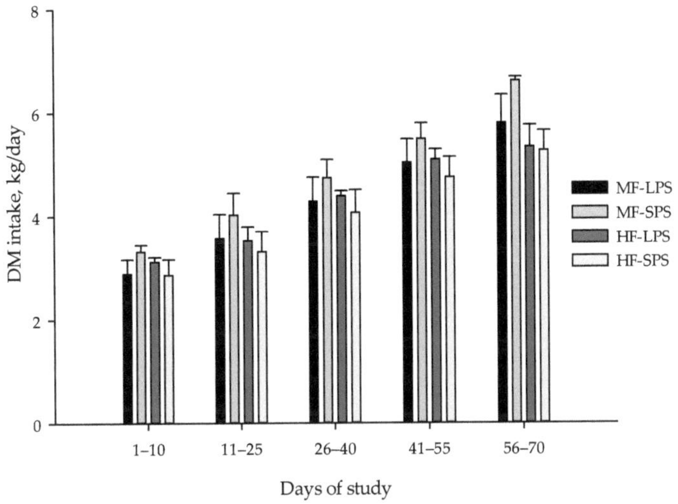

Figure 4. Calf DM intake (means ± SE) at different time points. The dietary treatments were as follows: (1) high-fiber (HF) diets with forage-to-concentrate ratio of 40:60 with long particle-sized alfalfa hay (LPS; HF-LPS); (2) HF diets with short particle-sized alfalfa hay (SPS; HF-SPS); (3) moderate-fiber (MF) diets with forage-to-concentrate ratio of 22.5:77.5 with LPS alfalfa hay (MF-LPS); and (4) MF diets with SPS alfalfa hay (MF-SPS).

Dietary fiber level and AH particle size interacted and affected heart girth and body length, with the smallest values observed in HF-SPS-fed calves. Calves offered MF diets had greater wither height, belly girth, and heart girth compared with calves fed HF diets. Fiber level and AH particle size interacted and affected BCS, with calves fed LPS AH having greater BCS compared with those fed SPS AH in HF diets, though an inverse response was observed with MF diets.

Table 5. Nutrient intake, performance, and structural growth measures of weaned calves fed diets differing in fiber level (F) and alfalfa hay particle size (PS) *.

Items	Moderate Fiber		High Fiber		SEM	*p* Values						
	LPS	SPS	LPS	SPS		F	PS	F × PS	Time (T)	F × T	PS × T	F × PS × T
Total DM intake, kg/d	4.31	4.83	4.29	4.01	0.18	0.03	0.43	0.04	<0.01	0.70	0.97	0.91
NDF intake, kg/d	1.24	1.37	1.44	1.31	0.06	0.24	0.97	0.02	<0.01	0.91	0.98	0.90
NFC intake, kg/d	1.82	2.08	1.53	1.47	0.07	<0.01	0.19	0.03	<0.01	0.31	0.97	0.94
Forage NDF intake, kg/d	0.46	0.51	0.81	0.77	0.03	<0.01	0.92	0.09	<0.01	0.36	0.97	0.96
Final body weight, kg	162.9	164.1	154.1	149.6	3.56	<0.01	0.65	0.44	–	–	–	–
ADG, kg/d	1.17	1.21	1.07	1.03	0.05	<0.01	0.94	0.42	<0.01	0.30	0.45	0.30
Feed conversion ratio [1]	3.68	3.99	4.00	3.88	0.31	0.08	0.22	0.79	<0.01	0.61	0.99	0.38
Structural growth measures												
Wither height, cm	100.2	100.3	99.7	99.2	0.27	<0.01	0.63	0.16	<0.01	0.55	0.86	0.89
Hip height, cm	101.9	102.6	102.2	101.5	0.28	0.13	0.81	0.01	<0.01	0.96	0.93	0.89
Hip width, cm	23.5	23.3	23.3	23.1	0.13	0.35	0.19	0.95	<0.01	0.10	0.75	0.96
Belly girth, cm	133.0	132.7	131.8	131.2	0.48	0.01	0.33	0.74	<0.01	0.21	0.84	0.97
Heart girth, cm	113.7	114.2	112.7	111.0	0.33	<0.01	0.04	<0.01	<0.01	0.11	0.99	0.94
Body length, cm	109.0	109.9	110.2	107.9	0.57	0.23	0.07	<0.01	<0.01	0.97	0.96	0.94
Body condition score	3.43	3.57	3.47	3.32	0.02	<0.01	0.98	<0.01	<0.01	0.59	0.87	0.91

* The dietary treatments were as follows: (1) high-fiber (HF) diets with forage-to-concentrate ratio of 40:60 with long particle-sized alfalfa hay (LPS; HF-LPS); (2) HF diets with short particle-sized alfalfa hay (SPS; HF-SPS); (3) moderate-fiber (MF) diets with forage-to-concentrate ratio of 22.5:77.5 with LPS alfalfa hay (MF-LPS); and (4) MF diets with SPS alfalfa hay (MF-SPS); NFC = non-fibrous carbohydrates. [1] Feed conversion ratio = total DM intake/ADG. SEM = standard error of mean.

3.4. Health Status Indicators

Respiration rate, rectal temperature, and eye/ear score as indicators of calf health are presented in Table 6. Weekly data on rectal temperature and respiration rate are also presented in Figures 5 and 6, respectively. Interactions existed between fiber level and forage particle size for respiration rate ($p < 0.01$), as feeding LPS vs. SPS AH increased respiration rate only in HF but not MF diets. Feeding HF vs. MF diets increased rectal temperature (38.99 vs. 39.04 °C; $p = 0.01$) and ear score ($p < 0.01$), while an opposite effect was noted for eye score. Respiration rate increased ($p = 0.01$) when calves were fed LPS vs. SPS AH.

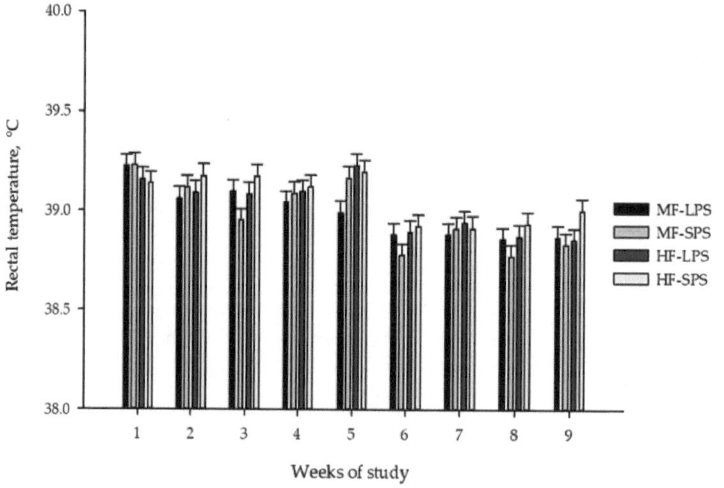

Figure 5. Calf rectal temperature (means ± SE) in weekly intervals. The dietary treatments were as follows: (1) high-fiber (HF) diets with forage-to-concentrate ratio of 40:60 with long particle-sized

alfalfa hay (LPS; HF-LPS); (2) HF diets with short particle-sized alfalfa hay (SPS; HF-SPS); (3) moderate-fiber (MF) diets with forage-to-concentrate ratio of 22.5:77.5 with LPS alfalfa hay (MF-LPS); and (4) MF diets with SPS alfalfa hay (MF-SPS).

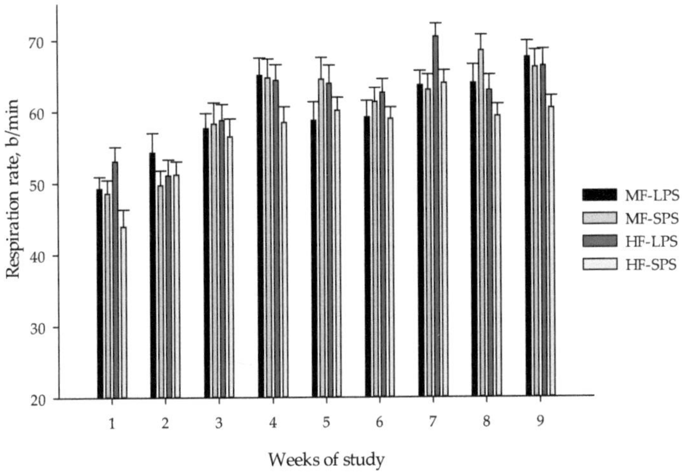

Figure 6. Calf respiration rate (means ± SE) in weekly intervals. The dietary treatments were as follows: (1) high-fiber (HF) diets with forage-to-concentrate ratio of 40:60 with long particle-sized alfalfa hay (LPS; HF-LPS); (2) HF diets with short particle-sized alfalfa hay (SPS; HF-SPS); (3) moderate-fiber (MF) diets with forage-to-concentrate ratio of 22.5:77.5 with LPS alfalfa hay (MF-LPS); and (4) MF diets with SPS alfalfa hay (MF-SPS).

Table 6. Health status indicators of weaned calves fed diets differing in fiber level (F) and alfalfa hay particle size (PS) *.

Items	Moderate Fiber		High Fiber		SEM	*p* Values						
	LPS	SPS	LPS	SPS		F	PS	F × PS	Time (T)	F × T	PS × T	F × PS × T
Respiration rate, breaths/min	59.8	60.4	61.4	56.9	0.761	0.20	0.01	<0.01	<0.01	0.58	0.91	0.98
Rectal temperature, °C	38.99	38.98	39.02	39.06	0.019	0.01	0.42	0.22	<0.01	0.50	0.89	0.41
Eye score	0.587	0.537	0.463	0.489	0.029	<0.01	0.67	0.18	<0.01	0.96	0.69	0.72
Ear score	0.371	0.314	0.404	0.491	0.029	<0.01	0.59	0.01	<0.01	0.60	0.99	0.06

* The dietary treatments were as follows: (1) high-fiber (HF) diets with forage-to-concentrate ratio of 40:60 with long particle-sized alfalfa hay (LPS; HF-LPS); (2) HF diets with short particle-sized alfalfa hay (SPS; HF-SPS); (3) moderate-fiber (MF) diets with forage-to-concentrate ratio of 22.5:77.5 with LPS alfalfa hay (MF-LPS); and (4) MF diets with SPS alfalfa hay (MF-SPS). SEM = standard error of the means.

3.5. Nutrient Digestibility and Rumen Fermentation

Nutrient digestibility and rumen fermentation variables are listed in Table 7. Nutrient digestibility remained unaffected by the interactive effect of dietary fiber level and AH particle size. Digestibility of DM, organic matter, and crude protein increased when calves were fed the MF diet rather than the HF diet. NDF and ADF digestibility were greater in calves fed SPS compared with those fed LPS AH.

Table 7. Nutrient digestion and rumen fermentation characteristics of weaned calves fed diets differing in fiber level (F) and forage particle size (PS) *.

Items	Moderate Fiber		High Fiber		SEM	*p* Values						
	LPS	SPS	LPS	SPS		F	PS	F × PS	Time (T)	F × T	PS × T	F × PS × T
Digestibility, %												
DM	72.2	77.1	68.6	69.8	2.62	0.03	0.19	0.40	0.60	0.37	0.20	0.42
Organic matter	74.2	78.8	71.2	72.0	2.11	0.04	0.24	0.38	0.59	0.41	0.22	0.40
Crude protein	72.8	77.4	68.9	66.6	2.77	0.02	0.69	0.27	0.91	0.87	0.31	0.46
NDF	49.4	60.3	49.4	53.6	2.19	0.12	<0.01	0.12	0.88	0.20	0.31	0.23
ADF	44.8	52.6	48.8	51.4	1.30	0.39	0.01	0.13	0.91	0.11	0.39	0.32
Rumen fermentation parameters												
pH	6.05	6.19	6.20	6.39	0.08	0.03	0.06	0.89	0.08	0.69	0.37	0.82
Ammonia-N, mg/dL	10.7	10.3	10.8	10.4	1.18	0.94	0.69	0.98	0.24	0.47	0.30	0.84
Total VFA, mM	83.8	82.5	85.6	85.0	4.03	0.60	0.82	0.92	0.83	0.14	0.71	0.91
Individual VFA, mol/100 mol												
Acetate	56.4	57.3	64.7	66.7	0.92	<0.01	0.13	0.55	0.29	0.73	0.48	0.73
Propionate	30.7	30.1	22.6	20.3	1.16	<0.01	0.23	0.44	0.33	0.44	0.57	0.66
Butyrate	9.68	9.44	9.84	9.96	0.35	0.35	0.82	0.58	0.76	0.52	0.61	0.63
Isobutyrate	0.55	0.48	0.55	0.65	0.04	0.03	0.71	0.02	0.12	0.08	0.29	0.43
Valerate	1.73	1.74	1.43	1.29	0.19	0.04	0.72	0.68	0.80	0.31	0.18	0.96
Isovalerate	0.94	0.91	0.92	1.12	0.08	0.26	0.34	0.16	0.12	0.09	0.83	0.99
Acetate: propionate	1.96	1.99	2.93	3.33	0.13	<0.01	0.09	0.15	0.22	0.67	0.55	0.95

* The dietary treatments were as follows: (1) high-fiber (HF) diets with forage-to-concentrate ratio of 40:60 with long particle-sized alfalfa hay (LPS; HF-LPS); (2) HF diets with short particle-sized alfalfa hay (SPS; HF-SPS); (3) moderate-fiber (MF) diets with forage-to-concentrate ratio of 22.5:77.5 with LPS alfalfa hay (MF-LPS); and (4) MF diets with SPS alfalfa hay (MF-SPS). SEM = standard error of the means.

There were no interactions between fiber level and AH particle size for rumen fermentation variables except for isobutyrate molar proportion, which was greatest in calves fed the HF-SPS diet. Calves fed HF diets had higher rumen pH compared with those fed MF diets. Ruminal pH tended to be greater in calves fed diets containing SPS AH than those fed LPS AH. Main dietary factors or their interaction had no effect on ruminal total VFA and ammonia-N concentrations. Feeding HF vs. MF diets resulted in greater acetate (65.7 vs. 56.9; $p < 0.01$), but lower propionate molar proportions (21.5 vs. 30.4; $p < 0.01$), which resulted in a greater acetate-to-propionate ratio. Valerate molar proportion also decreased (1.36 vs. 1.74; $p = 0.04$) when calves received HF rather than MF diets.

3.6. Behavioral Activities

Data on feeding behavior and sorting activity are presented in Table 8. An interaction between fiber content and particle size of AH was observed for standing ($p < 0.01$) and lying behaviors ($p < 0.01$); calves fed LPS AH spent more time standing with those fed SPS AH in MF diets, though an inverse response was observed with HF diets. Feeding SPS vs. LPS AH decreased lying behavior only in HF but not MF diets. Feeding SPS vs. LPS AH resulted in less lying time only in the HF diet. There was a time-by-AH particle size effect for drinking behavior, with calves fed SPS AH spending more time to drink compared with those fed LPS AH only on day 70 of the study. Additionally, calves receiving SPS AH spent more time eating on day 35 of the study, but this response was not observed on day 70 of the study. Calves fed MF vs. HF diets spent less time eating (205 vs. 275 min; $p < 0.01$) and ruminating (106 vs. 130 min; $p < 0.01$), but more time lying (953 vs. 891 min; $p < 0.01$).

Table 8. Feeding behaviors and sorting index of weaned calves fed diets differing in fiber level (F) and forage particle size (PS) *.

Items	Moderate Fiber		High Fiber		SEM	*p* Values						
	LPS	SPS	LPS	SPS		F	PS	F × PS	Time (T)	F × T	PS × T	F × PS × T
Feeding behavior												
Eating, min	198	211	262	288	5.53	<0.01	0.01	0.44	<0.01	0.80	<0.01	0.18
Rumination, min	103	109	126	133	2.77	<0.01	0.02	0.83	0.02	0.98	0.02	0.58
Lying, min	944	962	920	861	10.5	<0.01	0.05	<0.01	<0.01	0.51	0.97	0.22
Standing, min	250	209	207	240	9.23	0.54	0.64	<0.01	<0.01	0.93	0.65	0.77
Drinking, min	13.8	16.3	15.3	18.9	1.74	0.24	0.08	0.75	<0.01	0.40	0.02	0.04
NNOB, min	34.3	42.5	35.4	32.5	3.74	0.23	0.48	0.14	0.51	0.04	<0.01	0.37
Sorting index [1]												
19 mm	95.2 *	97.8	90.5 *	84.5 *	1.93	<0.01	0.37	0.02	0.01	0.40	0.15	0.87
8 mm	103 *	102 *	96.5 *	98.6 *	0.98	<0.01	0.61	0.05	0.56	0.19	0.88	0.86
1.18 mm	103 *	101 *	103 *	102 *	0.54	0.48	0.01	0.59	<0.01	0.99	0.09	0.97
<1.18 mm (pan)	93.7 *	93.6 *	101 *	97.6 *	1.18	<0.01	0.10	0.14	<0.01	0.52	0.60	0.98

* The dietary treatments were as follows: (1) high-fiber (HF) diets with forage-to-concentrate ratio of 40:60 with long particle-sized alfalfa hay (LPS; HF-LPS); (2) HF diets with short particle-sized alfalfa hay (SPS; HF-SPS); (3) moderate-fiber (MF) diets with forage-to-concentrate ratio of 22.5:77.5 with LPS alfalfa hay (MF-LPS); and (4) MF diets with SPS alfalfa hay (MF-SPS); NNOB = non-nutritive oral behavior, defined as oral activities such as licking, consuming non-eatable items, tongue rolling, and grooming, excluding eating or drinking behavior. [1] No sorting is indicated by a value equal to 100, sorting for is indicated by a value greater than 100, and sorting against is indicated by a value of less than 100. * $p < 0.05$; sorting values differ from 100. SEM = standard error of the means.

Fiber level and AH particle size of AH interacted ($p = 0.02$) and affected the sorting activity for long particles, as the extent of sorting against 19 mm particles was greatest in calves fed HF-SPS. All calves sorted against the 19 mm particles, the extent of which was greater in calves fed the HF than the MF diet. Regardless of AH particle size, calves fed MF sorted in favor of particles retained on the 8 mm sieve, but those fed the HF diet sorted against these particles (102.3 vs. 97.7; $p < 0.01$). All calves sorted in favor of particles retained on the 1.18 mm sieve, the extent of which was generally greater in calves fed LPS AH than in those fed SPS AH. All calves sorted against particles finer than 1.18 mm (pan) except those fed HF-LPS, which sorted for these particles.

4. Discussion

Studies on dairy cows have shown that an average THI above the threshold of 68 is a signal for the initiation of heat stress [48,49]. Rumination time or milk yield may decrease when THI reaches above this threshold. However, no criterion has been established for young calves, perhaps because it is impossible to measure performance parameters such as milk yield in growing calves [50]. Because of their greater surface area relative to body weight and lower heat production than dairy cows, calves may be better equipped to withstand warmer temperatures [5,51]. However, growing calves still suffer from heat stress as it can affect their physiology, growth, feed intake, rumen function, and wellbeing [6–8]. According to Dado-Senn et al. [50], THI in calves raised in a shaded, subtropical environment is the best indicator of heat stress, and it is important to keep a regular check on the calves when the THI exceeds 65 to 69.

Compared with adult cows, young calves have a narrower thermoneutral zone, with the upper limit of normal at about 29 °C, and heat stress generally occurs at temperatures above 32 °C and 60% humidity [52,53]. In this experiment, the average rectal temperature ranged from 38.98 to 39.06 °C, indicating that the temperature was below the upper limit of the normal physiological range (38.1 to 39.2 °C) [9,54]. Kovács et al. [55], who studied heat stress in 7-week-old dairy calves, found a poor correlation between rectal temperature and meteorological parameters, especially in calves kept in shaded environments. Kovács et al. [55] have suggested that other physiological measures, such as skin temperature at the ear, respiratory rate, and heart rate, may be more appropriate indicators for assessing heat stress in young calves. Dairy calves housed indoors appear to breathe normally, between

24 and 36 times per minute [56]. Moore et al. [57] have also found that the respiratory rate of dairy calves increased by 2 breaths per minute with each 1 °C increase in indoor barn temperature. In this experiment, the average respiration rate ranged from 56.9 to 61.4 breaths per minute. This indicates that respiration rate exceeded the upper limit of the normal range, likely implying that the calves experienced heat stress during the experiment.

Nemati et al. [26] studied the effects of AH content and particle size in calves aged 0 to 10 weeks. They observed a three-way interaction between the content of AH, particle size, and the time calves consumed the starter feed. Calves offered a high content of AH with moderate particle size consumed the greatest amount of starer feed from week 8 to 10 of age, in contrast with what we observed in the current experiment. Previously, Karami et al. [23] reported that DMI of weaned dairy calves increased linearly when forage-to-concentrate ratio was decreased from 50:50 to 20:80. Aragona et al. [22] have also demonstrated that DMI was greater in calves fed restricted, long-chopped hay than those fed free-choice, long-chopped hay. These results may support the hypothesis that post-weaned calves may benefit from a limited amount of forage in their diets and providing greater amounts of forage can compromise their feed intake. Studies with ruminant animals indicate that increasing the forage-to-concentrate ratio limits feed intake via the physical distension of the gastrointestinal tract [58]. Our results indicate that feeding weaned calves the MF diet containing SPS AH may have reduced gut fill and minimized its inhibitory effect on DMI. It is possible that chopping AH into smaller sizes may have reduced heat production during heat stress and encouraged greater feed consumption. Further research is needed to ascertain whether reducing the length of forage has an impact on body heat production.

Regardless of forage particle size, we observed that calves receiving MF diets gained more weight and were heavier at the end of the experiment than those offered HF diets. The enhanced growth efficiency of calves fed MF diet might be related to the combined effects of greater DMI and improved nutrient digestion. In support of our findings, previous research has also shown that diets with higher starch content and higher NFC/NDF ratios promote growth performance in calves older than 8 weeks [59–61]. Aragona et al. [22] have also reported that feeding a high starch diet (45%) instead of a low starch diet (8%) results in increased feed intake, growth, and feed efficiency in calves at 2 to 4 months of age. The lower rectal temperature in calves fed MF diets could possibly be explained by their comparatively lower whole-body heat production and higher tissue energy retention, as evidenced by increased ADG and lower feed conversion ratio. Similarly, Reynolds et al. [62] have found that feeding a diet with 75% concentrate resulted in growing beef heifers producing less heat energy and retaining more tissue energy compared with those fed a diet containing 75% alfalfa.

In support of our findings that feeding MF vs. HF diet had a positive effect on several structural growth parameters, Aragona et al. [22] have reported that feeding a high starch, low fiber starter diet (51% starch) results in a 14% increase in hip width in young weaned calves (aged 2 to 4 months) compared with a low starch, high fiber starter diet (11% starch). Karami et al. [23] have also reported that heart girth, withers height, and hip eight at d 120 of calf age decrease linearly as forage-to-concentrate ratio increases from 20:80 to 50:50. Previous studies have shown that increasing ME intake can improve the body size of dairy calves [63]. Previous studies suggest that, as forage proportion in calf diet increased, ADG and empty body weight gain decreased [64–66]. In addition, as forage intake increased, the reticulorumen weight and length, as well as papillae density, decreased, which could affect the absorption of VFA and ultimately decrease growth and development [66].

Dairy producers have two primary goals for a cost-effective heifer-raising program: achieving earlier age at first calving and maximizing milk production during the first lactation [67]. Previous research suggests that promoting prepubertal growth and musculoskeletal development may potentially help in achieving these goals [68]. For example, accelerated growth in the first six months of a calf's life has been shown to result in a lower age at first calving, which helps reduce heifer-raising costs [69]. Our preliminary data collected from this trial suggests the important role of the lower-fiber diet in promoting growth rate and in improving several skeletal growth measures of calves aged 3–6 months raised under challenging environmental conditions. Therefore, conducting long-term experiments is suggested in order to assess how this dietary factor might impact age at puberty and first calving, and the lactation performance of heifers as first-lactation cows.

Lower digestibility of DM and crude protein in calves receiving the HF than those receiving the MF diets could be explained by replacing lower-fiber, rapidly and extensively digestible components with medium-quality forage source with lower digestibility in HF diet. Feeding the HF diet may have exceeded the capacity of the undeveloped rumen and microbial community to digest nutrients, resulting in declined nutrient digestibility. Although AH particle size had no effect on digestibility of OM, DM, and crude protein, diets containing SPS vs. LPS AH increased NDF and ADF digestibility. Increased surface area of SPS AH may have resulted in increased digestibility of fibrous fractions by facilitating the attachment of ruminal cellulolytic bacteria to degrade cellulose and hemicellulose components [70]. Our results show that rumen pH tends to be higher in diets containing SPS vs. LPS AH, which is consistent with digestibility results. Lower ruminal pH may create an unfavorable environment for the optimal activity of cellulolytic bacteria in the rumen [71], compromising digestion of fibrous fractions.

Amat et al. [72] have reported that adult heifers fed diets with a lower forage-to-concentrate ratio had lower rumen pH than those fed diets with a higher forage-to-concentrate ratio. Although total rumen VFA production remained unaffected by dietary forage proportion, feeding HF diet increased eating and rumination time, which could promote salivary secretion. Previous studies have shown that forage supplementation in calf diets stimulates rumination activity [26,30], which is likely due to the necessary reduction in particle size of high-forage diets [73]. The longer duration of rumination in weaned calves fed HF compared with MF may be explained by the higher intake of peNDF, which primarily stimulates rumination activity [74].

Calves receiving HF vs. MF diets had greater acetate, but lower propionate molar proportions, resulting in acetate-to-propionate ratio to increase. Feeding high-fiber diets typically results in greater acetate proportions in the rumen [75]. In this experiment, greater acetate formation in HF-fed calves occurred at the expense of decreased propionate and valerate molar proportions. In agreement with our results, Nemati et al. [26] and Suarez-Mena et al. [76] have reported that feeding HF diets increased rumen acetate proportion, which could be related to higher pH and improved growth and activity of cellulolytic bacteria, resulting in higher acetate-to-propionate ratio [77]. Propionate is produced when NFC is fermented by amylolytic bacteria, while acetate results from the fermentation of structural carbohydrates by cellulolytic bacteria [78]. In support, heifers with body weights of 80 to 250 kg were fed diets with 12:88 vs. 30:70 forage-to-concentrate ratio, and the results suggest that the lower forage-to-concentrate ratio caused ruminal fermentation to become more amylolytic and proteolytic than cellulolytic, causing the propionate-to-acetate ratio and total VFA production to increase as heifer age progressed from 13 to 41 weeks [79]. Valerate may be used by cellulolytic microorganisms in the rumen [80], which could explain why its concentration decreased in HF-fed calves. In other words, the decrease in valerate concentration may be an indication of improved growth and activity of cellulolytic bacteria in HF-fed calves.

Although forage particle size did not affect the proportion of rumen VFA, the acetate-to-propionate ratio tended to be greater in calves receiving SPS vs. LPS AH. Nemati et al. [26] have also reported similar results. This response could be attributed to greater fiber digestibility, which shifts ruminal fermentation towards more acetate production, likely by stimulating cellulolytic microorganisms.

One of the key elements affecting the ruminating behavior in ruminants is the forage particle size. In this study, calves fed SPS AH spent longer eating and ruminating compared with calves fed LPS forage. Longer rumination time typically increases salivation during feeding, which further increases rumen pH, as seen in calves fed SPS AH. Studies in calves show contradictory results in relation to forage particle size and rumen pH. In contrast with our results, Nemati et al. [26] have reported that calves fed medium particle-sized AH (geometric mean = 3 mm) had higher rumen pH at both pre- and post-weaning periods (days 35 and 70) compared with those fed fine particle-sized AH (geometric mean = 1 mm). However, Mirzaei et al. [27] found that rumen pH was not different when calves were fed different particle sizes of AH (2.92 vs. 5.04 mm as geometrical mean). These discrepancies could be due to forage level, basal diet chemical composition or differences in calf stage of growth.

Our results indicate that the sorting behavior of growing calves against long particles may depend on dietary forage level as the greatest sorting against these particles occurred in calves fed the HF-SPS diet. Sorting against long particles of diet could also be an alleviating strategy for discomfort associated with the heat stress that occurred in the current study. As discussed earlier, greater forage proportion may increase heat increment in ruminants. Therefore, calves may prefer to consume less long particles in an effort to reduce the heat production that arises from physical and microbial digestion.

5. Conclusions

These findings provide valuable insight into the complex interplay between dietary fiber content, particle size, and performance of weaned calves (from approximate ages of 3 to 6 months). The findings suggest that the dietary fiber content has a greater impact on the performance of weaned calves under moderate heat stress, compared with the forage particle size. Weaned calves fed MF as compared with HF showed superior performance, as evidenced by greater feed intake and final body weight, improved nutrient digestion, and longer lying time, an indication of enhanced calf comfort. These results have practical implications for optimizing feed formulation for weaned calves in hot environments. Future research is needed to investigate the potential of manipulating dietary fiber content and forage particle size in order to optimize calf performance under different environmental conditions.

Author Contributions: Conceptualization, F.H., G.-R.G. and F.A.; methodology, M.-H.I., M.H. and F.A.; software, F.H.; validation, F.H., M.A., G.-R.G., M.K. and M.H.G.; formal analysis, F.H.; investigation, M.-H.I. and M.H.; resources, F.H., G.-R.G., M.A. and M.K.; data curation, F.H., M.H.G. and F.A.; writing—original draft preparation, M.-H.I. and M.H.; writing—review and editing, F.H., M.A., G.-R.G., M.H.G. and F.A.; supervision, F.H. and G.-R.G.; project administration, F.H. and G.-R.G. All authors have read and agreed to the published version of the manuscript.

Funding: This research was financially supported by Isfahan University of Technology (Isfahan, Iran).

Institutional Review Board Statement: All procedures involving animals were thoroughly reviewed and approved by the Animal Care and Use Committee of Isfahan University of Technology prior to the beginning of the study. The methods were conducted in accordance with the regulations of the Iranian Council of Animal Care (1995).

Informed Consent Statement: Not applicable.

Data Availability Statement: Data are available within the article.

Acknowledgments: The authors would like to extend their sincere gratitude to A. Zalbeik, the production manager of Namfar Agriculture and Animal Husbandry complex (Isfahan, Iran), for his valuable help and support in conducting this study. The authors would also like to thank the farm staff for providing the facilities used for this study. In addition, we thank H. Bagherpur for his assistance in conducting the study.

Conflicts of Interest: The authors declare no conflict of interest.

References

1. Khan, M.; Weary, D.; Von Keyserlingk, M. Invited review: Effects of milk ration on solid feed intake, weaning, and performance in dairy heifers. *J. Dairy Sci.* **2011**, *94*, 1071–1081. [CrossRef]
2. Cavallini, D.; Raspa, F.; Marliani, G.; Nannoni, E.; Martelli, G.; Sardi, L.; Valle, E.; Pollesel, M.; Tassinari, M.; Buonaiuto, G. Growth performance and feed intake assessment of Italian Holstein calves fed a hay-based total mixed ration: Preliminary steps towards a prediction model. *Vet. Sci.* **2023**, *10*, 554. [CrossRef]
3. Baumgard, L.H.; Rhoads, R.P. RUMINANT NUTRITION SYMPOSIUM: Ruminant production and metabolic responses to heat stress. *J. Anim. Sci.* **2012**, *90*, 1855–1865. [CrossRef] [PubMed]
4. Polsky, L.; von Keyserlingk, M.A.G. Invited review: Effects of heat stress on dairy cattle welfare. *J. Dairy Sci.* **2017**, *100*, 8645–8657. [CrossRef] [PubMed]
5. West, J.W. Effects of heat-stress on production in dairy cattle. *J. Dairy Sci.* **2003**, *86*, 2131–2144. [CrossRef] [PubMed]
6. Broucek, J.; Kisac, P.; Uhrincat, M.; Hanus, A.; Benc, F. Effect of high temperature on growth performance of calves maintained in outdoor hutches. *J. Anim. Feed Sci.* **2008**, *17*, 139. [CrossRef]
7. Hill, T.M.; Bateman, H.G.; Aldrich, J.M.; Schlotterbeck, R.L. Comparisons of housing, bedding, and cooling options for dairy calves. *J. Dairy Sci.* **2011**, *94*, 2138–2146. [CrossRef]
8. Dado-Senn, B.; Ouellet, V.; Lantigua, V.; Van Os, J.; Laporta, J. Methods for detecting heat stress in hutch-housed dairy calves in a continental climate. *J. Dairy Sci.* **2023**, *106*, 1039–1050. [CrossRef]
9. Roland, L.; Drillich, M.; Klein-Jöbstl, D.; Iwersen, M. Invited review: Influence of climatic conditions on the development, performance, and health of calves. *J. Dairy Sci.* **2016**, *99*, 2438–2452. [CrossRef]
10. Wang, J.; Li, J.; Wang, F.; Xiao, J.; Wang, Y.; Yang, H.; Li, S.; Cao, Z. Heat stress on calves and heifers: A review. *J. Anim. Sci. Biotechnol.* **2020**, *11*, 79. [CrossRef]
11. Laporta, J. ADSA Foundation Scholar Award: Early-life exposure to hyperthermia: Productive and physiological outcomes, costs, and opportunities. *J. Dairy Sci.* **2021**, *104*, 11337–11347. [CrossRef] [PubMed]
12. Yadav, B.; Singh, G.; Verma, A.; Dutta, N.; Sejian, V. Impact of heat stress on rumen functions. *Vet. World* **2013**, *6*, 992–996. [CrossRef]
13. Tajima, K.; Nonaka, I.; Higuchi, K.; Takusari, N.; Kurihara, M.; Takenaka, A.; Mitsumori, M.; Kajikawa, H.; Aminov, R.I. Influence of high temperature and humidity on rumen bacterial diversity in Holstein heifers. *Anaerobe* **2007**, *13*, 57–64. [CrossRef]
14. Nonaka, I.; Takusari, N.; Tajima, K.; Suzuki, T.; Higuchi, K.; Kurihara, M. Effects of high environmental temperatures on physiological and nutritional status of prepubertal Holstein heifers. *Livest. Sci.* **2008**, *113*, 14–23. [CrossRef]
15. Beiranvand, H.; Ghorbani, G.R.; Khorvash, M.; Nabipour, A.; Dehghan-Banadaky, M.; Homayouni, A.; Kargar, S. Interactions of alfalfa hay and sodium propionate on dairy calf performance and rumen development. *J. Dairy Sci.* **2014**, *97*, 2270–2280. [CrossRef]
16. Hill, S.R.; Knowlton, K.F.; Daniels, K.M.; James, R.E.; Pearson, R.E.; Capuco, A.V.; Akers, R.M. Effects of milk replacer composition on growth, body composition, and nutrient excretion in preweaned Holstein heifers. *J. Dairy Sci.* **2008**, *91*, 3145–3155. [CrossRef] [PubMed]
17. Diao, Q.; Zhang, R.; Fu, T. Review of strategies to promote rumen development in calves. *Animals* **2019**, *9*, 490. [CrossRef] [PubMed]
18. Suárez, B.J.; Van Reenen, C.G.; Stockhofe, N.; Dijkstra, J.; Gerrits, W.J.J. Effect of roughage source and roughage to concentrate ratio on animal performance and rumen development in veal calves. *J. Dairy Sci.* **2007**, *90*, 2390–2403. [CrossRef] [PubMed]
19. Suárez, B.J.; Van Reenen, C.G.; Gerrits, W.J.J.; Stockhofe, N.; van Vuuren, A.M.; Dijkstra, J. Effects of supplementing concentrates differing in carbohydrate composition in veal calf diets: II. rumen development. *J. Dairy Sci.* **2006**, *89*, 4376–4386. [CrossRef]
20. Hill, T.M.; Bateman, H.G.; Aldrich, J.M.; Schlotterbeck, R.L. Roughage amount, source, and processing for diets fed to weaned dairy calves. *Prof. Anim. Sci.* **2010**, *26*, 181–187. [CrossRef]
21. Dong, L.; Li, B.; Diao, Q. Effects of dietary forage proportion on feed intake, growth performance, nutrient digestibility, and enteric methane emissions of Holstein heifers at various growth stages. *Animals* **2019**, *9*, 725. [CrossRef] [PubMed]
22. Aragona, K.M.; Suarez-Mena, F.X.; Dennis, T.S.; Quigley, J.D.; Hu, W.; Hill, T.M.; Schlotterbeck, R.L. Effect of starter form, starch concentration, and amount of forage fed on Holstein calf growth from 2 to 4 months of age. *J. Dairy Sci.* **2020**, *103*, 2324–2332. [CrossRef]
23. Karami, A.; Alikhani, M.; Khorvash, M.; Hashemzadeh, F.; Sadeghi-Sefidmazgi, A.; Rafiee, H.; Ferraretto, L.F. Effects of different forage to concentrate ratios on performance, plasma metabolites, and feeding behaviour of weaned dairy calves from 70 to 120 days of age. *Ital. J. Anim. Sci.* **2021**, *20*, 1317–1327. [CrossRef]
24. Nampoothiri, V.M.; Mohini, M.; Malla, B.A.; Mondal, G.; Pandita, S. Growth performance, and enteric and manure greenhouse gas emissions from Murrah calves fed diets with different forage to concentrate ratios. *Anim. Nutr.* **2018**, *4*, 215–221. [CrossRef]
25. Montoro, C.; Miller-Cushon, E.K.; DeVries, T.J.; Bach, A. Effect of physical form of forage on performance, feeding behavior, and digestibility of Holstein calves. *J. Dairy Sci.* **2013**, *96*, 1117–1124. [CrossRef] [PubMed]
26. Nemati, M.; Amanlou, H.; Khorvash, M.; Moshiri, B.; Mirzaei, M.; Khan, M.A.; Ghaffari, M.H. Rumen fermentation, blood metabolites, and growth performance of calves during transition from liquid to solid feed: Effects of dietary level and particle size of alfalfa hay. *J. Dairy Sci.* **2015**, *98*, 7131–7141. [CrossRef] [PubMed]

27. Mirzaei, M.; Khorvash, M.; Ghorbani, G.R.; Kazemi-Bonchenari, M.; Riasi, A.; Nabipour, A.; Van Den Borne, J.J.G.C. Effects of supplementation level and particle size of alfalfa hay on growth characteristics and rumen development in dairy calves. *J. Anim. Physiol. Anim. Nutr.* **2015**, *99*, 553–564. [CrossRef]
28. Coverdale, J.A.; Tyler, H.D.; Quigley, J.D.; Brumm, J.A. Effect of various levels of forage and form of diet on rumen development and growth in calves. *J. Dairy Sci.* **2004**, *87*, 2554–2562. [CrossRef]
29. Khan, M.A.; Weary, D.M.; von Keyserlingk, M.A.G. Hay intake improves performance and rumen development of calves fed higher quantities of milk. *J. Dairy Sci.* **2011**, *94*, 3547–3553. [CrossRef]
30. Castells, L.; Bach, A.; Araujo, G.; Montoro, C.; Terré, M. Effect of different forage sources on performance and feeding behavior of Holstein calves. *J. Dairy Sci.* **2012**, *95*, 286–293. [CrossRef]
31. Suarez-Mena, F.X.; Hill, T.M.; Jones, C.M.; Heinrichs, A.J. Review: Effect of forage provision on feed intake in dairy calves. *Prof. Anim. Sci.* **2016**, *32*, 383–388. [CrossRef]
32. National Research Council (NRC). *Nutritional Requirements of Dairy Cattle*, 7th ed.; National Academies Press: Washington, DC, USA, 2001.
33. AOAC International. *Official Methods of Analysis*, 18th ed.; AOAC International: Gaithersburg, MD, USA, 2006.
34. American National Standards Institute. *Method of Determining and Expressing Particle Size of Chopped Forage Material by Screening*; ASAE: St Joseph, MI, USA, 1998.
35. Khan, M.A.; Lee, H.J.; Lee, W.S.; Kim, H.S.; Ki, K.S.; Hur, T.Y.; Suh, G.H.; Kang, S.J.; Choi, Y.J. Structural growth, rumen development, and metabolic and immune responses of Holstein male calves fed milk through step-down and conventional methods. *J. Dairy Sci.* **2007**, *90*, 3376–3387. [CrossRef]
36. Edmonson, A.J.; Lean, I.J.; Weaver, L.D.; Farver, T.; Webster, G. A body condition scoring chart for Holstein dairy cows. *J. Dairy Sci.* **1989**, *72*, 68–78. [CrossRef]
37. McGuirk, S.M. Disease management of dairy calves and heifers. *Vet. Clin. N. Am. Food Anim. Pract.* **2008**, *24*, 139–153. [CrossRef]
38. Van Keulen, J.; Young, B.A. Evaluation of acid-insoluble ash as a natural marker in ruminant digestibility studies. *J. Anim. Sci.* **1977**, *44*, 282–287. [CrossRef]
39. Hashemzadeh-Cigari, F.; Khorvash, M.; Ghorbani, G.R.; Kadivar, M.; Riasi, A.; Zebeli, Q. Effects of supplementation with a phytobiotics-rich herbal mixture on performance, udder health, and metabolic status of Holstein cows with various levels of milk somatic cell counts. *J. Dairy Sci.* **2014**, *97*, 7487–7497. [CrossRef] [PubMed]
40. Rahimi, A.; Ghorbani, G.R.; Hashemzadeh, F.; Mirzaei, M.; Saberipour, H.R.; Ahmadi, F.; Ghaffari, M.H. Impact of corn processing and weaning age on calf performance, blood metabolites, rumen fermentation, and behavior. *Sci. Rep.* **2023**, *13*, 10701. [CrossRef] [PubMed]
41. Leonardi, C.; Armentano, L.E. Effect of quantity, quality, and length of alfalfa hay on selective consumption by dairy cows. *J. Dairy Sci.* **2003**, *86*, 557–564. [CrossRef]
42. St-Pierre, N.R. Design and analysis of Pen Studies in the animal sciences. *J. Dairy Sci.* **2007**, *90*, E87–E99. [CrossRef]
43. Tempelman, R.J. Invited review: Assessing experimental designs for research conducted on commercial dairies. *J. Dairy Sci.* **2009**, *92*, 1–15. [CrossRef]
44. Bello, N.M.; Kramer, M.; Tempelman, R.J.; Stroup, W.W.; St-Pierre, N.R.; Craig, B.A.; Young, L.J.; Gbur, E.E. Short communication: On recognizing the proper experimental unit in animal studies in the dairy sciences. *J. Dairy Sci.* **2016**, *99*, 8871–8879. [CrossRef]
45. Lascano, G.; Koch, L.; Heinrichs, A. Precision-feeding dairy heifers a high rumen-degradable protein diet with different proportions of dietary fiber and forage-to-concentrate ratios. *J. Dairy Sci.* **2016**, *99*, 7175–7190. [CrossRef]
46. Kononoff, P.J.; Heinrichs, A.J.; Buckmaster, D.R. Modification of the Penn State forage and total mixed ration particle separator and the effects of moisture content on its measurements. *J. Dairy Sci.* **2003**, *86*, 1858–1863. [CrossRef] [PubMed]
47. Lammers, B.P.; Buckmaster, D.R.; Heinrichs, A.J. A simple method for the analysis of particle sizes of forage and total mixed rations. *J. Dairy Sci.* **1996**, *79*, 922–928. [CrossRef] [PubMed]
48. Armstrong, D.V. Heat stress interaction with shade and cooling. *J. Dairy Sci.* **1994**, *77*, 2044–2050. [CrossRef] [PubMed]
49. Bernabucci, U.; Lacetera, N.; Baumgard, L.H.; Rhoads, R.P.; Ronchi, B.; Nardone, A. Metabolic and hormonal acclimation to heat stress in domesticated ruminants. *Animal* **2010**, *4*, 1167–1183. [CrossRef] [PubMed]
50. Dado-Senn, B.; Ouellet, V.; Dahl, G.E.; Laporta, J. Methods for assessing heat stress in preweaned dairy calves exposed to chronic heat stress or continuous cooling. *J. Dairy Sci.* **2020**, *103*, 8587–8600. [CrossRef]
51. Beiranvand, H.; Khani, M.; Omidian, S.; Ariana, M.; Rezvani, R.; Ghaffari, M.H. Does adding water to dry calf starter improve performance during summer? *J. Dairy Sci.* **2016**, *99*, 1903–1911. [CrossRef]
52. Neuwirth, J.; Norton, J.; Rawlings, C.; Thompson, F.; Ware, G. Physiologic responses of dairy calves to environmental heat stress. *Int. J. Biometeorol.* **1979**, *23*, 243–254. [CrossRef]
53. Gebremedhin, K.; Cramer, C.; Porter, W. Predictions and measurements of heat production and food and water requirements of Holstein calves in different environments. *Trans. ASAE* **1981**, *24*, 715–0720. [CrossRef]
54. Moran, J. *Calf Rearing—A Practical Guide*, 2nd ed.; Landlinks Press: Collingwood, Australia, 2002.
55. Kovács, L.; Kézér, F.L.; Ruff, F.; Jurkovich, V.; Szenci, O. Assessment of heat stress in 7-week old dairy calves with non-invasive physiological parameters in different thermal environments. *PLoS ONE.* **2018**, *13*, e0200622.
56. Radostits, O.M.; Mayhew, I.G.; Houston, D.M. *Veterinary Clinical Examination and Diagnosis*; WB Saunders: Philadelphia, PA, USA, 2000.

57. Moore, D.; Duprau, J.; Wenz, J. Effects of dairy calf hutch elevation on heat reduction, carbon dioxide concentration, air circulation, and respiratory rates. *J. Dairy Sci.* **2012**, *95*, 4050–4054. [CrossRef] [PubMed]
58. Allen, M.S. Effects of diet on short-term regulation of feed intake by lactating dairy cattle. *J. Dairy Sci.* **2000**, *83*, 1598–1624. [CrossRef] [PubMed]
59. Hill, T.M.; Bateman, I.I.H.G.; Aldrich, J.M.; Schlotterbeck, R.L. Effects of the amount of chopped hay or cottonseed hulls in a textured calf starter on young calf performance. *J. Dairy Sci.* **2008**, *91*, 2684–2693. [CrossRef]
60. Hill, T.M.; Bateman, H.G.; Aldrich, J.M.; Schlotterbeck, R.L. High-starch, coarse-grain, low-fiber diets maximize growth of weaned dairy calves less than 4 months of age. *Prof. Anim. Sci.* **2012**, *28*, 325–331. [CrossRef]
61. Hill, T.M.; Quigley, J.D.; Bateman, I.I.H.G.; Aldrich, J.M.; Schlotterbeck, R.L. Source of carbohydrate and metabolizable lysine and methionine in the diet of recently weaned dairy calves on digestion and growth. *J. Dairy Sci.* **2016**, *99*, 2788–2796. [CrossRef]
62. Reynolds, C.K.; Tyrrell, H.F.; Reynolds, P.J. Effects of diet forage-to-concentrate ratio and intake on energy metabolism in growing beef heifers: Whole body energy and nitrogen balance and visceral heat production. *J. Nutr.* **1991**, *121*, 994–1003. [CrossRef] [PubMed]
63. Khan, M.A.; Bach, A.; Weary, D.M.; von Keyserlingk, M.A.G. Invited review: Transitioning from milk to solid feed in dairy heifers. *J. Dairy Sci.* **2016**, *99*, 885–902. [CrossRef]
64. Warner, R.G.; Flatt, W.P.; Loosli, J.K. Ruminant nutrition, dietary factors influencing development of ruminant stomach. *J. Agric. Food Chem.* **1956**, *4*, 788–792. [CrossRef]
65. Mitchell, L.K.; Heinrichs, A.J. Increasing grass hay inclusion level on weaned dairy calf growth, intake, digestibility, and ruminal fermentation. *J. Dairy Sci.* **2020**, *103*, 9012–9023. [CrossRef]
66. Stobo, I.; Roy, J.; Gaston, H.J. Rumen development in the calf: 1. The effect of diets containing different proportions of concentrates to hay on rumen development. *Br. J. Nutr.* **1966**, *20*, 171–188. [CrossRef]
67. Salte, R.; Storli, K.S.; Wærp, H.K.L.; Sommerseth, J.K.; Prestl, E.; Volden, H.; Klemetsdal, G. Designing a replacement heifer rearing strategy: Effects of growth profile on performance of Norwegian Red heifers and cows. *J. Dairy Sci.* **2020**, *103*, 10835–10849. [CrossRef] [PubMed]
68. Nor, N.M.; Steeneveld, W.; Van Werven, T.; Mourits, M.; Hogeveen, H. First-calving age and first-lactation milk production on Dutch dairy farms. *J. Dairy Sci.* **2013**, *96*, 981–992.
69. Brickell, J.; Bourne, N.; McGowan, M.; Wathes, D. Effect of growth and development during the rearing period on the subsequent fertility of nulliparous Holstein-Friesian heifers. *Theriogenology* **2009**, *72*, 408–416. [CrossRef]
70. Miron, J.; Ben-Ghedalia, D.; Morrison, M. Invited Review: Adhesion mechanisms of rumen cellulolytic bacteria. *J. Dairy Sci.* **2001**, *84*, 1294–1309. [CrossRef]
71. Osborne, J.K.; Mutsvangwa, T.; Alzahal, O.; Duffield, T.F.; Bagg, R.; Dick, P.; Vessie, G.; McBride, B.W. Effects of monensin on ruminal forage degradability and total tract diet digestibility in lactating dairy cows during grain-induced subacute ruminal acidosis. *J. Dairy Sci.* **2004**, *87*, 1840–1847. [CrossRef] [PubMed]
72. Amat, S.; McKinnon, J.J.; Penner, G.B.; Hendrick, S. Effects of dietary sulfur concentration and forage-to-concentrate ratio on ruminal fermentation, sulfur metabolism, and short-chain fatty acid absorption in beef heifers. *J. Anim. Sci.* **2014**, *92*, 712–723. [CrossRef] [PubMed]
73. Beauchemin, K.A.; Yang, W.Z.; Rode, L.M. Effects of barley grain processing on the site and extent of digestion of beef feedlot finishing diets. *J. Anim. Sci.* **2001**, *79*, 1925–1936. [CrossRef]
74. Mertens, D.R. Creating a system for meeting the fiber requirements of dairy cows. *J. Dairy Sci.* **1997**, *80*, 1463–1481. [CrossRef]
75. Yang, W.Z.; Beauchemin, K.A.; Rode, L.M. Effects of grain processing, forage to concentrate ratio, and forage particle size on rumen pH and digestion by dairy cows. *J. Dairy Sci.* **2001**, *84*, 2203–2216. [CrossRef]
76. Suarez-Mena, F.X.; Lascano, G.J.; Rico, D.E.; Heinrichs, A.J. Effect of forage level and replacing canola meal with dry distillers grains with solubles in precision-fed heifer diets: Digestibility and rumen fermentation. *J. Dairy Sci.* **2015**, *98*, 8054–8065. [CrossRef]
77. Castells, L.; Bach, A.; Aris, A.; Terré, M. Effects of forage provision to young calves on rumen fermentation and development of the gastrointestinal tract. *J. Dairy Sci.* **2013**, *96*, 5226–5236. [CrossRef] [PubMed]
78. Enjalbert, F.; Garrett, J.; Moncoulon, R.; Bayourthe, C.; Chicoteau, P. Effects of yeast culture (*Saccharomyces cerevisiae*) on ruminal digestion in non-lactating dairy cows. *Anim. Feed Sci. Technol.* **1999**, *76*, 195–206. [CrossRef]
79. Rotger, A.; Ferret, A.; Calsamiglia, S.; Manteca, X. Changes in ruminal fermentation and protein degradation in growing Holstein heifers from 80 to 250 kg fed high-concentrate diets with different forage-to-concentrate ratios. *J. Anim. Sci.* **2005**, *83*, 1616–1624. [CrossRef]
80. Cline, J.H.; Hershberger, T.V.; Bentley, O.G. Utilization and/or synthesis of valeric acid during the digestion of glucose, starch and cellulose by rumen microorganisms in vitro. *J. Anim. Sci.* **1958**, *17*, 284–292. [CrossRef]

animals

MDPI

Article

Comparable Evaluation of Nutritional Benefits of *Lactobacillus plantarum* and *Bacillus toyonensis* Probiotic Supplementation on Growth, Feed Utilization, Health, and Fecal Microbiota in Pre-Weaning Male Calves

Mohamed S. Ayyat [1,*], Hamdy A. El-Nagar [2], Wael M. Wafa [2], Khaled M. Abd El-Latif [3], Samir Mahgoub [4] and Adham A. Al-Sagheer [1,*]

1 Department of Animal Production, Faculty of Agriculture, Zagazig University, Zagazig 44511, Egypt
2 Animal Production Research Institute, Agricultural Research Center, Giza 12619, Egypt
3 Specialized Hospital, Ain Shams University, Cairo 11588, Egypt
4 Department of Agricultural Microbiology, Faculty of Agriculture, Zagazig University, Zagazig 44511, Egypt
* Correspondence: ayyatm@yahoo.com (M.S.A.); adham_alsaht@zu.edu.eg (A.A.A.-S.)

Simple Summary: Pre-weaning nutrition plays a crucial role in the growth of male calves, as it shapes their future productivity and overall health. The administration of probiotics to calves during this period can help in maintaining a healthy gastrointestinal tract, establishing a strong immune system, enhancing nutrient absorption, and ultimately leading to optimal growth rates and a successful transition to a productive adult life. This study aims to investigate and compare the blood biochemical, hematological, and fecal microbiota of pre-weaned Holstein male calves supplemented with a mixture of *Lactobacillus plantarum* and *Bacillus toyonensis* versus each individual strain alone. The current study found that using the probiotic mixture displayed superior positive effects compared to each individual strain alone on growth, feed intake, feed efficiency, blood constituents, and modulation of fecal microbiota composition. This approach represents a novel and potentially more effective strategy for enhancing the growth and health of pre-weaned Holstein male calves.

Abstract: This study was conducted to investigate the impact of probiotic supplementation using *Lactobacillus plantarum* DSA 20174 and/or *Bacillus toyonensis* ATCC 55050 on growth performance, blood parameters, hematological measures, and fecal microbiota in pre-weaning Holstein calves. Thirty-two four-day-old male calves with a similar genetic background, weighing an average of 38.27 ± 0.12 kg, were randomly assigned to four groups. The groups consisted of a control group (CON) without supplementation, a group receiving *B. toyonensis* (BT) at 3×10^9 cfu/calf/day, a group receiving *L. plantarum* (LP) at 1×10^{10} cfu/calf/day, and a group receiving a combination of LP and BT (LP + BT) at half the dosage for each. The study found that calves supplemented with LP and LP + BT experienced significant improvements in average daily gain and final body weight compared to the control group. The LP + BT group showed the most positive effects on TDMI, starter intake, and CP intake. RBC counts tended to be higher in the probiotic groups, with the LP + BT group having the highest values. The LP + BT group also had higher total protein, albumin, globulin, and hematocrit concentrations. All probiotic groups showed higher serum IgG concentrations. Probiotic supplementation led to increased total bacterial count and decreased levels of *E. coli*, salmonella, and clostridium. The LP + BT group had a significant decrease in coliform count, while both LP and LP + BT groups had increased *Lactobacillus* populations. In conclusion, LP + BT probiotic supplement showed the most beneficial effects on growth, feed efficiency, blood constituents, and modulation of fecal microbiota composition.

Keywords: dairy calf; probiotic; *Lactobacillus plantarum*; Bacillus toyonensis; growth; hematology; immunity; fecal bacteria

1. Introduction

Pre-weaned calves have a higher risk of morbidity and mortality during their first few days of life [1–3]. Neonatal calves have immature immune and antioxidant systems and low disease resistance. Consequently, these calves are susceptible to respiratory and intestinal ailments, which compromises their subsequent growth and overall health [4]. Antibiotics are used to reduce diarrhea-related morbidity and mortality, but stricter regulations in the livestock sector have led to microbial resistance risks. The EU banned antibiotics and ionophores, while the USA introduced the Veterinary Feed Directive [5]. Therefore, it is essential to offer scientific techniques, like probiotics and natural feed additives, to guarantee optimal growth, development, and health in the first few years of life, as they could have a favorable effect on how the livestock business will evolve in the future [6].

Increased scientific interest in probiotics has persisted for decades due to their potential health advantages. There exist probiotics that are specifically formulated for ruminant animals, which encompass yeast strains such as *Y. lipolytica* and *S. cerevisiae*, as well as bacterial genera including *Bacillus*, *Propionibacterium*, *Bifidobacterium*, *Lactobacillus*, *Enterococcus*, and other noteworthy species including *Megasphaera elsdenii* [7,8]. *Lactobacillus plantarum* has been identified in various silages and fermented foods [9,10]. *L. plantarum* is well known for its ability to endure the process of gastric transit, enabling it to colonize the intestines and positively affect the host [11]. *L. plantarum* has demonstrated beneficial effects in promoting nutrient digestibility, boosting immunological function and mitigating pathogenic colonization across several animal species [12,13]. Moreover, its robustness, resistance to bile and acid, along with the ability to generate antimicrobial compounds [14] render it a highly suitable contender for probiotic supplementation in pre-weaning calves.

Bacillus species, including *Bacillus toyonensis*, exhibit potential as probiotics due to their capacity to produce endospores, which allows them to endure adverse environmental circumstances and maintain viability throughout storage and transit within the gastrointestinal system [15]. *Bacillus toyonensis* is a strain of *Bacillus cereus* group that occurs naturally and is characterized by its non-toxigenic and non-pathogenic properties. It has been recognized as a safe additive since it has been found to have no detrimental effects on many animal species [16]. *B. toyonensis* has exhibited probiotic characteristics, including the synthesis of antimicrobial compounds, activation of the immune response, and enhancement of gut health [17,18]. The aforementioned attributes make it a potential additive for augmenting the overall health and performance of pre-weaned calves.

Both investigated probiotics exert their mode of action within the intestines by establishing an unfavorable environment for the pathogenic colonization while simultaneously facilitating the growth of beneficial commensal bacteria. In response, organic acids are produced, which elevate the pH level and stimulates the synthesis of antimicrobial compounds that hinders the growth of pathogenic bacteria. Moreover, probiotics promote the proliferation of favorable bacteria, thus altering the equilibrium of microorganisms within the gut. In addition, they bolster immune function and stimulate immune cells, supporting protection against pathogen-induced or stress-induced intestinal diseases [7,19,20].

Studies have shown that probiotics containing multiple bacterial species and strains are superior to those containing only one strain or species in improving calf health and performance [21–23]. However, there has been a lack of research examining the combined effect of supplementary *L. plantarum* and *B. toyonensis* on the performance of pre-weaned dairy calves. Therefore, we hypothesized that supplementing pre-weaning calves with *L. plantarum* and/or *B. toyonensis* probiotics would have a beneficial effect on overall health and performance. The objective of this study was to examine the combination impact of *L. plantarum* and *B. toyonensis*, compared with their individual impacts, on pre-weaning calf growth performance, blood biochemical, hematological, and fecal microbiota.

2. Materials and Methods

2.1. Probiotic Bacterial and Inoculum Preparation

The *L. plantarum* DSA 20174 and *B. toyonensis* ATCC 55050 strains were acquired from The Center of Bacterial Collection (MERCIN) located at Ain Shams University in Cairo, Egypt. *L. plantarum* was cultivated overnight at a temperature of 37 °C in de Man Rogosa Sharpe broth, as described by De Man et al. [24]. Prior to its use, *L. plantarum* was subcultured three times. *B. toyonensis* was cultivated for a duration of 48 h at a temperature of 37 °C in nutrient broth (NB), which consists of the following components per liter: 5.0 g peptone, 1.0 g beef extract, 2.0 g yeast extract, 5.0 g sodium chloride, with a final pH of 6.8 ± 0.2. The stock cultures of *L. plantarum and B. toyonensis* were kept at −20 °C in 20% glycerol and −80 °C in 10% skim milk with cryoprotectant, respectively. To enumerate *B. toyonensis* and *L. plantarum*, diluted samples were plated on nutrient agar and MRS agar, respectively. The plates were subjected to incubation at a temperature of 37 °C for 48 h. The cell density, as determined through plate counting, was ~10^8 CFU/mL. This inoculum ratio was used in all experiments of dosing animals.

2.2. Experimental Design, Animals, and Treatment Diets

The current study was conducted at the El-Gemmiza Research Station located in El-Santa, Gharbia Governorate, Egypt, which is a part of the Agriculture Research Center's Animal Production Research Institute in Giza. The research protocol that involved animals was in accordance with the experimental recommendations of the Animal Ethics Committee of Zagazig University (ZU-IACUC). This study was performed using a completely randomized design involving 32 male Holstein calves with a similar genetic background, approximately four days old. The calves were allocated into four different treatment groups, each consisting of eight calves. The overall mean weight of the calves was 38.27 ± 0.12 kg at the start of the study. Immediately after birth, the calves were given colostrum amounting to 10% of their birth body weight, followed by transition milk until the third day of life. On the third day, they were transported to the experimental facility where they were housed for a duration of 84 days.

The experimental groups consisted of a control group (CON) without supplementation, a group receiving *B. toyonensis* (BT) at 3 × 10^9 cfu/calf/day, a group receiving *L. plantarum* (LP) at 1 × 10^{10} cfu/calf/day, and a group receiving a combination of LP and BT (LP + BT) at half the dosage for each. The tested probiotics dosage was selected based on the findings of Jiang, et al. and Novak, et al. [25,26]. During the period of 4 to 60 days of age, the milk was supplemented with the tested probiotics before morning feeding. Subsequently, from 61 to 88 days of age, the probiotics were added to the starter mixture that was fed to calves every morning.

Calves were raised in individual pens (1.5 m length × 1.0 m width) with rice straw bedding for the duration of the trial. Fresh water was made available to the calves at all times through buckets. Throughout the experiment, the calves were provided with starter mixture ad libitum, which was formulated based on the recommendations of NRC [27]. Table 1 presents the ingredients and chemical composition the starter mixture.

Table 1. Ingredients chemical composition of the starter mixture fed to calves.

Item	Value
Ingredient, g/kg	
Yellow corn	350
Soybean meal, 44% CP	250
Alfalfa hay	150
Wheat bran	95
Barley grain	100
Molasses	30
Iodized sodium chloride	8
Calcium phosphate	10
Mineral and vitamin premix [1]	7
Total	1000
Chemical analysis, g/kg of air dry basis	
Dry matter	904.8
Organic matter	939.6
Crude protein	190.8
Ether extract	29.9
Ash	60.3
Neutral detergent fiber	248.5
Acid detergent fiber	128.6

[1] Each kilogram of premix contains: vitamin A, 50,000 IU; vitamin D3, 10,000 IU; vitamin E, 2900 IU; Ca, 196 g; P, 96 g; $MgSO_4$, 19,000 mg; $FeSO_4$, 9000 mg; $CuSO_4$, 5000 mg; Mn, 6000 mg; $ZnSO_4$, 8000 mg; Co, 20 mg; I, 227 mg; Se, 67 mg.

The chemical composition of the starter mixture was estimated by AOAC [28] procedures using representative samples that were milled to pass a 1 mm screen. The starter samples were analyzed for crude ash (method no. 954.01), DM contents (method no. 925.10), ether extract content (method no. 920.39) using petroleum ether in a Soxhlet apparatus, and crude protein (N × 6.25) using the Kjeldahl procedure (method no. 954.01). The quantification of neutral detergent fiber (NDF) and acid detergent fiber (ADF) was done using the procedures described by Van Soest, et al. and the AOAC [28,29] method 973.18), respectively. All the chemical analyses were performed on DM basis.

2.3. Growth Performance and Feed Intake

The calves' weights were recorded before morning feedings on d 4 of age (initial weight) and thereafter every 28 days (d 32, d 60, and d 88 of age). Data of starter mixture intake were collected for each calf every four weeks. The difference between the amount of starter mixture offered and the orts was calculated and recorded as an average on a grams-per-day basis for each calf. The total dry matter intake (DMI) included both starter mixture and milk. The calculation of the average daily gain (ADG) involved dividing the difference between the final and initial body weight by the duration in days. The calculation of feed efficiency (FE) was done by dividing ADG in kg by the amount of TDMI in kg. These measurements were taken for four different time periods: d 4–32, d 33–60, d 61–88, and d 4–88 (overall period).

2.4. Hematological Study and Biochemical Assay

Blood specimens were drawn 3 h ± 30 min after morning feeding from the jugular vein of 20 calves (n = 5 calves per group) at 32, 60, and 88 days of age. Blood was drawn into tubes with K3-EDTA for hematological profile or without anticoagulant (for serum collection to measure biochemical and immunological parameters). Samples without anticoagulant were centrifuged at 3000× g at 4 °C for 15 min. The resulting supernatant was carefully collected and kept in a deep freezer at −20 °C for subsequent serum measurements. The hematological profiles were assessed using an automated hematology analyzer (Medonic CA620 VET, Stockholm, Sweden). The following parameters were measured: red blood cell (RBC) count, hemoglobin (Hb) level, hematocrit (Hct) value, mean corpuscular hemoglobin concentration (MCHC), platelet count, mean corpuscular hemoglobin (MCH), mean corpuscular volume (MCV), white blood cell (WBC) count, and proportion of basophils, neutrophils, lymphocytes, monocytes, and eosinophils.

The contents of total cholesterol (TC), triglyceride (TG), total protein, albumin, urea, high-density lipoprotein (HDL), blood urea nitrogen (BUN), very low-density lipoprotein (VLDL), and low-density lipoprotein (LDL) were estimated from the sera samples using commercial kits (Diamond Diagnostic, Dokki, Giza, Egypt) following the manufacturer's instructions. To determine the globulin levels, the total protein levels were subtracted from the serum albumin levels. Additionally, the serum concentration of immunoglobulin G (IgG) was determined using commercial ELISA kits according to the manufacturer's directives.

2.5. *Enumeration of Probiotic and Associated Bacteria in Calf Feces and Cloacal Swabs*

For detection of total bacterial count (TBC), coliform count (CFC), Lactobacilli count (LC), spore-forming bacteria (SFB: *Bacillus* and *Clostridium*), *Salmonella*, and *Escherichia coli* in animal samples, the calf feces, and cloacal swabs were mixed into buffered peptone water (8.5 g sodium chloride/L and 1.0 g peptone/L) and incubated at 35–37 °C for 18–24 h. There are two steps for detecting the tested bacteria. The first step: before incubation, total cultivable bacteria TBC, CFC, Lactobacilli, and spore-forming bacteria count (SFB) *Escherichia coli* and *Salmonella* were counted. TBC was counted onto plate count agar (Merck, 1.05463, Darmstadt, Germany) for 48 h at 30 °C. *Bacillus* spp. count was determined after pasteurizing the samples at 80 °C for 15 min onto plate count agar (Merck, 1.05463, Darmstadt, Germany) for 48 h at 30 °C. Coliform counts were taken following incubation on Violet Red Bile Agar (VRB, Biolife, Italy) for 24 h at 37 °C. After 24 h of incubation at 37 °C, the number of *E. coli* colonies on eosin-methylene blue agar plates (Merck, Darmstadt, Germany) was recorded. After 24 h of incubation at 37 °C, *Salmonella* spp. was counted on Xylose Lysine Deoxycholate agar (XLD: Biolife, Italy). *Clostridium* was counted onto CHROMagar™, following 24 h of anaerobic incubation at 35–37 °C, the colonies could be read under UV light. Excel 2010 was used to convert the bacterial count to a Log_{10} before performing statistical analysis. The second step: after incubation, the samples at 37 °C for 18–24 h, 100 µL were streaked onto eosin-methylene blue agar plates, Xylose Lysine Deoxycholate agar plates and CHROMagar™ for *E. coli*, *Salmonella* spp. and *Clostridium*, respectively. Common colony types and morphological features were analyzed in the plates. *E. coli* and *Salmonella* colonies were identified using citrate, Voges–Proskauer, methyl red, and indole reaction biochemical assays, and their identities were confirmed using the standards set out by ISO 16654 and ISO 6579, respectively [30,31].

2.6. *Statistical Analysis*

The data of growth performance, starter intake, DMI, FE, hematological parameters, and serum measures were subjected to analysis using the general linear model method of SPSS statistical software (version, 20) by considering the period as the repeated measure. The repeated measures model contained fixed effects of treatment, period, and interaction of period × treatment, and the calf identity was considered as a random effect. Each individual calf served as an independent experimental unit. The following model was adopted:

$$Y_{ijkl} = \mu + Calf_i + P_j + T_k + (P \times T)_{jk} + e_{ijkl} \quad (1)$$

where Y_{ijkl} is the observation; µ is the overall mean; P_j is the effect of period; T_k is the effect of treatment; $(P \times T)_{jk}$ is the effect of interaction between treatment and period; e_{ijkl} is the error term. The data concerning fecal microbial populations were analyzed using the same model, excluding the effect of the experimental period as follows:

$$Y_{ijkl} = \mu + Calf_i + T_k + e_{ijkl} \quad (2)$$

The data are reported as the least squares means along with their standard error of the mean. The comparison of the means of treatment groups was conducted using Tukey's multiple range test, and statistical differences were evaluated at a significance level of $p < 0.05$. Statistical trends were identified in the range of p-values from 0.05 to 0.10.

3. Results

3.1. Growth Performance and Feed Intake

As presented in Table 2, calves supplemented with LP and LP+ BT exhibited a significant increase in both ADG and final weight compared to the control group ($p < 0.01$). Throughout the experimental period, all calves received an equal daily allowance of milk, while the starter mixture was made available to them ad libitum. However, the treatments had a significant impact on TDMI, starter intake, and CP intake ($p < 0.01$). Notably, the calves supplemented with the probiotic mixture (LP + BT group) displayed the most prominent positive effects when compared to the control group. The individual addition of LP and BT did not affect the TDMI, the starter intake, or the CP intake. There was a significant reduction in the values of TDMI/$W^{0.75}$ in the LP and LP+ BT groups compared to the other groups ($p < 0.01$). The LP+ BT group showed a higher FE compared to the other groups ($p < 0.01$; Table 2). Compared to the other groups, the LP and LP+ BT treatment groups showed a significant increase in ADG ($p = 0.012$) and FE ($p < 0.01$) during the periods from d 4 to d 32 and 33 d to 60 d (Figure 1A,B). From d 4 to d 32, the TDMI/$W^{0.75}$ values of LP + BT calves were higher, but from d 32 to d 60 and d 61 to d 88, all probiotic groups had lower TDMI/$W^{0.75}$ values ($p < 0.01$) compared to the CON group (Figure 1C).

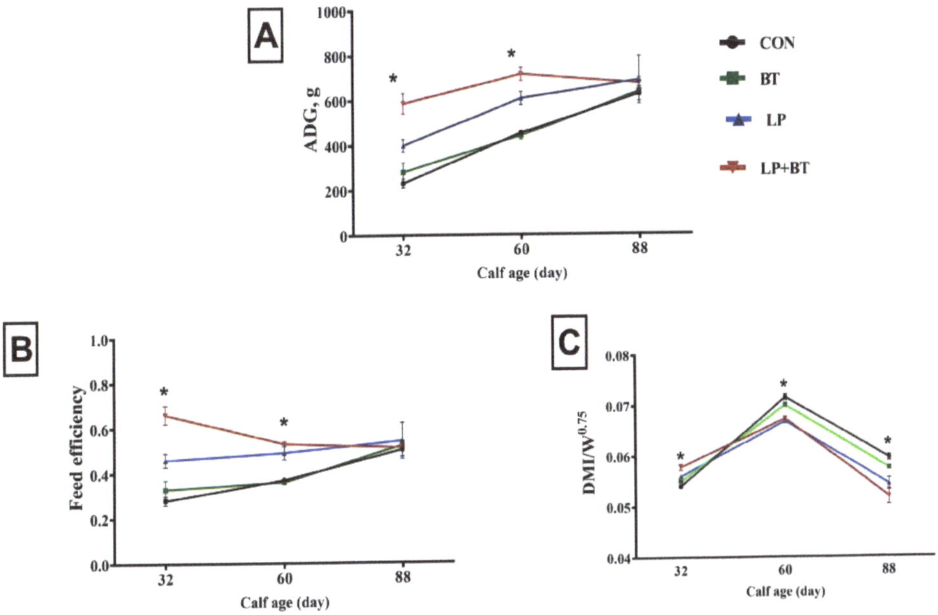

Figure 1. Changes in average daily gain (ADG, **A**), feed efficiency (**B**), and dry matter intake per kilogram of $BW^{0.75}$ (**C**) in pre-weaned calves supplemented without (CON group) or with *Bacillus toyonensis* (BT) at a dosage of 3×10^9 cfu/calf/day, *Lactobacillus plantarum* (LP) at a dosage of 1×10^{10} cfu/calf/day, and their combination (LP + BT) at half the dosage for each from day 4 to day 88 of age. Error bars indicate the SEM; *, $p < 0.05$.

Table 2. The feed intake and growth performance of non-supplemented calves and calves supplemented with *Bacillus toyonensis* (BT), *Lactobacillus plantarum* (LP), and their combination (LP + BT).

	Treatment (Trt)				SEM	p-Value		
	CON	BT	LP	LP + BT		Trt	Period	Trt × Period
Initial BW (kg)	38.20	38.47	38.17	38.23	0.12	0.850	-	-
Final BW (kg)	74.80 [c]	76.43 [c]	85.53 [b]	93.47 [a]	2.11	<0.01	-	-
ADG; (g/d)	435.7 [c]	452.0 [c]	563.9 [b]	657.5 [a]	24.91	<0.01	<0.01	0.012
TDMI [2] (g/d)	1099.5 [b]	1101.7 [b]	1121.8 [b]	1183.2 [a]	28.93	<0.01	<0.01	0.183
Starter DMI (g/d)	416.8 [b]	419.1 [b]	439.0 [b]	500.6 [a]	40.31	<0.01	<0.01	0.184
CP intake (g/d)	244.8 [b]	245.2 [b]	249.1 [b]	260.8 [a]	5.16	<0.01	<0.01	0.183
DMI/$W^{0.75}$	0.062 [a]	0.061 [a]	0.059 [b]	0.059 [b]	0.001	<0.01	<0.01	<0.01
Feed efficiency [3]	0.384 [c]	0.403 [c]	0.497 [b]	0.567 [a]	0.017	<0.01	<0.01	<0.01

[a–c] Means within the row with different letters differed significantly ($p < 0.05$). SEM: standard error of the mean; ADG, average daily gain; TDMI, total dry matter intake; Starter DMI, starter dry matter intake; CP intake, crude protein intake. [2] TDMI: total dry matter intake of milk replacer and starter feed, g/d. [3] Feed efficiency: g of ADG/g of TDMI.

3.2. Hematological Parameters

The effects of LP and/or BT supplementation on the hematological components of pre-weaned calves throughout the experimental period are shown in Table 3. There was no effect of treatment on most hematological parameters including Hb, MCV, MCH, MCHC, platelet, WBCs, lymphocyte, neutrophils monocytes, eosinophils, and basophils ($p > 0.05$). However, a trend ($p = 0.072$) towards an increase in RBC counts was observed in the probiotic-supplemented groups, with the highest values recorded in the mixture group (LP+ BT). The calves in the LP+ BT group exhibited significantly elevated Hct concentrations ($p = 0.028$) compared to the calves in the CON group.

Table 3. Changes in hematological parameters of non-supplemented calves (CON) and calves supplemented with *Bacillus toyonensis* (BT), *Lactobacillus plantarum* (LP), and their combination (LP + BT).

	Treatment (Trt)				SEM	p Value		
	CON	BT	LP	LP + BT		Trt	Period	Trt × Period
Erythrogram								
RBCs, $10^6 \times \mu L$	4.74	5.18	5.51	5.54	0.13	0.072	0.026	0.677
Hb, g/dL	7.83	8.66	9.03	9.45	0.30	0.120	0.003	0.146
Hct, %	28.19 [b]	29.26 [b]	30.02 [ab]	32.59 [a]	0.85	0.028	<0.001	0.118
MCV, fL	59.61	56.44	54.59	59.63	1.27	0.105	<0.001	0.343
MCH, pg/dL	16.37	16.70	16.37	17.00	0.28	0.785	0.055	0.130
MCHC, g/dL	27.96	29.81	30.14	29.97	0.70	0.507	<0.001	0.458
Platelet, $10^3/\mu L$	342.22	321.22	350.44	378.78	21.18	0.643	<0.001	0.131
Leukogram								
WBCs, $10^3/\mu L$	9.84	9.90	10.89	10.65	0.33	0.657	0.540	0.900
Lymphocyte, %	58.22	55.54	56.83	59.43	0.95	0.529	0.389	0.538
Neutrophils, %	36.64	39.26	37.82	35.21	1.02	0.575	0.354	0.574
Monocytes, %	3.24	3.36	3.49	3.34	0.10	0.862	0.442	0.659
Eosinophils, %	1.71	1.65	1.66	1.83	0.06	0.442	<0.001	0.133
Basophils, %	0.18	0.19	0.20	0.20	0.01	0.774	<0.001	0.193

[a,b] Means within the row with different letters differed significantly ($p < 0.05$). SEM stands for standard error of the mean; Hb, hemoglobin; RBCs: red blood cells; Hct: the hematocrit; MCV: mean corpuscular volume; MCH: mean corpuscular hemoglobin; MCHC: mean corpuscular hemoglobin concentration; WBCs: white blood cells.

3.3. Blood Biochemical Parameters

As shown in Table 4, serum total protein, albumin, and globulin values significantly increased in the group receiving LP + BT ($p < 0.05$). All treatment groups that received probiotic supplementation experienced a significant elevation in serum concentration of

IgG in comparison to the CON group ($p = 0.013$). The treatment did not have a significant effect ($p > 0.05$) on serum concentrations of, TC, TG, HDL, LDL, VLDL, and albumin to globulin ratio (Table 4). There was a tendency for higher serum concentrations of urea and BUN in the LP + BT group compared to the other group on days 60 and 88 ($p = 0.067$; Figure 2).

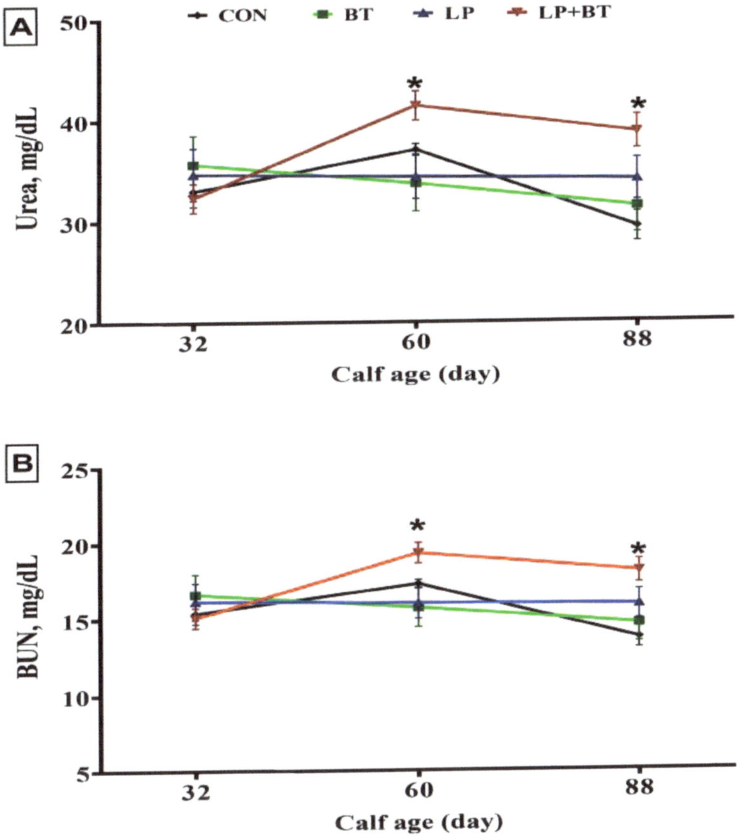

Figure 2. Changes in concentrations of blood urea (**A**) and urea nitrogen (BUN, (**B**)) in pre-weaned calves supplemented without (CON group) or with *Bacillus toyonensis* (BT) at a dosage of 3×10^9 cfu/calf/day, *Lactobacillus plantarum* (LP) at a dosage of 1×10^{10} cfu/calf/day, and their combination (LP + BT) at half the dosage for each from day 4 to day 88 of age. Error bars indicate the SEM; *, $p = 0.066$.

Table 4. Changes in selected measures in the serum of non-supplemented calves (CON) and calves supplemented with *Bacillus toyonensis* (BT), *Lactobacillus plantarum* (LP), and their combination (LP + BT).

	Treatment (Trt)				SEM	*p*-Value		
	CON	BT	LP	LP + BT		Trt	Period	Trt × Period
TP, g/dL	5.39 [b]	5.60 [b]	5.64 [b]	6.18 [a]	0.09	0.003	0.002	0.508
ALB, g/dL	2.94 [b]	2.93 [b]	2.92 [b]	3.27 [a]	0.05	0.003	0.001	0.411
GLOB, g/dL	2.45 [b]	2.68 [ab]	2.72 [ab]	2.91 [a]	0.05	0.016	0.046	0.757
A/G	1.21	1.10	1.09	1.12	0.02	0.112	0.339	0.884
TC, mg/dL	81.56	83.11	80.00	79.44	1.11	0.458	0.001	0.185
TG, mg/dL	82.89	87.44	87.67	88.11	1.67	0.700	0.539	0.574
HDL, mg/dL	42.22	39.78	41.22	45.33	1.07	0.188	0.009	0.198
LDL, mg/dL	21.49	23.29	19.96	18.09	1.30	0.515	0.158	0.305
VLDL, mg/dL	17.84	20.04	18.82	16.02	1.20	0.690	0.155	0.563
Urea, mg/dL	33.11	33.56	34.33	37.44	0.72	0.065	0.076	0.067
BUN, mg/dL	15.45	15.66	16.02	17.47	0.33	0.065	0.076	0.066
IgG, mg/mL	21.86 [b]	29.34 [a]	30.17 [a]	31.64 [a]	1.17	0.013	0.950	0.362

[a,b] Means within the row with different letters differed significantly ($p < 0.05$). SEM stands for standard error of the mean; TP: total protein; ALB: albumin; GLOB: globulin; A/G: albumin/globulin ratio; TC: total cholesterol; TG: total triglycerides; HDL: high-density lipoprotein; LDL: low-density lipoprotein; VLDL: very low-density lipoprotein; BUN: blood urea nitrogen; IgG: immunoglobulin G.

3.4. Fecal Bacterial Microbiota

Table 5 summarizes the results of fecal microbial population in pre-weaned calves supplemented with LP and/or BT. Significant increases in TBC ($p < 0.001$) and decreases in the count of *E. coli* ($p < 0.001$), Salmonella ($p < 0.001$), and Clostridium ($p = 0.005$) were observed in all probiotic groups. The LP + BT group had a significant decrease in coliform count compared to the other groups ($p = 0.005$). The count of *Lactobacillus* populations was significantly increased in both the LP and LP + BT groups when compared to the other groups ($p < 0.001$). Additionally, the BT group had a significantly higher fecal *bacillus* count, while the LP + BT group had a lower count compared to the CON group ($p = 0.001$).

Table 5. Fecal bacterial microbiota (Log cfu/g) of non-supplemented calves (CON) and calves supplemented with *Bacillus toyonensis* (BT), *Lactobacillus plantarum* (LP), and their combination (LP + BT).

	Treatment				SEM	*p*-Value
	CON	BT	LP	LP + BT		
Total bacterial count	8.24 [d]	10.71 [b]	9.56 [c]	10.90 [a]	0.321	<0.001
Coliforms	5.79 [a]	5.25 [a]	5.37 [a]	4.39 [b]	0.176	0.008
Lactobacillus	5.69 [b]	6.31 [b]	7.89 [a]	7.94 [a]	0.308	<0.001
Bacillus	5.66 [b]	7.87 [a]	5.92 [b]	4.62 [c]	0.379	0.001
E. coli	5.87 [a]	4.26 [b]	4.11 [c]	3.12 [d]	0.298	<0.001
Salmonella spp.	5.56 [a]	3.41 [b]	3.21 [b]	3.18 [b]	0.309	<0.001
Clostridium	5.03 [a]	3.76 [b]	3.67 [b]	3.40 [b]	0.214	0.005

[a–d] Means within the row with different letters differed significantly ($p < 0.05$). SEM: standard error of the mean.

4. Discussion

In the present work, calves received milk or starter supplemented with LP and LP + BT displayed a significantly enhanced ADG and final weight than those fed non-supplemented diets. Likewise, in a study by Casper et al. [32], it was observed that neonatal Holstein calves receiving *L. plantarum* GB LP-1 supplementation at a dosage of 8 g/d exhibited a significant increase of over 14% in ADG compared to the CON group during the 0-to-56-day period of the experiment. According to Cangiano et al. [33], 5 of the 11 studies that investigated the benefits of probiotics demonstrated significant improvements in ADG. Additionally, Timmerman et al. [23] demonstrated that feeding pre-weaning Holstein Friesian calves a combination of five *Lactobacillus* strains and *Enterococcus faecium* resulted in enhanced ADG. LP-enhancing effect on ADG is thought to occur through enhanced gut health by limiting pathogen invasion [33]. It could additionally be attributable to the decrease in intestinal viscosity and the increase in crypt depth and villus height [34]. Previous research using LP has demonstrated substantial nutritional advantages of greater weight gains and energy and protein retention while reducing *Salmonella* and *Clostridia counts* when fed to stressed swine [35]. In accordance with the present study, the analysis of the fecal microbial population in pre-weaned calves fed whole milk or starter fortified with LP, BT, and LP + BT showed decreases in the count of *E. coli*, *Salmonella*, and *Clostridium* with the maximum reduction of coliform count in the LP + BT group. In this regard, LP has been demonstrated to exhibit a wide range of antibacterial activities across several hosts [36,37]. LP has been reported to exert inhibitory effects on the proliferation of the prevalent intestinal pathogen. This is achieved by its antibacterial metabolism, creation of a lower pH condition through hydrogen peroxide production, and its ability to effectively cling to intestinal cells [35]. The growth-promoting activity in LP and LP + BT could be also related the enhanced FE in both groups and increased TDMI, starter intake, and CP intake in the LP + BT. Similarly, a significant enhancement was recorded in the feed conversion in veal calves fed milk replacers fortified with five strains of *Lactobacillus* and *Enterococcus faecium* as a probiotic [23]. In this respect, Cangiano et al. [33] reported that feeding microbial-based probiotics promoted stable gut microbiota in dairy calf, thereby enhancing digestive efficiency. Interestingly, the pronounced growth-enhancing effect in the calves that received the combination of the two probiotic LP and BT could be highly related to the increased TDMI. In this regard, Raza et al. [38] reported that increased TDMI in the pre-weaned calves could potentially stimulate rumen development and subsequently enhance growth performance.

Blood metabolites, including blood total protein, albumin, globulin, urea, and BUN are indicators of fermentative changes and microbial protein synthesis occurring in the developing rumen in pre-weaned calves [39]. The concentration of BUN can serve as a useful tool for assessing the efficacy of dietary protein utilization in young calves [40]. The efficient functioning of the rumen has also been linked to elevated levels of BUN [38]. Increased levels of BUN in the group that received LP + BT indicate improved utilization and metabolism of proteins. Increased BUN levels indicate an enhanced rate of protein catabolism and utilization. The potential impact of probiotics on protein metabolism depends on their ability to enhance the efficient breakdown of dietary protein into amino acids, which can then be utilized for a range of physiological functions such as tissue growth and repair [41]. In the current investigation, it was shown that the pre-weaned calves fed milk or starter supplemented with LP + BT had higher levels of serum total protein, albumin, and BUN than the control calves and higher serum albumin levels of albumin in LP group. The aforementioned findings could be explained by an increased intake of starter and CP, leading to larger hydrolysis of dietary proteins or the conversion of amino acids into ammonia through deamination [42]. Importantly, the other blood indices including TC, TG, HDL, LDL, VLDL, and albumin to globulin ratio as well as the erythrogram and leukogram components exhibited values within the average range, as previously documented in pre-weaning calves [40,43]. This reflects the safety of this probiotic combination. Notably, an increasing trend in RBC counts was observed in the probiotic-supplemented groups, with the highest values recorded in the mixture group (LP + BT). The latter group also exhibited significantly elevated Hct concentrations than the calves in the CON group. The favorable effect of LP+ BT on the gut microbiota, which has a significant influence on the metabolism and absorption of iron, the key component of erythrocytes Hct, could be responsible for the improvement of RBC count and Hct concentration [44]. In line with this, LP treatment has been reported to increase the iron absorption in the body [45].

The neonatal phase of calf growth is a critical stage characterized by a decreased immune response, which can lead to serious health issues such as respiratory and intestinal distress leading to high morbidity and mortality [38]. The results of our study showed that adding LP, BT, or their combination in whole milk or starter improved newborn health by increasing serum IgG and globulin levels compared to the CON group. In previous studies, multispecies probiotic supplementation in pre-weaning dairy calves was associated with an increase in IgA levels [46], which is in line with our results. Also, Hong et al. [47] reported that the supplementation of probiotics in milk fortification was associated with elevated IgG levels, indicating an enhanced immune response against spores.

In this regard, Cangiano et al. [33] proposed that the supplementing of lactic acid bacteria may have a greater effect throughout the lower gastrointestinal tract, as it has the potential to influence systemic immune function. This is achieved by enhancing both humoral and cell-mediated immunity through the promotion of B and T cell activity, as well as the reduction of cortisol levels in the bloodstream. Lactic acid bacteria have been shown to interact with intestinal resident microflora and immunological and epithelial cells, as well as initiate and promote immune function, resulting in antibody formation [48]. Moreover, the positive effects of *Bacillus* probiotic strains in stimulating immunity in calves have been reported [49].

Notably, the pre-weaned calves supplemented with the probiotic mixture (LP + BT group) displayed the most prominent positive effects when compared to the groups fed the individual probiotic. The beneficial impact on calf health may have resulted from the synergistic effects observed when a combination of probiotics is added to milk and then starter mixture, as observed in the present study. This could possibly lead to increased starter consumption, hence potentially inducing favorable alterations in the rumen microbial population. Comparably, Laarman and Oba [50] confirmed that ruminal fermentation parameters and microbial population were enhanced by the increased consumption of milk replacer and starter feed, as in our study, in pre-weaned calves. Another important issue to note is some earlier reports proposed that using a probiotic blend of anaerobic (LP) and aerobic (BT) strains of bacteria may have synergistic beneficial effects on animal health, growth, and welfare [51].

5. Conclusions

This study demonstrated that the supplementation of pre-weaning Holstein calves with a probiotic combination of *Lactobacillus plantarum* and *Bacillus toyonensis* had significant positive effects on growth, feed efficiency, blood constituents, and modulation of fecal microbiota composition. These findings support the potential use of LP + BT as a beneficial probiotic strategy to enhance calf performance and health during the critical pre-weaning period. Still, more research with direct assessments of rumen development in calves given a probiotic combination is required.

Author Contributions: Conceptualization, M.S.A. and A.A.A.-S.; Data curation, M.S.A., H.A.E.-N., W.M.W., S.M. and A.A.A.-S.; Formal analysis, K.M.A.E.-L., S.M. and A.A.A.-S.; Investigation, H.A.E.-N., W.M.W. and A.A.A.-S.; Methodology, H.A.E.-N., W.M.W., S.M. and A.A.A.-S.; Resources, M.S.A., K.M.A.E.-L., H.A.E.-N. and W.M.W.; Software, M.S.A.; Writing—original draft, S.M. and A.A.A.-S.; Writing—review and editing, M.S.A. and K.M.A.E.-L. All authors have read and agreed to the published version of the manuscript.

Funding: This research received no external funding.

Institutional Review Board Statement: The animal study protocol was approved by the Institutional animal care and Use committee of Zagazig University (approval code: ZU-IACUC/2/F/269/2023).

Informed Consent Statement: Not applicable.

Data Availability Statement: The data that support the findings of this study are available upon reasonable request.

Conflicts of Interest: The authors declare no conflict of interest.

References

1. Jiang, X.; Ma, G.; Cui, Z.; Li, Y.; Zhang, Y. Effects of fermented corn gluten meal on growth performance, plasma metabolites, rumen fermentation and bacterial community of holstein calves during the pre-weaning period. *Livest. Sci.* **2020**, *231*, 103866. [CrossRef]
2. Mee, J.F. Why do so many calves die on modern dairy farms and what can we do about calf welfare in the future? *Animals* **2013**, *3*, 1036–1057. [CrossRef]
3. Hordofa, D.; Abunna, F.; Megersa, B.; Abebe, R. Incidence of morbidity and mortality in calves from birth to six months of age and associated risk factors on dairy farms in Hawassa city, southern Ethiopia. *Heliyon* **2021**, *7*, e08546. [CrossRef] [PubMed]
4. Chester-Jones, H.; Heins, B.; Ziegler, D.; Schimek, D.; Schuling, S.; Ziegler, B.; De Ondarza, M.; Sniffen, C.; Broadwater, N. Relationships between early-life growth, intake, and birth season with first-lactation performance of holstein dairy cows. *J. Dairy Sci.* **2017**, *100*, 3697–3704. [CrossRef]
5. Stefańska, B.; Sroka, J.; Katzer, F.; Goliński, P.; Nowak, W. The effect of probiotics, phytobiotics and their combination as feed additives in the diet of dairy calves on performance, rumen fermentation and blood metabolites during the preweaning period. *Anim. Feed Sci. Technol.* **2021**, *272*, 114738. [CrossRef]
6. Wang, L.; Zhang, C.; Zhang, Q.; Xu, H.; Feng, G.; Zhang, G.; Zhang, Y. Effects of feeding different doses of 25-hydroxyvitamin D_3 on the growth performance, blood minerals, antioxidant status and immunoglobulin of preweaning calves. *Anim. Feed Sci. Technol.* **2022**, *285*, 115220. [CrossRef]
7. Seo, J.K.; Kim, S.-W.; Kim, M.H.; Upadhaya, S.D.; Kam, D.K.; Ha, J.K. Direct-fed microbials for ruminant animals. *Asian Australas. J. Anim. Sci.* **2010**, *23*, 1657–1667. [CrossRef]
8. Stefańska, B.; Komisarek, J.; Stanisławski, D.; Gąsiorek, M.; Kasprowicz-Potocka, M.; Frankiewicz, A.; Nowak, W. The effect of *Yarrowia lipolytica* culture on growth performance, ruminal fermentation and blood parameters of dairy calves. *Anim. Feed Sci. Technol.* **2018**, *243*, 72–79. [CrossRef]
9. Busconi, M.; Reggi, S.; Fogher, C. Evaluation of biodiversity of lactic acid bacteria microbiota in the calf intestinal tracts. *Antonie Leeuwenhoek* **2008**, *94*, 145–155. [CrossRef]
10. Goel, A.; Halami, P.M.; Tamang, J.P. Genome analysis of *lactobacillus plantarum* isolated from some indian fermented foods for bacteriocin production and probiotic marker genes. *Front. Microbiol.* **2020**, *11*, 40. [CrossRef]
11. Le, B.; Yang, S.H. Efficacy of *lactobacillus plantarum* in prevention of inflammatory bowel disease. *Toxicol. Rep.* **2018**, *5*, 314–317. [CrossRef]
12. Wang, J.; Ji, H.; Wang, S.; Liu, H.; Zhang, W.; Zhang, D.; Wang, Y. Probiotic lactobacillus plantarum promotes intestinal barrier function by strengthening the epithelium and modulating gut microbiota. *Front. Microbiol.* **2018**, *9*, 1953. [CrossRef] [PubMed]
13. Ding, S.; Wang, Y.; Yan, W.; Li, A.; Jiang, H.; Fang, J. Effects of *lactobacillus plantarum* 15-1 and fructooligosaccharides on the response of broilers to pathogenic *Escherichia coli* O78 challenge. *PLoS ONE* **2019**, *14*, e0212079.
14. Ahire, J.J.; Jakkamsetty, C.; Kashikar, M.S.; Lakshmi, S.G.; Madempudi, R.S. In vitro evaluation of probiotic properties of *Lactobacillus plantarum* UBLP40 isolated from traditional indigenous fermented food. *Probiotics Antimicrob. Proteins* **2021**, *13*, 1413–1424. [CrossRef]
15. Casula, G.; Cutting, S.M. *Bacillus* probiotics: Spore germination in the gastrointestinal tract. *Appl. Environ. Microbiol.* **2002**, *68*, 2344–2352. [CrossRef]
16. EFSA, E.F.S.A. Scientific opinion on the safety and efficacy of toyocerin® (*Bacillus toyonensis*) as a feed additive for chickens for fattening, weaned piglets, pigs for fattening, sows for reproduction, cattle for fattening and calves for rearing and for rabbits for fattening. *EFSA J.* **2014**, *12*, 3766.
17. Pechrkong, T.; Incharoen, T.; Hwanhlem, N.; Kaewkong, W.; Subsoontorn, P.; Tartrakoon, W.; Numthuam, S.; Jiménez, G.; Charoensook, R. Effect of *Bacillus toyonensis* BCT-7112^{T} supplementation on growth performance, intestinal morphology, immune-related gene expression, and gut microbiome in barbary ducks. *Poult. Sci.* **2023**, *102*, 102991. [CrossRef] [PubMed]
18. Abd El-Hack, M.E.; El-Saadony, M.T.; Shafi, M.E.; Qattan, S.Y.A.; Batiha, G.E.; Khafaga, A.F.; Abdel-Moneim, A.-M.E.; Alagawany, M. Probiotics in poultry feed: A comprehensive review. *J. Anim. Physiol. Anim. Nutr.* **2020**, *104*, 1835–1850. [CrossRef]
19. Alagawany, M.; Bassiony, S.S.; El-Kholy, M.S.; El-Naggar, K.; El-Metwally, A.E.; Al-Sagheer, A.A. Comparison of the effects of probiotic-based formulations on growth, feed utilization, blood constituents, cecal fermentation, and duodenal morphology of rabbits reared under hot environmental conditions. *Ann. Anim. Sci.* **2023**, *23*, 777–787. [CrossRef]
20. Sanz, Y.; De Palma, G. Gut microbiota and probiotics in modulation of epithelium and gut-associated lymphoid tissue function. *Int. Rev. Immunol.* **2009**, *28*, 397–413. [CrossRef]
21. Timmerman, H.M.; Koning, C.J.; Mulder, L.; Rombouts, F.M.; Beynen, A.C. Monostrain, multistrain and multispecies probiotics—A comparison of functionality and efficacy. *Int. J. Food Microbiol.* **2004**, *96*, 219–233. [CrossRef] [PubMed]
22. Lambo, M.T.; Chang, X.; Liu, D. The recent trend in the use of multistrain probiotics in livestock production: An overview. *Animals* **2021**, *11*, 2805. [CrossRef]
23. Timmerman, H.M.; Mulder, L.; Everts, H.; van Espen, D.C.; van der Wal, E.; Klaassen, G.; Rouwers, S.M.G.; Hartemink, R.; Rombouts, F.M.; Beynen, A.C. Health and growth of veal calves fed milk replacers with or without probiotics. *J. Dairy Sci.* **2005**, *88*, 2154–2165. [CrossRef] [PubMed]
24. De Man, J.; Rogosa, D.; Sharpe, M.E. A medium for the cultivation of lactobacilli. *J. Appl. Microbiol.* **1960**, *23*, 130–135. [CrossRef]

25. Jiang, X.; Xu, H.J.; Cui, Z.Q.; Zhang, Y.G. Effects of supplementation with *Lactobacillus plantarum* 299v on the performance, blood metabolites, rumen fermentation and bacterial communities of preweaning calves. *Livest. Sci.* **2020**, *239*, 104120. [CrossRef]
26. Novak, K.N.; Davis, E.; Wehnes, C.A.; Shields, D.R.; Coalson, J.A.; Smith, A.H.; Rehberger, T.G. Effect of supplementation with an electrolyte containing a *Bacillus*-based direct-fed microbial on immune development in dairy calves. *Res. Vet. Sci.* **2012**, *92*, 427–434. [CrossRef]
27. NRC. *National Research Council. Nutrient Requirements of Dairy Cattle*, 7th ed.; National Academy Press: Washington, DC, USA, 2001.
28. AOAC. *Official Methods of Analysis*, 20th ed.; AOAC: Gaithersburg, MD, USA, 2016.
29. Van Soest, P.V.; Robertson, J.; Lewis, B. Methods for dietary fiber, neutral detergent fiber, and nonstarch polysaccharides in relation to animal nutrition. *J. Dairy Sci.* **1991**, *74*, 3583–3597. [CrossRef]
30. *6579:1993*; Microbiology—General Guidance on Methods for the Detection of Salmonella Rev. 2nd ed. International Organization for Standardization: Geneva, Switzerland, 1991.
31. *EN ISO 16654:2002*; Microbiology—General Guidance on Methods for the Detection of Escherichia, Rev. International Organization for Standardization: Geneva, Switzerland, 2001.
32. Casper, D.P.; Hultquist, K.M.; Acharya, I.P. *Lactobacillus plantarum* GB LP-1 as a direct-fed microbial for neonatal calves. *J. Dairy Sci.* **2021**, *104*, 5557–5568. [CrossRef]
33. Cangiano, L.R.; Yohe, T.T.; Steele, M.A.; Renaud, D.L. Invited review: Strategic use of microbial-based probiotics and prebiotics in dairy calf rearing. *Appl. Anim. Sci.* **2020**, *36*, 630–651. [CrossRef]
34. Meyer, M.M.; Fries-Craft, K.A.; Bobeck, E.A. Composition and inclusion of probiotics in broiler diets alter intestinal permeability and spleen immune cell profiles without negatively affecting performance. *J. Anim. Sci.* **2020**, *98*, skz383. [CrossRef] [PubMed]
35. Wang, J.; Zeng, Y.; Wang, S.; Liu, H.; Zhang, D.; Zhang, W.; Wang, Y.; Ji, H. Swine-derived probiotic *Lactobacillus plantarum* inhibits growth and adhesion of enterotoxigenic *Escherichia coli* and mediates host defense. *Front. Microbiol.* **2018**, *9*, 1364. [CrossRef]
36. Guerra-Ordaz, A.; González-Ortiz, G.; La Ragione, R.; Woodward, M.; Collins, J.; Pérez, J.; Martín-Orúe, S. Lactulose and *Lactobacillus plantarum*, a potential complementary synbiotic to control postweaning colibacillosis in piglets. *Appl. Environ. Microbiol.* **2014**, *80*, 4879–4886. [CrossRef]
37. El Halfawy, N.M.; El-Naggar, M.Y.; Andrews, S.C. Complete genome sequence of *Lactobacillus plantarum* 10CH, a potential probiotic lactic acid bacterium with potent antimicrobial activity. *Genome Announc.* **2017**, *5*, e01398-17. [CrossRef] [PubMed]
38. Raza, M.; Yousaf, S.M.; Ahmad, J.; Rashid, A.M.; Majeed, A.K.; Tahir, K.S.; Ashraf, S.; Numan, M.; Khalid, A.; Rehman, U.H. Prebiotics supplementation modulates pre-weaning stress in male cattle calves by improving growth performance, health scores and serum biomarkers. *Czech J. Anim. Sci.* **2022**, *67*, 102–113. [CrossRef]
39. Zhang, J.; Shang, J.; Hao, Y.; Wang, Y.; Cao, Z.; Yang, H.; Wang, W.; Li, S. Growth performance, blood metabolites, ruminal fermentation, and bacterial community in preweaning dairy calves fed corn silage-included starter and total mixed ration. *J. Dairy Sci.* **2023**, *106*, 4545–4558. [CrossRef]
40. Liu, S.; Ma, J.; Zhou, J.; Wu, J.; Li, J.; Alugongo, G.; Xiao, J.; Wang, J.; Wang, Y.; Wang, W. Tributyrin supplementation in pasteurized waste milk: Effects on growth performance, health, and blood parameters of dairy calves. *J. Dairy Sci.* **2021**, *104*, 12496–12507. [CrossRef]
41. Wang, J.; Ji, H. Influence of probiotics on dietary protein digestion and utilization in the gastrointestinal tract. *Curr. Protein Pept. Sci.* **2019**, *20*, 125–131. [CrossRef]
42. Quigley, J.; Wolfe, T.; Elsasser, T. Effects of additional milk replacer feeding on calf health, growth, and selected blood metabolites in calves. *J. Dairy Sci.* **2006**, *89*, 207–216. [CrossRef]
43. Zou, Y.; Wang, Y.; Deng, Y.; Cao, Z.; Li, S.; Wang, J. Effects of feeding untreated, pasteurized and acidified waste milk and bunk tank milk on the performance, serum metabolic profiles, immunity, and intestinal development in Holstein calves. *J. Anim. Sci. Biotechnol.* **2017**, *8*, 53. [CrossRef]
44. Rusu, I.G.; Suharoschi, R.; Vodnar, D.C.; Pop, C.R.; Socaci, S.A.; Vulturar, R.; Istrati, M.; Moroșan, I.; Fărcaș, A.C.; Kerezsi, A.D.; et al. Iron supplementation influence on the gut microbiota and probiotic intake effect in iron deficiency-A literature-based review. *Nutrients* **2020**, *12*, 1993. [CrossRef] [PubMed]
45. Hoppe, M.; Önning, G.; Berggren, A.; Hulthén, L. Probiotic strain *Lactobacillus plantarum* 299v increases iron absorption from an iron-supplemented fruit drink: A double-isotope cross-over single-blind study in women of reproductive age. *Br. J. Nutr.* **2015**, *114*, 1195–1202. [CrossRef] [PubMed]
46. Wu, Y.; Wang, L.; Luo, R.; Chen, H.; Nie, C.; Niu, J.; Chen, C.; Xu, Y.; Li, X.; Zhang, W. Effect of a multispecies probiotic mixture on the growth and incidence of diarrhea, immune function, and fecal microbiota of pre-weaning dairy calves. *Front. Microbiol.* **2021**, *12*, 681014. [CrossRef]
47. Hong, H.A.; Duc, L.H.; Cutting, S.M. The use of bacterial spore formers as probiotics. *FEMS Microbiol. Rev.* **2005**, *29*, 813–835. [CrossRef]
48. Kaburagi, T.; Yamano, T.; Fukushima, Y.; Yoshino, H.; Mito, N.; Sato, K. Effect of *Lactobacillus johnsonii* La1 on immune function and serum albumin in aged and malnourished aged mice. *Nutrition* **2007**, *23*, 342–350. [CrossRef] [PubMed]
49. Yao, J.; Wang, L.; Zhang, W.; Liu, M.; Niu, J. Effects of *Bacillus megaterium* on growth performance, serum biochemical parameters, antioxidant capacity, and immune function in suckling calves. *Open Life Sci.* **2020**, *15*, 1033–1041. [CrossRef] [PubMed]

50. Laarman, A.; Oba, M. Effect of calf starter on rumen ph of Holstein dairy calves at weaning. *J. Dairy Sci.* **2011**, *94*, 5661–5664. [CrossRef]
51. Abou-Kassem, D.E.; Elsadek, M.F.; Abdel-Moneim, A.E.; Mahgoub, S.A.; Elaraby, G.M.; Taha, A.E.; Elshafie, M.M.; Alkhawtani, D.M.; Abd El-Hack, M.E.; Ashour, E.A. Growth, carcass characteristics, meat quality, and microbial aspects of growing quail fed diets enriched with two different types of probiotics (*Bacillus toyonensis* and *Bifidobacterium bifidum*). *Poult. Sci.* **2021**, *100*, 84–93. [CrossRef]

animals

MDPI

Article

Growth Performance and Fecal Microbiota of Dairy Calves Supplemented with Autochthonous Lactic Acid Bacteria as Probiotics in Mexican Western Family Dairy Farming

José Martín Ruvalcaba-Gómez [1,*], Fernando Villaseñor-González [2], Mario Alfredo Espinosa-Martínez [3], Lorena Jacqueline Gómez-Godínez [1], Edith Rojas-Anaya [1], Zuamí Villagrán [4], Luis Miguel Anaya-Esparza [4], Germán Buendía-Rodríguez [5] and Ramón Ignacio Arteaga-Garibay [1,*]

[1] Centro Nacional de Recursos Genéticos, Instituto Nacional de Investigaciones Forestales, Agrícolas y Pecuarias, Boulevard de la Biodiversidad 400, Tepatitlán de Morelos 47600, Jalisco, Mexico; gomez.lorena@inifap.gob.mx (L.J.G.-G.); rojas.edith@inifap.gob.mx (E.R.-A.)

[2] Campo Experimental Centro Altos de Jalisco, Instituto Nacional de Investigaciones Forestales, Agrícolas y Pecuarias, Av. Biodiversidad 2470, Tepatitlán de Morelos 47600, Jalisco, Mexico; villasenor.fernando@inifap.gob.mx

[3] Centro Nacional de Investigación Disciplinaria en Fisiología y Mejoramiento Animal, Instituto Nacional de Investigaciones Forestales, Agrícolas y Pecuarias, Ajuchitlán Colón 76280, Querétaro, Mexico; espinosa.mario@inifap.gob.mx

[4] Centro Universitario de los Altos, Universidad de Guadalajara, Av. Rafael Casillas Aceves 1200, Tepatitlán de Morelos 47600, Jalisco, Mexico; blanca.villagran@academicos.udg.mx (Z.V.); luis.aesparza@academicos.udg.mx (L.M.A.-E.)

[5] Sitio Experimental Hidalgo, Campo Experimental Valle de México, Instituto Nacional de Investigaciones Forestales, Agrícolas y Pecuarias, Carr. Pachuca-Tulancingo 104ª, Pachuca de Soto 42090, Hidalgo, Mexico; buendia.german@inifap.gob.mx

[*] Correspondence: ruvalcaba.josemartin@inifap.gob.mx (J.M.R.-G.); arteaga.ramon@inifap.gob.mx (R.I.A.-G.)

Simple Summary: *Lactobacillus* is one of the bacterial genera frequently used as probiotics. Although probiotic effects are strain dependent, several *Lactobacillus* strains have been used to improve the growth rate in dairy calves, and the health or productivity parameters of dairy cows, mainly through the modulation of rumen microbiota. This study explains the effects observed on the growth parameters and fecal microbiota of dairy calves supplemented with three different *Lactobacillus*-based probiotic formulations supplied between days 5 and 19 after birth.

Abstract: Probiotic supplementation in dairy cattle has achieved several beneficial effects (improved growth rate, immune response, and adequate ruminal microbiota). This study assessed the effects on the growth parameters and gut microbiota of newborn dairy calves supplemented with two *Lactobacillus*-based probiotics, individually (6BZ or 6BY) or their combination (6BZ + 6BY), administrated with the same concentration (1×10^9 CFU/kg weight) at three times, between days 5 and 19 after birth. The control group consisted of probiotic-unsupplemented calves. Growth parameters were recorded weekly until eight weeks and at the calves' ages of three, four, and five months. Fecal microbiota was described by high-throughput sequencing and bioinformatics. Although no significant effects were observed regarding daily weight and height gain among probiotic-supplemented and non-supplemented calves, correlation analysis showed that growth rate was maintained until month 5 through probiotic supplementation, mainly when the two-strain probiotics were supplied. Modulation effects on microbiota were observed in probiotic-supplemented calves, improving the Bacteroidota: Firmicutes and the Proteobacteria ratios. Functional prediction by PICRUSt also showed an increment in several pathways when the two-strain probiotic was supplemented. Therefore, using the three-administration scheme, the two-strain probiotic improved the growth rate and gut microbiota profile in newborn dairy calves. However, positive effects could be reached by applying more administrations of the probiotic during the first 20 days of a calf's life.

Keywords: *Lactobacillus*; PICRUSt; nutraceuticals; microbiota modulation; ion torrent

1. Introduction

The intestinal microbiota has been recognized as a key organ in animals and humans due to its associated functions related to harvesting and absorbing nutrients involved in energy expenditure [1,2]. Specifically, the bovine gastrointestinal tract, considered sterile at animal birth, is rapidly colonized by bacteria, and it involves a very complex microbiota in response to several factors such as sex, genetics, and the exposome (feeding, environmental conditions, stress levels, among others) [3,4]; therefore, perturbations in gut microbiota have been related to the appearance or development of several pathologies or even to losses in metabolic and immunological abilities in the animal [3,5]. For this reason, microbiota modulation has become one of the main strategies to keep a healthy gut microbiota in animals to ensure its correct performance in obtaining energy from nutrients and modulating the immune response to improve tolerance and resistance to pathogenic agents [6,7].

In this context, it has been reported that correct colonization of gut microbiota at birth positively increases the success in the development of the immune system, contributing to keeping a good health status in animals [8]. Probiotics, which refer to the use of live microorganisms as a nutritional supplement to confer a health benefit to the host, have been one of the main microbiota modulation-based strategies to improve the health status and performance in several productive animal species, including dairy cows, alone or in combination with other strategies such as prebiotics [9–11]. Remarkably, dairy cow-associated rumen microbiota is mainly composed of members of the Firmicutes and Bacteroidetes phyla [12], strongly represented by *Clostridium*, *Lactococcus*, *Flavobacterium*, and *Prevotella* genera, and some other members of the Ruminococcaceae family [12]. The rumen-associated bacterial communities are acquired from the first days of life, before rumen development. These communities are mainly defined by the interactions of the newborn to environmental factors such as contact with the mother during the delivery, colostrum, and milk, and the establishment of the bacterial communities occurs the first 12 days after birth [13]. For this reason, it has been demonstrated that the intervention with bacteria-based probiotics is more efficient when performed in young calves [2,14].

Lactic acid bacteria (LAB), considered a natural resident of the gastrointestinal tract (GIT) and numerous fermented foods, have been extensively used as probiotics, mainly several members of the *Lactobacillus* genus, which was recently reclassified into 25 genera, due to its versatility in the use of substrates, colonization capacity, and beneficial effects [15–18]. This study aimed to assess the effects of using LAB-based probiotics on establishing GIT microbiota and growth parameters of calves allocated in Mexican western family dairy farming.

2. Materials and Methods

2.1. LAB Strains and Probiotic Preparation

Two LAB strains, previously isolated from the feces of dairy cows and characterized according to the FAO/WHO guidelines [19], were selected [20]. Strains were recovered according to the procedure described by Ruvalcaba et al. [21]. Strains were identified as *Lactobacillus* 6BZ and *Lactobacillus* 6BY. A pre-inoculum of each strain was prepared by inoculating 100 mL of MRS broth, followed by incubation at 37 °C for 24 h. Then, 24 h pre-inoculum was used for growth in a stirred-batch mode using a Biostat® B Twin 5 L double-wallet, round-bottom glass bioreactor (Sartorius BBI Systems, Melsungen, DE). Temperature (35 °C), pH (6.5), and stirring (120 rpm) were automatically controlled. Next, 48 h bioreactor-produced cultures were mixed with sterile skim milk (10%, 1:1 *v/v*), distributed into 5 mL sterile serum bottles provided with two-leg Stoppers, and freeze-dried using a Labconco FreeZone® 12 L Freeze Dry system equipped with a Stoppering Tray

Dryer (Labconco, Kansas, MO, USA). The final concentration of each reconstituted probiotic dose was around 5×10^9 CFU/mL.

2.2. Animals, Treatments and Feeding

A total of 42 Holstein newborn calves were included in the study. Calves were allocated into three dairy family farms in the Los Altos de Jalisco region. At the age of 5 days, calves were aleatorily assigned to four treatments identified as the 6BZ group (*Lactobacillus* 6BZ supplemented; $n = 8$), 6BY group (*Lactobacillus* 6BY supplemented; $n = 8$), and 6BZ + 6BY group (supplemented with both 6BZ and 6BY strains; $n = 17$); and CG consisting of a control group (probiotic unsupplemented; $n = 9$). The calves' feeding was based on a milk replacer (protein: 20%, fat: 19.5%, fiber: 0.9%) and then complemented with a calf starter concentrate that covered the nutrient requirements suggested by the Nutrient Requirements of Dairy Cattle of 2001 [22]. Water was offered at libitum from day two. Probiotics were reconstituted using the milk replacer and were supplemented at days 5, 12, and 19 after birth (1×10^9 CFU/kg weight).

2.3. Growth Performance

From week two up to week eight of age, the calves' weight was estimated weekly using a heart girth tape designed for Holstein's calves [23]. The calves' height was also recorded using a somatometric ruler (Nasco, Whitewater, WI, USA) [24]. Later, weight and height were recorded monthly up to month five. Data from weight and height were used to estimate daily weight gain and daily height gain.

2.4. Feces Sampling and DNA Extraction

At days 1, 5, 15, 30, 45, and 60 after birth, 10 g of fecal samples was obtained from each calf. Samples were obtained aseptically before morning feeding, placed into sterile cryogenic vials (DNase-RNase free; Corning®, Glendale, AZ, USA), and brought to the laboratory. Samples corresponding to the same treatment, the same sampling day, and the same dairy farm were pooled, and 0.2 g of each pooled sample was used for DNA extraction. Metagenomic DNA was extracted and purified using the Quick-DNA™ Fecal/soil Microbe Miniprep system (Zymo Research, Irvine, CA, USA) according to the manufacturer´s instructions. DNA integrity was verified by electrophoresis in 1% agarose gel. DNA was stored at −20 °C until sequencing procedures.

2.5. Construction of the 16S rRNA Libraries and Sequencing Procedures

Libraries of 16S DNA were constructed based on PCR amplification of 7 of the 9 hypervariable regions of the 16S rDNA gene (V2, V3, V4, V6–V9). Amplification was performed in two independent reactions using the 16S Metagenomics™ system according to the manufacturer's instructions (Thermo Fisher Scientific, Waltham, MA, USA) in a Verity™ thermal cycler (Thermo Fisher Scientific, Waltham, MA, USA). An equimolar mixture using the amplification products was prepared, and 50 nanograms were used to construct the 16S rDNA libraries with the Ion Plus Fragment Library commercial system and the Ion Xpress barcode adapters (Thermo Fisher Scientific). Library purification was carried out using the Agentcourt AMPure XP system according to the manufacturer's instructions (Beckman Coulter, Brea, CA, USA) and quantified with a highly sensitive DNA commercial system and the Bioanalyzer 2100 (Agilent Technologies, Santa Clara, CA, USA). Library concentration was adjusted to 26 pM followed by PCR amplification of the PCR emulsion using a volume of 25 µL of the equimolar mixture for all samples (One-Touch 2, Thermo Fisher Scientific, Waltham, MA, USA) and enriched with the OneTouch Enrichment system (Thermo Fisher Scientific, Waltham, MA, USA). Sequencing was carried out using the Ion S5™ system (Thermo Fisher Scientific, Waltham, MA, USA).

2.6. Bioinformatics

Sequences were quality assessed and trimmed to remove low-quality regions using the Trimmomatic tool [25] and visualized in MultiQC [26]. Remotion of chimeric and low-quality sequences using the DADA2 module and ASV assignation using the SILVA v-132 database as a reference for the taxonomy assignation (16S rRNA gene sequences clustered at 99% of similarity) [27] were performed using the nf-core ampliseq pipeline v2.3.2 [28,29]. Alpha diversity (observed features, evenness, faith, and Shannon) was calculated using the QIIME2 [30] module and distances based on ASV abundance between samples (beta diversity) were measured using the method of "weighted UniFrac" and visualized using principal coordinate analysis (PCoA) plot. Relative abundance graphs and PCA graphics were generated using the abundance tables in the Origin v. 2022 software (OriginLab, Northampton, MA, USA).

2.7. Functional Prediction

The software PICRUSt2 (v. 2.4.1) was used to conduct a deeper exploration of the possible metabolic mechanism associated with bacterial communities in the fecal samples, and obtain the functional profile [31]. The predictions were made by corresponding the marker gene data and the reference genomes in databases, including MetaCyc [32].

2.8. Statistical Analysis

The experimental design corresponded to a completely randomized design. Data corresponding to the growth performance were analyzed by REML ANOVA using the PROC MIXED procedure of SAS 9.4 v (SAS Institute Inc., Cary, NC, USA) for the repeated measures of weight and height obtained during the milk-feeding period (first eight weeks) and monthly records (0 to 5 months). Treatment, age, and treatment × age were considered fixed effects in the model, and the random effect of the block (production unit). The covariance structure was autoregressive. Significance was determined at $p < 0.05$. Pearson correlations were also calculated from weight and wither height data ($\alpha = 0.05$).

3. Results

3.1. Growth Parameters

Lactobacillus-based probiotics were successfully supplemented to dairy calves, and the effects on growth parameters (weight and wither height) were recorded. Table 1 shows the weight and wither height of Holstein calves supplemented with probiotics during the milk-feeding period (first eight weeks of a calf's life). Regarding the weight, no statistical differences were observed by probiotic treatment during the first eight weeks ($p = 0.4308$). Nonetheless, daily weight gain in the probiotic-supplemented calves was not significantly different ($p = 0.456$, Table S1) in comparison with the non-supplemented calves. Regarding wither height scores, no statistical differences ($p = 0.8233$) were observed by probiotic supplementation. On the other hand, a significant effect was observed for both weight and wither height by age of calves ($p < 0.001$). Additionally, no interactions were observed between probiotic supplementation and age of calves for weight ($p = 0.7198$) and wither height ($p = 0.4377$) during the milk-feeding period.

Table 2 shows the weight and wither height of Holstein calves from birth to 5 months of age. In this context, during the first five months of a calf's life, a significant effect was detected in the calf's weight by treatment ($p = 0.0480$), month ($p < 0.0001$), and their interaction (treatment × age, $p < 0.0001$). In general, the calves treated with 6BZ (200.28 kg) and 6BY + 6BZ (202.00 kg) exhibited higher weights than the 6BY (184.66 kg) and control group (186.66 kg). On the other hand, no statistical effects were observed by probiotic supplementation ($p = 0.3541$) on wither height, but a significant effect associated with the calves' age ($p < 0.0001$) was observed; moreover, no interactive effects (treatment group × age) were detected ($p = 0.5826$) for such parameter.

Table 1. Body weight and wither height of Holstein calves during milk-feeding period.

Parameter/Age (Week)	Calves' Groups (Teatment)				Significance		
	Control	6BZ	6BY	6BY + 6BZ	T	A	T × A
			Body weight (kg)				
0	39.55 ± 0.37	42.12 ± 0.76	40.62 ± 1.22	40.23 ± 0.64	0.4308	<0.0001	0.7198
1	41.11 ± 0.78	42.87 ± 1.02	42.50 ± 1.42	42.23 ± 0.92			
2	43.33 ± 1.05	45.62 ± 1.06	44.00 ± 1.80	44.29 ± 1.18			
3	47.11 ± 1.33	48.87 ± 1.43	45.75 ± 2.20	47.58 ± 1.15			
4	50.88 ± 1.52	54.75 ± 1.67	49.25 ± 2.45	52.76 ± 1.47			
5	54.88 ± 1.78	59.50 ± 2.28	55.00 ± 3.21	57.47 ± 1.67			
6	61.11 ± 2.07	64.12 ± 2.18	58.87 ± 3.51	62.82 ± 2.01			
7	67.77 ± 3.13	71.00 ± 2.64	66.25 ± 2.51	69.82 ± 2.51			
8	72.55 ± 3.01	78.25 ± 3.78	73.12 ± 3.30	77.52 ± 2.92			
			Wither height (cm)				
0	78.66 ± 0.91	80.50 ± 0.53	80.50 ± 0.90	78.70 ± 0.98	0.8233	<0.0001	0.4377
1	80.11 ± 1.05	80.62 ± 0.94	81.12 ± 1.07	79.58 ± 0.74			
2	81.88 ± 1.00	82.50 ± 0.88	82.50 ± 0.80	81.47 ± 0.73			
3	84.22 ± 0.79	84.25 ± 0.95	84.12 ± 0.85	83.41 ± 0.71			
4	85.77 ± 0.74	86.50 ± 0.84	84.87 ± 0.93	85.70 ± 0.64			
5	87.88 ± 0.77	87.87 ± 1.04	86.75 ± 1.17	87.52 ± 0.79			
6	89.22 ± 0.86	90.62 ± 1.01	88.50 ± 1.21	89.23 ± 0.82			
7	91.88 ± 1.47	92.00 ± 1.19	89.25 ± 1.91	91.11 ±0.75			
8	93.77 ± 0.87	93.50 ± 1.23	92.25 ± 1.91	93.29 ± 0.98			

Average value ± standard error. T: treatment group; A: age in weeks, T × A: treatment group × age.

Table 2. Body weight and wither height of Holstein's calves from birth to 5 months of age.

Parameter/Age (Months)	Calves' Groups (Teatment)				Significance		
	Control	6BZ	6BY	6BY + 6BZ	T	A	T × A
			Body weight (kg)				
0	39.55 ± 0.78	42.12 ± 1.02	40.62 ± 1.22	40.23 ± 0.64	0.0480	<0.0001	<0.0001
1	101.11 ± 4.48	107.8 ± 4.20	99.50 ± 2.85	96.61 ± 2.61			
2	123.55 ± 5.77	135.4 ± 3.16	130.37 ± 5.02	119.94 ± 3.46			
3	145.66 ± 4.98	155.28 ± 3.16	157.37 ± 7.97	152.33 ± 5.15			
4	166.55 ± 4.92	180.14 ± 4.65	195.87 ± 17.63	159.66 ± 4.44			
5	186.66 ± 4.57	200.28 ± 5.97	202.00 ± 12.73	184.66 ± 5.90			
			Wither height (cm)				
0	78.66 ± 0.91	80.50 ± 0.53	80.50 ± 0.90	78.70 ± 0.98	0.3541	<0.0001	0.5826
1	98.33 ± 1.01	96.42 ± 2.10	97.62 ± 1.22	95.38 ± 0.74			
2	102.00 ± 1.01	101.42 ± 1.19	102.62 ± 1.71	95.38 ± 0.63			
3	107.88 ± 0.88	107.57 ± 1.26	108.00 ± 2.35	100.72 ± 0.74			
4	111.33 ± 0.84	111.28 ± 1.40	113.00 ± 2.14	107.44 ± 0.76			
5	116.55 ± 1.00	118.28 ± 1.24	117.50 ± 1.76	110.38 ± 0.66			

Average value ± standard error. T: treatment group; A: age in months, T × A: treatment group × age.

Aiming to corroborate the possible effects of probiotic supplementation on the maintenance of growth rate in calves, regression analysis of the body weight in week one, with the body weight recorded in the following seven weeks and months 3, 4, and 5 of age, was performed and the results are shown in Table 3. According to their p-values, correlations were considered null (0–0.01), weak (0.1–0.4), medium (0.4–0.6), or strong (0.6–1.0), when they were significant ($p < 0.05$). In the case of the control group, there was a positive correlation ($p < 0.05$) between the weight in week one and week 2. Subsequently, it was only significant in month 5. The groups of calves that received single-strain-based probiotics generally had a positive correlation only during the first 6 to 8 weeks of age, while in the group that received the two-strain-based probiotic, the correlation was maintained until five months of age ($p < 0.05$).

Table 3. Pearson's correlation coefficients and probability values (p values), among first week and different ages, for the weight of both probiotic-supplemented and non-supplemented calves during the study.

		Age									
		Weeks							Months		
Treatment		**2**	**3**	**4**	**5**	**6**	**7**	**8**	**3**	**4**	**5**
Control	P	0.8955	0.590	0.590	0.514	0.380	0.382	0.498	0.528	0.601	0.681
	p	0.001	0.094	0.094	0.156	0.313	0.310	0.172	0.143	0.086	0.04
6BZ	P	0.889	0.881	0.872	0.735	0.724	0.695	0.858	0.255	0.217	0.149
	p	0.003	0.003	0.004	0.037	0.042	0.055	0.006	0.542	0.606	0.724
6BY	P	0.907	0.655	0.766	0.782	0.743	0.612	0.609	0.190	0.519	0.277
	p	0.001	0.077	0.026	0.021	0.034	0.106	0.109	0.652	0.187	0.507
6BY + 6BZ	P	0.914	0.813	0.631	0.633	0.652	0.488	0.526	0.579	0.653	0.801
	p	<0.0001	<0.0001	0.006	0.006	0.004	0.046	0.03	0.014	0.004	0.0001

P = Pearsons' correlation coefficients; p = probability values. p values $\leq$ 0.05 are statistically significant.

Similar results were observed when regression analysis was performed to wither height scores at different times during the study and the wither height values at the first week of age of the calves (Table 4). In general, positive and significant correlations ($p < 0.05$) were observed during the first six weeks for the calves in the control group. Meanwhile, positive and significant correlations were registered until week eight in the calves supplemented with the 6BZ strain probiotic. On the other hand, when the two-strain probiotic was supplemented, positive and statistically significant correlation values were observed up to month 3 of age.

Table 4. Pearson's´ correlation coefficients and probability values (p values), among first week and different ages, for the wither height of both probiotic-supplemented and non-supplemented calves during the study.

		Age									
		Weeks							Months		
Treatment		**2**	**3**	**4**	**5**	**6**	**7**	**8**	**3**	**4**	**5**
Control	P	0.939	0.639	0.782	0.579	0.741	0.492	0.481	0.547	0.478	0.465
	p	0.000	0.064	0.013	0.102	0.022	0.178	0.190	0.127	0.193	0.207
6BZ	P	0.879	0.231	0.269	0.455	0.413	0.428	0.581	0.502	0.689	0.527
	p	0.004	0.583	0.520	0.257	0.309	0.290	0.131	0.205	0.059	0.180
6BY	P	0.822	0.725	0.795	0.683	0.872	0.712	0.786	0.358	0.320	0.430
	p	0.012	0.042	0.018	0.062	0.005	0.047	0.021	0.384	0.440	0.287
6BY + 6BZ	P	0.886	0.820	0.664	0.661	0.601	0.541	0.626	0.538	0.472	0.457
	p	<0.000	<0.000	0.004	0.004	0.011	0.025	0.007	0.026	0.056	0.065

P = Pearsons' correlation coefficients; p = probability values. p values $\leq$ 0.05 are statistically significant.

3.2. *Fecal Microbiota*

The observed feature vector values differed between the probiotic-supplemented and non-supplemented calves ($p = 0.018$). In addition, based on the evenness vector, the abundance distribution of bacterial groups in fecal microbiota was different in the probiotic-supplemented calves compared to the non-supplemented group ($p = 0.021$). Both Shannon diversity index and faith diversity index values were higher in samples from the probiotic-supplemented calves ($p < 0.05$). Differences in relative abundances at different bacterial taxa were observed in feces from the probiotic-supplemented and non-supplemented dairy calves (Table 5).

Table 5. Relative abundance of main bacterial taxa that exhibited differences in feces from *Lactobacillus*-based probiotic-supplemented or non-supplemented dairy calves.

	Treatment			
	Average Relative Abundance (%, at Day 60)			
	Control	6BZ	6BY	6BY + 6BZ
Phylum				
Bacteroidota	59.3	46.5	48.6	44.1
Firmicutes	34.0	45.3	40.6	42.2
Actinobacteriota	1.7	2.5	2.6	3.1
Proteobacteria	3.6	4.1	7.0	9.1
Class				
Bacteroidia	59.3	46.5	48.6	44.1
Bacilli	3.6	3.5	6.5	12.3
Clostridia	26.9	37.5	32.1	26
Negativicutes	3.3	4.0	1.8	3.2
Desulfovibrionia	0.0	0.0	0.1	0.1
Coriobacteriia	1.6	2.4	1.7	1.7
Gammaproteobacteria	3.4	3.9	6.6	8.7
Family				
Enterobacteriaceae	0.0	0.0	3.85	0.09
Prevotellaceae	22.7	31.9	24.8	16.7
Bacteroidaceae	2.9	2.3	5.6	2.6
Lachnospiraceae	8.9	17.1	13.2	7.7
Muribaculaceae	26.4	6.1	9.1	9.8
Succinivibrionaceae	2.3	2.1	1.2	5.0
Carnobacteriaceae	0.0	0.1	1.3	3.5
Acholeplasmataceae	0.3	0.8	0.8	3.9
Genus				
Bacteroides	2.9	2.3	5.6	2.6
Alistipes	0.3	0.1	0.6	0.5
Prevotella	17.7	21.7	18.5	10.3
Alloprevotella	3.7	9.1	5.8	5.0
Ruminococcus	1.7	2.0	2.5	2.2
Faecalibacterium	0.7	1.0	1.1	1.1
Pseudobutyrivibrio	0.07	0.34	0.23	0.15

In general, at least 19 Phyla in both types of samples were observed, with Firmicutes, Proteobacteria, and Bacteroidota being the most representative (Figure 1a). Bacteroidia, Bacilli, Clostridia, Negativicutes, Desulfovibrionia, Coriobacteriia, and Gammaproteobacteria were the main classes observed in the samples, together representing more than 90% of bacteria in the feces samples. Bacteroidia exhibited higher relative abundance in the probiotic-unsupplemented calves compared to the probiotic-supplemented groups. Bacteroidia's relative abundance was higher in the non-supplemented calves in comparison to the probiotic-supplemented groups; meanwhile, Bacilli's relative abundance was superior in samples from the probiotic-supplemented calves, mainly when the combination of probiotic strains (6BY + 6BZ) was used. The Clostridia proportion in feces was similar between the non-supplemented and 6BY + 6BZ-supplemented calves but remained higher when probiotic strains were used individually. Gammaproteobacteria's relative abundance was higher in the non-supplemented calves during the first 45 days of evaluation compared to all probiotic-supplemented groups. However, at day 60, this abundance was similar to the individual strain groups but not to 6BY + 6BZ-supplemented calves that exhibited higher values.

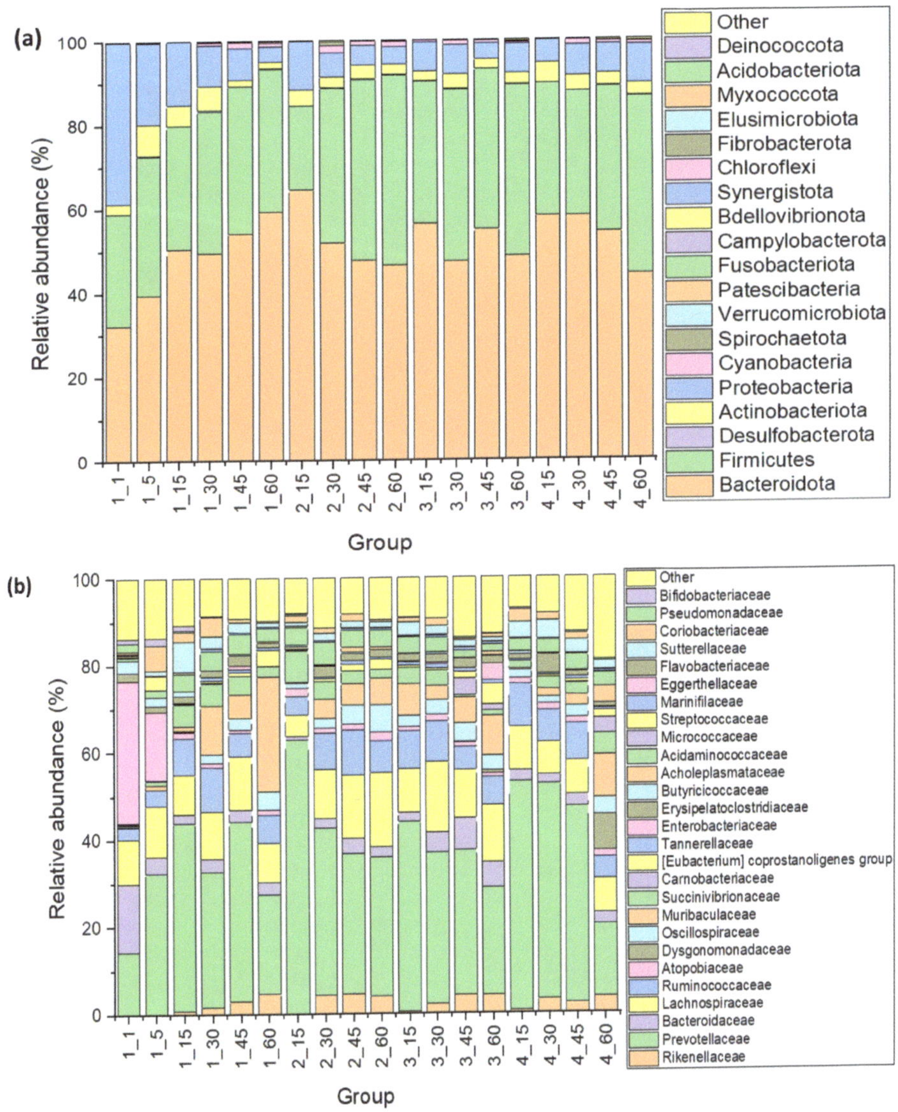

Figure 1. Relative abundance of main bacterial (**a**) Phyla and (**b**) families in feces from probiotic-supplemented and non-supplemented dairy calves' groups: (1: non-supplemented calves; 2: 6BZ-strain probiotic; 6BY-strain probiotic, 6BZ + 6BY probiotic) at different times after birth (1, 5, 15, 30, 45, and 60 days).

Enterobacteriaceae was the most abundant family in calves´ feces at the beginning of the experiment but rapidly decreased, mainly in the probiotic-supplemented calves, and, at day 60, was practically undetectable, except in the 6BY-supplemented calves (Figure 1b). The Prevotellaceae family exhibited a significant increase in the four groups but, at day 60, remained higher in the 6BZ-supplemented calves, followed by the 6BY-supplemented associated samples and the non-supplemented calves; however, the 6BY + 6BZ-supplemented calves registered the lowest values for the relative abundance of this bacterial family. On the other hand, the Bacteroidaceae family remained under 3% of relative abundance at day 60, except in the 6BY-supplemented group that registered values closer to 6%. Additionally, on day 60, the Lachnospiraceae family exhibited higher relative abundance values in samples from the 6BZ- and 6BY-supplemented calves (13 and 17%, respectively) at day 60, meanwhile lower values were observed in feces from the non-supplemented and 6BY + 6BZ-supplemented calves (9 and 8%, respectively).

The Muribaculaceae family was mainly observed at day 60 in feces from the non-supplemented calves (26% of relative abundance), followed by the 6BY + 6BZ-supplemented calves (10% of relative abundance). The Succinivibrionaceae family seemed to be increased by the use of the 6BY + 6BZ probiotic since relative abundance was higher in this group at day 60 (5%) compared to values observed in the rest of the groups (2% or less) as well as the Carnobacteriaceae family (3.5% of relative abundance in 6BY + 6BZ-supplemented calves vs. less than 1% in the rest of the groups), and the Acholeplasmataceae family (3.9% of relative abundance in 6BY + 6BZ-supplemented calves vs. less than 1% in the rest of the groups). The remaining bacterial families detected were found in low relative abundances (less than 1%). However, values were higher in feces from the 6BY + 6BZ-supplemented calves (11.5%) compared to the 6BZ-supplemented (7.2%), 6BY-supplemented (8.2%), or non-supplemented calves (6.9%), possibly indicating a greater bacterial diversity in that type of samples.

Finally, at genus level, the 6BY-based probiotic maintained the relative abundance of *Bacteroides* in the supplemented calves better than when the 6BZ-based or the 6BY + 6BZ-based probiotic was supplied. The *Alistipes* genus was also observed in all the samples, but reached its highest relative abundance at day 60 in samples from the 6BY- and 6BY + 6BZ-supplemented calves. The highest relative abundance of the *Prevotella* genus at day 60 was recorded in the samples from the 6BZ-supplemented calves, followed by samples from the 6BY-supplemented, non-supplemented, and 6BY + 6BZ-supplemented calves, respectively.

On the other hand, the *Alloprevotella* genus exhibited superior relative abundances in samples from the probiotic-supplemented calves in comparison with the non-supplemented calves, being higher in the 6BZ-supplemented calves, followed by the 6BY-supplemented and 6BY + 6BZ-supplemented calves. The *Ruminococcus* genus was observed in all the samples but, at day 60, the relative abundance of this genus was higher in samples from the 6BY + 6BZ-supplemented calves. The relative abundance of the *Faecalibacterium* genus was also improved at day 60 when probiotics were supplemented, as well as observed for the *Pseudobutyrivibrio* genus. Differences among treatments, in terms of diversity at genus level, were corroborated through a three-component PCA graphic, explaining the 52% of the variance between samples (Figure 2).

3.3. Functional Prediction of Bacterial Communities

A total of 405 METACYC pathways were predicted to establish the functionality of fecal bacterial communities associated with probiotic-supplemented and non-supplemented dairy calves. The frequency of certain pathways was increased when probiotic supplementation was used, such as the L-lysine biosynthesis III, L-isoleucine biosynthesis II, starch degradation V, L-isoleucine biosynthesis IV, glycogen degradation I (bacterial), aromatic amino acid biosynthesis, coenzyme A biosynthesis I, glycolysis I (from glucose 6-phosphate), pyruvate fermentation to isobutano, glycolysis II (from fructose 6-phosphate), L-glutamate and L-glutamine biosynthesis, pyrimidine deoxyribonucleotides de novo biosynthesis III, tetrapyrrole biosynthesis II (from glycine), 4-deoxy-L-threo-hex-4-enopyranuronate degradation, L-methionine biosynthesis III, D-galacturonate degradation I, peptidoglycan biosynthesis IV (*Enterococcus faecium*), pyruvate fermentation to acetone, TCA cycle IV (2-oxoglutarate decarboxylase), glutaryl-CoA degradation, toluene degradation II (aerobic) (via 4-methylcatechol), and biotin biosynthesis II pathways (Figure 3).

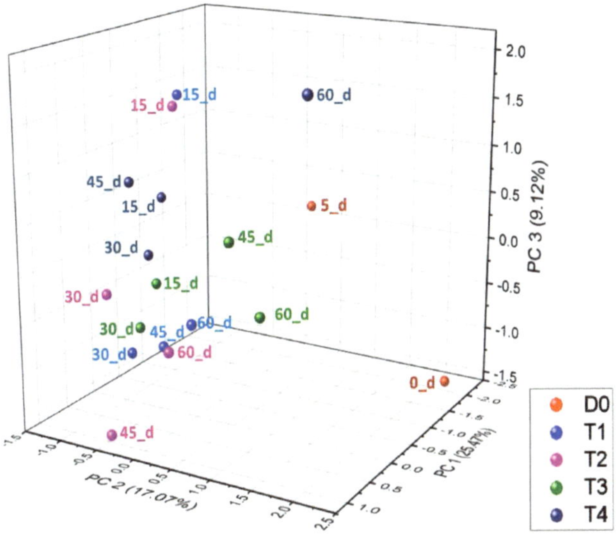

Figure 2. Principal component analysis graphic calculated from relative abundance of bacterial genera in feces samples from probiotic-supplemented and non-supplemented dairy calves' groups: (T1: non-supplemented calves; T2: 6BZ-strain probiotic; T3: 6BY-strain probiotic, and T4: 6BZ + 6BY probiotic) at different times after birth (1, 5, 15, 30, 45, and 60 days).

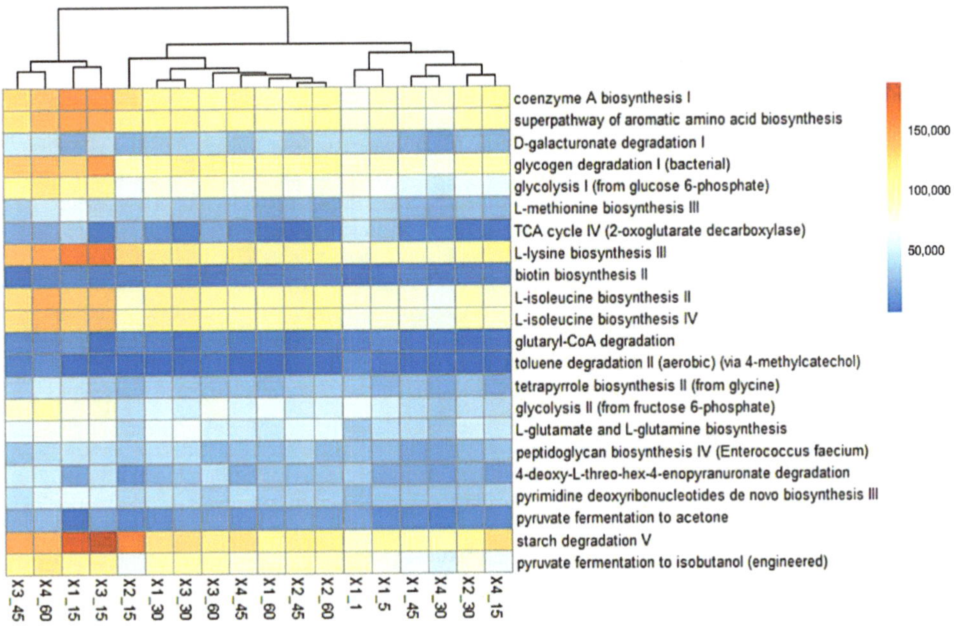

Figure 3. Heatmap representation of the main PICRUSt-predicted MetaCyc pathways incremented in feces from probiotic-supplemented calves. X1: non-supplemented calves, X2: 6BZ-supplemented group, X3: 6BY-supplemented group, X4: 6BZ + 6BY-supplemented group at different times after birth (1, 5, 15, 30, 45, and 60 days).

4. Discussion

Lactobacillus is a bacterial genus commonly used in the probiotic formulation for humans and animals [17]. Particularly for animal production purposes, probiotic strains are generally isolated from animal feces and characterized to establish the safety and probiotic potential of the microbial strains, following well-validated protocols for this purpose, which has allowed strains with promising results for their use as probiotics to be obtained, as observed for various *Lactobacillus* strains obtained from buffalo or calf feces [33,34]. In this study, we evaluated the effects on dairy calves' growth and fecal microbiota supplemented with three probiotic formulations using autochthonous lactic acid bacteria isolated from dairy cows´ feces. Two *Lactobacillus* strains were used as single-strain or two-strain probiotics, where no significant effects were observed on weight and wither height in dairy calves. Several studies have been conducted to assess the effects of different probiotic supplementation on the growth of dairy calves, reporting similar results, such as the study performed by Karamzadeh-Dehaghani et al. [35]. The authors in that study used a commercial probiotic containing dextrose, six different strains of lactic acid bacteria, and one *Bifidobacterium* strain, with no effects on body weight and growth parameters, including wither height after 28 days of evaluation. However, the authors reported improved diarrhea prevalence and some immune response indicators. Feed intake, body weight, and daily gain were neither improved by the concomitant use of a *Bacillus*-based probiotic and nucleotides derived from yeast supplementation through milk replacer in calves, but nucleotide supplementation seemed to reduce the *Lactobacillus* concentration in feces [36]. Nonetheless, some probiotic strains induce increases in calf daily weight gain, as reported by Jiang et al. [37] using a *Lactiplantibacillus plantarum* 299v strain as probiotic Holstein calves. They mentioned that the probiotic significantly increased the feed starter intake, as well as the average daily gain in the supplemented calves, at the time, reduced the diarrhea incidence, and increased the glucose, IgG, IgA, Interferon-gamma, and soluble CD4+ concentrations in plasma.

In this study, results suggested that probiotic supplementation significatively improved the maintenance weight gain of calves during their first five months of life, mainly after the milk-feeding period (first eight weeks of life), possibly by the microbiota modulation in calves that induces a medium- and long-term effect on the growth parameters, mainly when the two-strain probiotic was used [38]. Effects on feed efficiency, body weight gain, and reduced incidence of diarrhea have been commonly assessed when LAB-based probiotic supplementation is implemented in calves, but the observed results may depend on the bacterial strains. Some meta-analyses have intended to summarize the main effects of LAB-based probiotic supplementation. Through a meta-analysis of randomized controlled trials regarding LAB supplementation as a probiotic to young calves, including nine studies, Signorini et al. [39] observed that only when multi-strain probiotics were used the diarrhea incidence in the calves was reduced. Meanwhile, another meta-analysis published by Wang et al. [40], which included 49 studies of probiotic supplementation to pre-weaning dairy calves, reported that probiotics improved the growth performance but decreased digestibility and feed efficiency by increasing the dry matter intake and concluded that effects are indexed to the probiotic strain, supplementation dosage, and methods. On the other hand, Dehghan et al. [41] mentioned that, after analyzing the results of eight articles related to the probiotic supplementation to dairy calves, no significant effects on dry matter intake were observed as well as for feed efficiency; and Frizzo et al. [42] indicated, as a result of the meta-analysis of 21 articles and 14 studies, that growth of calves did not change when a LAB-based probiotic was supplemented in whole milk; however, beneficial effects were achieved when they were added to a milk replacer. The authors also highlighted that the number of supplemented strains seemed not to affect the results, but conclusions could be related to the number of calves included in each experiment.

In general, microbiota modulation is the main wanted effect when probiotics are used [43]. In this study, we proposed an administration scheme that only considered the administration of the probiotic, after the colostrum at day five after birth, on three occasions with a time between administrations of seven days. Following the administration scheme, we observed differences in fecal microbiota composition and structure as an indicator of changes in intestinal microbiota which, eventually, will promote the rumen microbiota establishment in the calves. Differences in the relative abundance of the different taxa were observed since day 15 and generally maintained until day 60.

Regarding phylum, the structure of the fecal microbiota observed in the samples from probiotic-supplemented calves was comparable to microbiota reported for healthy calves [44], which is mainly characterized by the maintenance of the proportion 1:1 or closed to between phyla Firmicutes and Bacteroidota [45] that are considered as a biomarker for metabolic potential of the gut microbiota [41]. Similar results were reported by Chang et al. [46] after the calves' supplementation with galacto-oligosaccharides, which facilitated the increase in the relative abundance of beneficial bacterial in the rumen, promoting the growth of the calves and reducing the incidence of diarrhea. Additionally, Actinobacteria and Proteobacteria phyla tended to increase in supplemented calves compared to the control. Although these phyla used to predominate in fecal microbiota of diarrheic calves [47], the observed values of relative abundance remained lower than those reported to cause dysbiosis in calves, particularly considering the Proteobacteria: Firmicutes + Bacteroidetes ratio, as reported previously [48] in beef cattle. On the other hand, Bacteroidia and Clostridia were the most abundant classes in all the treatments. It has been reported that Clostridia could be considered a marker of gastrointestinal dysbiosis [44]; in contrast, Bacilli and Gammaproteobacteria have been observed as predominant classes in healthy calves [43]; moreover, abundance of Bacteroidia plays an essential role in nutrition mainly for its contribution to carbon degradation [49].

Under day 60 after birth, the leading bacterial families in samples from the probiotic-supplemented calves were Prevotellaceae, Muribaculaceae, Succinivibrionaceae, and Acholeplasmataceae. It has been reported that the Prevotellaceae family usually increases with age in neonatal dairy calves, particularly after the first eight weeks of life [50], and is considered an essential family involved in carbohydrate degradation such as starch, xylan, pectin, and hemicellulose to produce propionate, succinate, and acetate [46,51], as well as in protein degradation to obtain peptides in the rumen [52]. The abundance of this family of microorganisms helps maintain calves' normal digestive function [49]. Prevotellaceae has been reported to be the most abundant family in fecal microbiota from Holstein calves fed with *Lacticaseibacillus rhamnosus* CG during the preweaning stage [49]. Additionally, the presence of Muribaculaceae family members has been related to a lower incidence of diarrhea in neonatal calves [53]; besides, they can degrade complex polysaccharides and potentially produce short-chain fatty acids such as acetate, propionate, and butyrate [54]. According to Kodithuwakku et al. [55], Muribaculaceae in the rumen contribute to efficient feed conversion into energy sources for milk production in dairy cows. Furthermore, the increase in the Succinivibrionaceae members' family has been reported as promoters for establishing the early rumen microbiota of young calves [56]. Similar trends were observed in the Acholeplasmataceae family, which was previously reported as part of the microbial taxon reported for rumen fluid, maintaining relative abundances of $\approx$5% in adult cows [57].

Regarding Genus taxa, *Bacteroides*, *Alistipes*, *Prevotella*, *Alloprevotella*, *Faecalibacterium*, and *Pseudobutyrivibrio* were the genera that exhibited the most notable differences in terms of relative abundance, depending on the study group at the age of 60 days. *The Bacteroides* genus is essential in absorbing amino acids in the intestine [49]. Furthermore, *Alistipes* contain a polysaccharide-degrading enzyme that facilitates the complex oligosaccharides degradation, and it has been reported with higher prevalence in the development of mature rumen of dairy calves [58]. A high abundance of *Bacteroides* and *Alistipes* in the feces of *Lactiplantibacillus plantarum* 299v-supplemented preweaning calves has been reported compared to non-supplemented calves [37]. *Prevotella* has been reported as a predominant genus in newborn calves during the first two weeks of age [50]; moreover, it is a key rumen microbial genus in calves [46], recognized as responsible for maintaining the normal digestion process of the calf's rumen [49] and protein degradation [50]. The *Prevotella* genus was frequently identified in the feces of beef cattle fed high-energy diets as a component of the core microbiome in the absence of probiotics [59,60]. Furthermore, it has been reported that a high abundance of *Prevotella* and *Alloprevotella* in dairy calves could exert a probiotic effect by reducing the incidence of diarrhea in the early stage [53]. On the other hand, *the Faecalibacterium* genus increased in feces from probiotic-supplemented calves compared to the control group. It has been reported that the fecal microbiota of pre-weaned dairy calves was dominated by microorganisms of this genus, which exhibited a lower incidence of diarrhea during the first four weeks [58]. Moreover, this genus has been associated with butyrate production, which promotes a greater weight gain during the first weeks of calves' lives due to butyrate enhancing the integrity of the intestinal epithelial barrier that may reduce susceptibility to intestinal infections in calves [61]. Virginio et al. [62] reported that both genera *Allopevotella* and *Faecalibacterium* increased in abundance in dairy calves supplemented with β-glucans until week eight. Finally, *Pseudobutyrivibrio* increased its relative abundance in feces from probiotic-supplemented calves. Some *Faecalibacterium* species have been isolated from various ruminants, and reports indicate that they can use complex polysaccharides as substrates for growth, including xylan and hemicellulose. However, the *Pseudobutyrivibrio* proportion has been reported to be decreased in the rumen of older calves [63].

The current results suggest that probiotic supplementation regulates the bacterial community composition in the intestine of calves, particularly in 6BY + 6BZ-supplemented calves (mainly in terms of diversity at the genus level) that may promote the absorption of nutrients in diets [37]. It has been reported that feeding lactic acid bacteria preparations for one day affects the microbial species composition of the feces of calves through β diversity data analysis using multivariate statistical tools such as principal component analysis [64]. Additionally, Fan et al. [65] reported that the β diversity of fecal microbiota of Holstein dairy calves was influenced by the supplementation with milk replaced with ethoxyquin, improving early rumen microbial development; this suggests that the interactions among different bacteria genera might play an essential role in maintaining intestinal homeostasis in the gut of newborn calves; moreover, the major impact on microbiota composition and diversity is directly related to the combined probiotic supplementation [65]. The changes observed in the structure of the bacterial community from feces were reflected in the predicted functional profile of samples. At least 18 predicted MetaCyc pathways showed incremented frequencies compared with non-supplemented calves. Samples from calves that were supplemented with the single-strain probiotic exhibited changes in the frequency of the selected pathways at day 15, but it seemed those frequency levels were not maintained at day 30 and later; on the other hand, when the 6BZ + 6BY-based probiotic was supplemented, the feces samples exhibited a sustained increment in the frequency of these predicted pathways.

5. Conclusions

This study demonstrated that using two *Lactobacillus* strains for calves' supplementation, considering three administrations during the two weeks after colostrum, resulted in higher weights after the milk-feeding period and at least during the first five months of life of supplemented calves mainly when the combined two strains were used. This could be related to the gut microbiota modulation, expressed as changes in composition and structure of the fecal microbiota of the probiotic-supplemented calves, promoting a balance between the most representative bacterial phyla. Nonetheless, exploring a different administration scheme that considers a more extended exposition of calves to the probiotic would be essential to improve the daily weight and height gain and increase, to a greater extent, the relative abundance of groups of microorganisms considered as probiotics in the rumen.

Supplementary Materials: The following supporting information can be downloaded at: https://www.mdpi.com/article/10.3390/ani13182841/s1, Table S1. Weight and height gain at eight weeks of Holstein calves treated with probiotics during the milk-feeding period.

Author Contributions: Conceptualization, J.M.R.-G. and R.I.A.-G.; methodology, J.M.R.-G., F.V.-G., M.A.E.-M., E.R.-A. and G.B.-R.; validation, J.M.R.-G., M.A.E.-M. and R.I.A.-G.; formal analysis, J.M.R.-G., M.A.E.-M. and L.J.G.-G.; investigation, J.M.R.-G., M.A.E.-M., E.R.-A., F.V.-G. and G.B.-R.; resources, J.M.R.-G. and R.I.A.-G.; data curation, J.M.R.-G.; writing—original draft preparation, J.M.R.-G., M.A.E.-M., Z.V. and L.M.A.-E.; writing—review and editing, J.M.R.-G., Z.V. and L.M.A.-E.; supervision, J.M.R.-G. and R.I.A.-G.; project administration, J.M.R.-G.; funding acquisition, J.M.R.-G., R.I.A.-G. and E.R.-A. All authors have read and agreed to the published version of the manuscript.

Funding: This research was funded by the National Institute of Forestry, Agricultural and Livestock Research through the project 1034034810.

Institutional Review Board Statement: The animal study protocol was approved by The bioethics and biosafety committee of the University of Guadalajara (CUALTOS; CEI-05-2022-09, CBIO-05-2022-05/02).

Informed Consent Statement: Informed consent was obtained from the client-owned animals involved in the study.

Data Availability Statement: The derived 16S rRNA gene sequences are available at the NCBI under the Bioproject ID: PRJNA991611.

Acknowledgments: The authors thank the milk producers of the Los Altos de Jalisco region who participated and facilitated the completion of the study. Thanks to the Subcommittee of Microbial and Invertebrate Genetic Resources for Food and Agriculture, especially to the macro-networks of the livestock and food industry.

Conflicts of Interest: The authors declare no conflict of interest.

References

1. Krajmalnik-Brown, R.; Ilhan, Z.E.; Kang, D.W.; DiBaise, J.K. Effects of Gut Microbes on Nutrient Absorption and Energy Regulation. *Nutr. Clin. Pract.* **2012**, *27*, 201–214. [CrossRef]
2. Zommiti, M.; Ferchichi, M. Probiotics and Prebiotics in Animal Feed. In *Probiotics and Prebiotics in Foods: Challenges, Innovations, and Advances*; Academic Press: Cambridge, MA, USA, 2021. [CrossRef]
3. Zeineldin, M.; Barakat, R.; Elolimy, A.; Salem, A.Z.M.; Elghandour, M.M.Y.; Monroy, J.C. Synergetic Action between the Rumen Microbiota and Bovine Health. *Microb. Pathog.* **2018**, *124*, 106–115. [CrossRef] [PubMed]
4. Shiels, P.G.; Painer, J.; Natterson-Horowitz, B.; Johnson, R.J.; Miranda, J.J.; Stenvinkel, P. Manipulating the Exposome to Enable Better Ageing. *Biochem. J.* **2021**, *478*, 2889–2898. [CrossRef]
5. Al Omran, Y.; Aziz, Q. The Brain-Gut Axis in Health and Disease. In *Microbial Endocrinology: The Microbiota-Gut-Brain Axis in Health and Disease*; Lyte, M., Cryan, J.F., Eds.; Advances in Experimental Medicine and Biology; Springer: New York, NY, USA, 2014; Volumn 817. [CrossRef]
6. Chang, M.N.; Wei, J.Y.; Hao, L.Y.; Ma, F.T.; Li, H.Y.; Zhao, S.G.; Sun, P. Effects of Different Types of Zinc Supplement on the Growth, Incidence of Diarrhea, Immune Function, and Rectal Microbiota of Newborn Dairy Calves. *J. Dairy Sci.* **2020**, *103*, 6100–6113. [CrossRef] [PubMed]
7. Varada, V.V.; Kumar, S.; Tyagi, N.; Tyagi, A.K. Effects of Compound Lyophilized Probiotics on Selected Faecal Microbiota, Immune Response, and Antioxidant Status in Newborn Buffalo Calves. *Curr. Res. Biotechnol.* **2022**, *4*, 493–502. [CrossRef]
8. Golić, N.; Đokić, J. Farm Animals and Pets—Impact on Gut Microbiota. In *Comprehensive Gut Microbiota*; Elsevier: Amsterdam, The Netherlands, 2022; pp. 125–138. [CrossRef]
9. Kimminau, E.A.; Karnezos, T.P.; Berghaus, R.D.; Jones, M.K.; Baxter, J.A.; Hofacre, C.L. Combination of Probiotic and Prebiotic Impacts *Salmonella* Enteritidis Infection in Layer Hens. *J. Appl. Poult. Res.* **2021**, *30*, 100200. [CrossRef]
10. Liao, S.F.; Nyachoti, M. Using Probiotics to Improve Swine Gut Health and Nutrient Utilization. *Anim. Nutr.* **2017**, *3*, 331–343. [CrossRef]
11. Xu, H.; Huang, W.; Hou, Q.; Kwok, L.; Sun, Z.; Ma, H.; Zhao, F.; Lee, Y.-K.; Zhang, H. The Effects of Probiotics Administration on the Milk Production, Milk Components and Fecal Bacteria Microbiota of Dairy Cows. *Sci. Bull.* **2017**, *62*, 767–774. [CrossRef]
12. Dankwa, A.S.; Humagain, U.; Ishaq, S.L.; Yeoman, C.J.; Clark, S.; Beitz, D.C.; Testroet, E.D. Bacterial Communities in the Rumen and Feces of Lactating Holstein Dairy Cows Are Not Affected When Fed Reduced-Fat Dried Distillers' Grains with Solubles. *Animal* **2021**, *15*, 100281. [CrossRef]
13. Liu, K.; Zhang, Y.; Yu, Z.; Xu, Q.; Zheng, N.; Zhao, S.; Huang, G.; Wang, J. Ruminal Microbiota–Host Interaction and Its Effect on Nutrient Metabolism. *Anim. Nutr.* **2021**, *7*, 49–55. [CrossRef]

14. Rey, M.; Enjalbert, F.; Combes, S.; Cauquil, L.; Bouchez, O.; Monteils, V. Establishment of Ruminal Bacterial Community in Dairy Calves from Birth to Weaning Is Sequential. *J. Appl. Microbiol.* **2014**, *116*, 245–257. [CrossRef] [PubMed]
15. Fuller, R. Probiotics for Farm Animals. In *Probiotics: A Critical Review*; Tannock, G.W., Ed.; Horizon Scientific Press: Wymondham, UK, 1999; pp. 15–22.
16. Terzić-Vidojević, A.; Veljović, K.; Tolinački, M.; Živković, M.; Lukić, J.; Lozo, J.; Fira, Đ.; Jovčić, B.; Strahinić, I.; Begović, J.; et al. Diversity of Non-Starter Lactic Acid Bacteria in Autochthonous Dairy Products from Western Balkan Countries—Technological and Probiotic Properties. *Food Res. Int.* **2020**, *136*, 109494. [CrossRef] [PubMed]
17. Das, T.K.; Pradhan, S.; Chakrabarti, S.; Mondal, K.C.; Ghosh, K. Current Status of Probiotic and Related Health Benefits. *Appl. Food Res.* **2022**, *2*, 100185. [CrossRef]
18. Gopal, P.K. Bacteria, Beneficial: Probiotic Lactic Acid Bacteria: An Overview. In *Encyclopedia of Dairy Sciences*; Elsevier: Amsterdam, The Netherlands, 2022; pp. 32–33.
19. FAO/WHO. *Guidelines for the Evaluation of Probiotics in Food. Report of a Joint FAO/WHO Working Group on Drafting Guidelines for the Evaluation of Probiotics in Food*; FAO/WHO: London, ON, Canada, 2002.
20. Reyes-Leos, J.R. Identificación de Cepas de *Lactobacillus* a Partir de Muestras de Estiércol Bovino de Granjas Productoras de Leche de La Región Altos Sur Del Estado de Jalisco. Bachelor Thesis, Centro Universitario Lagos de Moreno, Universidad de Guadalajara, Lagos de Moreno, Mexico, 2020.
21. Ruvalcaba-Gómez, J.M.; Ruiz-Espinosa, H.; Méndez-Robles, M.D.; Arteaga-Garibay, R.I.; Anaya-Esparza, L.M.; Villagrán, Z.; Delgado-Macuil, R.J. Use of Autochthonous Lactic Acid Bacteria as Starter Culture of Pasteurized Milk Adobera Cheese. *Fermentation* **2022**, *8*, 234. [CrossRef]
22. National Research Council. *Nutrient Requirements of Dairy Cattle*; National Academies Press: Washington, DC, USA, 2001; ISBN 978-0-309-06997-7.
23. Dingwell, R.T.; Wallace, M.M.; McLaren, C.J.; Leslie, C.F.; Leslie, K.E. An Evaluation of Two Indirect Methods of Estimating Body Weight in Holstein Calves and Heifers. *J. Dairy Sci.* **2006**, *89*, 3992–3998. [CrossRef]
24. Villaseñor González, F.; Estrada Cortés, E.; Montes Oceguera, L.D.R.; Vera Ávila, H.R.; Montiel Olguín, L.J.; Jiménez Severiano, H.; Espinosa Martínez, M.A. Factores Asociados a Indicadores de Crianza de Reemplazos Bovinos Durante El Periodo de Lactancia En Unidades Lecheras de Pequeña Escala. *Rev. Mex. Cienc. Pecu.* **2022**, *13*, 64–81. [CrossRef]
25. Bolger, A.M.; Lohse, M.; Usadel, B. Trimmomatic: A Flexible Trimmer for Illumina Sequence Data. *Bioinformatics* **2014**, *30*, 2114–2120. [CrossRef]
26. Ewels, P.; Magnusson, M.; Lundin, S.; Käller, M. MultiQC: Summarize Analysis Results for Multiple Tools and Samples in a Single Report. *Bioinformatics* **2016**, *32*, 3047–3048. [CrossRef]
27. Quast, C.; Pruesse, E.; Yilmaz, P.; Gerken, J.; Schweer, T.; Yarza, P.; Peplies, J.; Glöckner, F.O. The SILVA Ribosomal RNA Gene Database Project: Improved Data Processing and Web-Based Tools. *Nucleic Acids Res.* **2012**, *41*, D590–D596. [CrossRef]
28. Ewels, P.A.; Peltzer, A.; Fillinger, S.; Patel, H.; Alneberg, J.; Wilm, A.; Garcia, M.U.; Di Tommaso, P.; Nahnsen, S. The Nf-Core Framework for Community-Curated Bioinformatics Pipelines. *Nat. Biotechnol.* **2020**, *38*, 276–278. [CrossRef]
29. Straub, D.; Blackwell, N.; Langarica-Fuentes, A.; Peltzer, A.; Nahnsen, S.; Kleindienst, S. Interpretations of Environmental Microbial Community Studies Are Biased by the Selected 16S RRNA (Gene) Amplicon Sequencing Pipeline. *Front. Microbiol.* **2020**, *11*, 550420. [CrossRef]
30. Caporaso, J.G.; Kuczynski, J.; Stombaugh, J.; Bittinger, K.; Bushman, F.D.; Costello, E.K.; Fierer, N.; Peña, A.G.; Goodrich, J.K.; Gordon, J.I.; et al. QIIME Allows Analysis of High-Throughput Community Sequencing Data. *Nat. Methods* **2010**, *7*, 335–336. [CrossRef] [PubMed]
31. Douglas, G.M.; Maffei, V.J.; Zaneveld, J.R.; Yurgel, S.N.; Brown, J.R.; Taylor, C.M.; Huttenhower, C.; Langille, M.G.I. PICRUSt2 for Prediction of Metagenome Functions. *Nat. Biotechnol.* **2020**, *38*, 685–688. [CrossRef] [PubMed]
32. Caspi, R.; Billington, R.; Keseler, I.M.; Kothari, A.; Krummenacker, M.; Midford, P.E.; Ong, W.K.; Paley, S.; Subhraveti, P.; Karp, P.D. The MetaCyc Database of Metabolic Pathways and Enzymes—A 2019 Update. *Nucleic Acids Res* **2020**, *48*, D445–D453. [CrossRef] [PubMed]
33. Lin, W.-C.; Ptak, C.P.; Chang, C.-Y.; Ian, M.-K.; Chia, M.-Y.; Chen, T.-H.; Kuo, C.-J. Autochthonous Lactic Acid Bacteria Isolated From Dairy Cow Feces Exhibiting Promising Probiotic Properties and in Vitro Antibacterial Activity Against Foodborne Pathogens in Cattle. *Front. Vet. Sci.* **2020**, *7*, 239. [CrossRef] [PubMed]
34. Singh, A.; Kumar, S.; Vinay, V.V.; Tyagi, B.; Choudhury, P.K.; Rashmi, H.M.; Banakar, P.S.; Tyagi, N.; Tyagi, A.K. Autochthonous *Lactobacillus* spp. Isolated from Murrah Buffalo Calves Show Potential Application as Probiotic. *Curr. Res. Biotechnol.* **2021**, *3*, 109–119. [CrossRef]
35. Karamzadeh-Dehaghani, A.; Towhidi, A.; Zhandi, M.; Mojgani, N.; Fouladi-Nashta, A. Combined Effect of Probiotics and Specific Immunoglobulin Y directed against *Escherichia coli* on Growth Performance, Diarrhea Incidence, and Immune System in Calves. *Animal* **2021**, *15*, 100124. [CrossRef]
36. Górka, P.; Budzińska, K.; Budziński, W.; Jankowiak, T.; Kehoe, S.; Kański, J. Effect of Probiotic and Nucleotide Supplementation in Milk Replacer on Growth Performance and Fecal Bacteria in Calves. *Livest. Sci.* **2021**, *250*, 104556. [CrossRef]
37. Jiang, X.; Xu, H.J.; Cui, Z.Q.; Zhang, Y.G. Effects of Supplementation with *Lactobacillus plantarum* 299v on the Performance, Blood Metabolites, Rumen Fermentation and Bacterial Communities of Preweaning Calves. *Livest. Sci.* **2020**, *239*, 104120. [CrossRef]

38. Huuki, H.; Tapio, M.; Mäntysaari, P.; Negussie, E.; Ahvenjärvi, S.; Vilkki, J.; Vanhatalo, A.; Tapio, I. Long-Term Effects of Early-Life Rumen Microbiota Modulation on Dairy Cow Production Performance and Methane Emissions. *Front. Microbiol.* **2022**, *13*, 983823. [CrossRef]
39. Signorini, M.L.; Soto, L.P.; Zbrun, M.V.; Sequeira, G.J.; Rosmini, M.R.; Frizzo, L.S. Impact of probiotic administration on the health and fecal microbiota of young calves: A meta-analysis of randomized controlled trials of lactic acid bacteria. *Res. Vet. Sci.* **2012**, *93*, 250–258. [CrossRef] [PubMed]
40. Wang, L.; Sun, H.; Gao, H.; Xia, Y.; Zan, L.; Zhao, C. A meta-analysis on the effects of probiotics on the performance of pre-weaning dairy calves. *J. Anim. Sci. Biotechnol.* **2023**, *14*, 3. [CrossRef] [PubMed]
41. Dehghan, B.M.; Rajaei-Sharifabadi, H.; Vazirigohar, M. A Meta-Analysis of the Effect of Probiotics Administration on Growth Performance of Suckling Calves in Iran. *Iran. J. Appl. Anim. Sci.* **2020**, *10*, 213–219.
42. Frizzo, L.S.; Zbrun, M.V.; Soto, L.P.; Signorini, M.L. Effects of probiotics on growth performance in young calves: A meta-analysis of randomized controlled trials. *Anim. Feed. Sci. Technol.* **2011**, *169*, 147–156. [CrossRef]
43. Uyeno, Y.; Shigemori, S.; Shimosato, T. Effect of Probiotics/Prebiotics on Cattle Health and Productivity. *Microbes Environ.* **2015**, *30*, 126–132. [CrossRef]
44. Zeineldin, M.; Aldridge, B.; Lowe, J. Dysbiosis of the Fecal Microbiota in Feedlot Cattle with Hemorrhagic Diarrhea. *Microb. Pathog.* **2018**, *115*, 123–130. [CrossRef]
45. Xiang, K.; Hu, X.; Mu, R.; Animal, L.; Bureau, H.; Li, S.; Wang, Y.; Zhao, C.; Zhang, N.; Fu, Y. Rumen Microbiota Alterations During Ketosis is Associated with the development of Mastitis in Dairy Cows. *Res. Sq.* **2021**, *preprint*. [CrossRef]
46. Chang, M.; Wang, F.; Ma, F.; Jin, Y.; Sun, P. Supplementation with Galacto-Oligosaccharides in Early Life Persistently Facilitates the Microbial Colonization of the Rumen and Promotes Growth of Preweaning Holstein Dairy Calves. *Anim. Nutr.* **2022**, *10*, 223–233. [CrossRef]
47. Gomez, D.E.; Arroyo, L.G.; Costa, M.C.; Viel, L.; Weese, J.S. Characterization of the Fecal Bacterial Microbiota of Healthy and Diarrheic Dairy Calves. *J. Vet. Intern. Med.* **2017**, *31*, 928–939. [CrossRef] [PubMed]
48. Auffret, M.D.; Dewhurst, R.J.; Duthie, C.A.; Rooke, J.A.; John Wallace, R.; Freeman, T.C.; Stewart, R.; Watson, M.; Roehe, R. The Rumen Microbiome as a Reservoir of Antimicrobial Resistance and Pathogenicity Genes Is Directly Affected by Diet in Beef Cattle. *Microbiome* **2017**, *5*, 159. [CrossRef]
49. Wang, L.; Zhang, G.; Xu, H.; Xin, H.; Zhang, Y. Metagenomic Analyses of Microbial and Carbohydrate-Active Enzymes in the Rumen of Holstein Cows Fed Different Forage-to-Concentrate Ratios. *Front. Microbiol.* **2019**, *10*, 649. [CrossRef]
50. Kim, E.-T.; Lee, S.-J.; Kim, T.-Y.; Lee, H.-G.; Atikur, R.M.; Gu, B.-H.; Kim, D.-H.; Park, B.-Y.; Son, J.-K.; Kim, M.-H. Dynamic Changes in Fecal Microbial Communities of Neonatal Dairy Calves by Aging and Diarrhea. *Animals* **2021**, *11*, 1113. [CrossRef]
51. Poudel, P.; Froehlich, K.; Casper, D.P.; St-Pierre, B. Feeding Essential Oils to Neonatal Holstein Dairy Calves Results in Increased Ruminal Prevotellaceae Abundance and Propionate Concentrations. *Microorganisms* **2019**, *7*, 120. [CrossRef] [PubMed]
52. Zhang, L.; Jiang, X.; Liu, X.; Zhao, X.; Liu, S.; Li, Y.; Zhang, Y. Growth, Health, Rumen Fermentation, and Bacterial Community of Holstein Calves Fed *Lactobacillus rhamnosus* GG during the Preweaning Stage. *J. Anim. Sci.* **2019**, *97*, 2598–2608. [CrossRef] [PubMed]
53. Chen, H.; Liu, Y.; Huang, K.; Yang, B.; Zhang, Y.; Yu, Z.; Wang, J. Fecal Microbiota Dynamics and Its Relationship to Diarrhea and Health in Dairy Calves. *J. Anim. Sci. Biotechnol.* **2022**, *13*, 132. [CrossRef] [PubMed]
54. Wang, Y.; Nan, X.; Zhao, Y.; Jiang, L.; Wang, H.; Hua, D.; Zhang, F.; Wang, Y.; Liu, J.; Yao, J.; et al. Dietary Supplementation with Inulin Improves Lactation Performance and Serum Lipids by Regulating the Rumen Microbiome and Metabolome in Dairy Cows. *Anim. Nutr.* **2021**, *7*, 1189–1204. [CrossRef]
55. Kodithuwakku, H.; Maruyama, D.; Owada, H.; Watabe, Y.; Miura, H.; Suzuki, Y.; Hirano, K.; Kobayashi, Y.; Koike, S. Alterations in Rumen Microbiota via Oral Fiber Administration during Early Life in Dairy Cows. *Sci. Rep.* **2022**, *12*, 10798. [CrossRef]
56. Bu, D.; Zhang, X.; Ma, L.; Park, T.; Wang, L.; Wang, M.; Xu, J.; Yu, Z. Repeated Inoculation of Young Calves With Rumen Microbiota does not Significantly Modulate the Rumen Prokaryotic Microbiota Consistently but Decreases Diarrhea. *Front. Microbiol.* **2020**, *11*, 1403. [CrossRef] [PubMed]
57. Wang, X.; Li, X.; Zhao, C.; Hu, P.; Chen, H.; Liu, Z.; Liu, G.; Wang, Z. Correlation between Composition of the Bacterial Community and Concentration of Volatile Fatty Acids in the Rumen during the Transition Period and Ketosis in Dairy Cows. *Appl. Environ. Microbiol.* **2012**, *78*, 2386–2392. [CrossRef]
58. Oikonomou, G.; Teixeira, A.G.V.; Foditsch, C.; Bicalho, M.L.; Machado, V.S.; Bicalho, R.C. Fecal Microbial Diversity in Pre-Weaned Dairy Calves as Described by Pyrosequencing of Metagenomic 16S RDNA. Associations of *Faecalibacterium* Species with Health and Growth. *PLoS ONE* **2013**, *8*, e63157. [CrossRef]
59. Rice, W.C.; Galyean, M.L.; Cox, S.B.; Dowd, S.E.; Cole, N.A. Influence of Wet Distillers Grains Diets on Beef Cattle Fecal Bacterial Community Structure. *BMC Microbiol.* **2012**, *12*, 25. [CrossRef] [PubMed]
60. Shanks, O.C.; Kelty, C.A.; Archibeque, S.; Jenkins, M.; Newton, R.J.; McLellan, S.L.; Huse, S.M.; Sogin, M.L. Community Structures of Fecal Bacteria in Cattle from Different Animal Feeding Operations. *Appl. Environ. Microbiol.* **2011**, *77*, 2992–3001. [CrossRef]
61. Du, W.; Wang, X.; Hu, M.; Hou, J.; Du, Y.; Si, W.; Yang, L.; Xu, L.; Xu, Q. Modulating Gastrointestinal Microbiota to Alleviate Diarrhea in Calves. *Front. Microbiol.* **2023**, *14*, 1181545. [CrossRef]

62. Virginio, G.F.; Reis, M.E.; da Silva, A.P.; de Toledo, A.F.; Cezar, A.M.; Mendes, L.W.; Greco, L.; Montenegro, H.; Coutinho, L.L.; Bittar, C.M.M. Does Algae β-Glucan Affect the Fecal Bacteriome in Dairy Calves? *PLoS ONE* **2021**, *16*, e0258069. [CrossRef] [PubMed]
63. Guo, W.; van Niekerk, J.K.; Zhou, M.; Steele, M.A. PSIX-32 Assessment of Mucosa-Associated Microbiota in the Colon and Rumen of Dairy Calves Fed High Plane of Milk and during Weaning Transition. *J. Anim. Sci.* **2020**, *98* (Suppl. S4), 311. [CrossRef]
64. Dong, C.; Wei, M.; Sun, F.; Bao, H.; Bao, M.; Ju, J.; Du, L. Effect of Lactic Acid Bacteria Preparations on Calf Fecal Flora. *Rev. Bras. Zootec.* **2023**, *52*, e20210199. [CrossRef]
65. Wei, X.; Zou, J.; Zhang, Y.; Yang, J.; Wang, J.; Wang, Y.; Wang, C. Effects of Milk, Milk Replacer, and Milk Replacer plus Ethoxyquin on the Growth Performance, Weaning Stress, and the Fecal Microbiota of Holstein Dairy Calves. *Front. Microbiol.* **2023**, *14*, 1113518. [CrossRef]

animals

MDPI

Article

Assessment of Herd, Calf, and Colostrum Management Practices on Austrian Dairy Farms Using a Scoring System

Nicole Hechenberger [1], Katharina Lichtmannsperger [2,*], Daniela Klein-Jöbstl [3], Alexander Tichy [4] and Thomas Wittek [2]

1 Animal Health Service (Tiergesundheitsdienst) Salzburg, Bundesstraße 6, 5071 Wals-Siezenheim, Austria; nicole.hechenberger@salzburg.gv.at

2 Clinical Unit of Ruminant Medicine, University Clinic for Ruminants, Department for Farm Animals and Veterinary Public Health, University of Veterinary Medicine Vienna, Veterinärplatz 1, 1210 Vienna, Austria; thomas.wittek@vetmeduni.ac.at

3 Clinical Unit for Herd Health Management in Ruminants, University Clinic for Ruminants, Department for Farm Animals and Veterinary Public Health, University of Veterinary Medicine Vienna, Veterinärplatz 1, 1210 Vienna, Austria; daniela.klein@vetmeduni.ac.at

4 Bioinformatics and Biostatistics Unit, University of Veterinary Medicine Vienna, Veterinärplatz 1, 1210 Vienna, Austria; alexander.tichy@vetmeduni.ac.at

* Correspondence: katharina.lichtmannsperger@vetmeduni.ac.at

Simple Summary: As calves are born with a naïve immune system, they depend on the transfer of immunoglobulins via colostrum. If calves do not receive a sufficient supply of high-quality colostrum (>50 g/L of immunoglobulins; 100–300 g Immunoglobulins in total) within the first hours after birth, they are likely to suffer from Failure of Transfer of Passive Immunity. The objective of the present study was to evaluate herd, calf, and colostrum management on Austrian dairy farms, focusing on challenges and possibilities for improvement. A scoring system was implemented to compare and classify management practices. Farms in foothills/flatland regions of Austria, conventional producing farms, and full-time operated farms overall received a higher and therefore better score rating than farms in alpine regions of Austria, organic producing farms, and part-time operated farms.

Abstract: The objectives of the study were to describe colostrum management on Austrian dairy farms and to explore differences between regions (alpine/flatlands), organic and conventional producing farms, and full-time or part-time operated farms. An online survey (24 questions) on general farm characteristics and herd and calf management was sent to 16,246 farmers. In total, 2328 farmers (response rate 14.3%) answered the questionnaire. To allow an objective comparison, a scoring system was implemented. Farm size is, on average, smaller in the alpine regions than in the foothills/flatlands regions of Austria. Small farms were more often organic-producing farms (81.6%) and operated part-time (93.8%). In foothills/flatland regions, 70.0% of farms have a separate calving area, and in the alpine regions, it is solely 42.8%. Colostrum testing is still mostly done by visual appraisal (63.7%); only a few farmers use a colostrometer (8.8%), brix-refractometer (18.3%), or ColostroCheck® (9.2%, a cone-shaped device to rate the flow velocity of colostrum). The results of the present study using the scoring system showed differences in herd and calf management practices in all sectors. In the future, the findings and especially the scoring system can support Austrian dairy farmers or veterinarians to better assess areas of improvement on farms in order to prevent calves from suffering from Failure of Transfer of Passive Immunity.

Keywords: calf management; herd management; survey; colostrum

1. Introduction

Cattle have an epitheliochorial placenta type. The maternal uterine tissue layers remain intact, resulting in a separation between the maternal and the fetal blood circulation [1,2].

Therefore, an intrauterine transfer of immunoglobulins is almost impossible, and calves are born with a naïve immune system [3]. Unless an adequate amount of immunoglobulins is provided via colostrum, calves have an increased likelihood of succumbing to infections [4]. The process of transferring maternal immunoglobulins via colostrum to the calf is called Transfer of Passive Immunity (TPI). The amount of maternal colostrum needed depends on the concentration of immunoglobulins, the ingested volume of colostrum, and the ability of the calf's gut to absorb the immunoglobulins (apparent efficiency of immunoglobulin absorption). If calves are fed with >2.5 L colostrum (>50 g/L IgG) within the first three hours after birth, they are less likely to suffer from FTPI [5]. During the first twelve hours after birth, the calf's ability to absorb maternal immunoglobulins decreases substantially [6]. The reasons for this are not yet fully understood, and some findings are conflicting [7]. Some studies showed the presence of a tubular vesicle-vacuolar mechanism in neonatal enterocytes. The vacuoles that transport the immunoglobulins from the intestine to the blood decrease over time as the fetal intestinal cells mature [8]. The insufficient supply of the calf with immunoglobulins via a low amount of immunoglobulins in maternal colostrum or insufficient amount of colostrum is termed a Failure of Transfer of Passive Immunity (FTPI). Reschke and coworkers (2017) investigated 373 mother-dam pairs, where 162 (43.50%) of the calves showed an FTPI [9]. A sufficient maternal colostrum quality is not only defined by the amount of immunoglobulins but also by the level of bacterial contamination [10]. Colostral immunoglobulins, which are bound to bacteria in the colostrum, are deactivated before they can be absorbed [11]. Furthermore, bacteria present in colostrum compete with immunoglobulins on the unspecific receptors, which are necessary for absorbing immunoglobulins from the intestine into the bloodstream, resulting in a decreased apparent efficiency of immunoglobulin absorption [12]. It is essential to use adequate management methods for collecting, storing, and feeding colostrum to ensure calves get an appropriate amount of colostral immunoglobulins after birth and, in consequence, do not suffer from FTPI [13]. Additionally, cow-related factors such as the number of lactations, genetic parameters, dry period length, antepartum milk leakage, colostrum quantity, metabolic status of the cow, and udder health can influence colostrum quality [9,14–18]. Furthermore, colostrum and nutritional management of calves (quantities, occurrence of FTPI, etc.) have an effect on the future adult dairy cow (epigenetic programming) [19–21]. Calves suffering from FTPI have higher mortality and morbidity rates and reduced daily growth rates [20]. They are more likely to suffer from diseases such as diarrhea, respiratory diseases, navel infections, and omphalitis [13,22–24]. Calves suffering from FTPI result in substantial economic losses for farmers, and additionally, FTPI poses a major animal welfare issue [25].

In 2012, calf management practices have been evaluated in Austria [26]. The study described calf management practices, estimated differences in disease incidences on Austrian dairy farms depending on the farm structure (small farms $\leq$ 20 cows versus large farms > 20 cows), and different management practices [18]. Investigations carried out in countries such as the Netherlands [27], Brazil [28], Canada [29], and the USA [30] are only partially comparable to studies conducted in Austria since the farm structure is significantly different. In the aforementioned studies, the average number of cattle per farm is significantly higher, with more than 50 to 100 dairy cows per farm. In Austria, there are primarily small family-owned farms with an average size of 19 dairy cows per dairy farm [31]. This structure leads to the fact that many farms are run as part-time farms [32]. They are similarly structured like full-time farms. The only difference is that the farm owners have an additional income besides dairy farming (e.g., from agritourism, employment, wage work, and forestry). The on-farm produced raw milk is either collected using a dairy company and further processed in the dairy plant and/or processed directly on the dairy farm and sold as cheese and/or milk at the local farmer's store. Due to the different geographical structures, Austria has diverse dairy farming structures [32]. Cows calf all year round, but especially in the alpine region, block calving is conducted, and the cows' calves from around September to January. This is carried out due to the fact that alpine transhumance is carried out, which means that the dairy farmers with their respective cows and young stock move to mountainous areas during summertime. Additionally, Austria has a high number of organic-producing farms, especially in Salzburg, and therefore, the cows are on the pasture from May until October. In the western and southwestern parts of Austria, there are mostly small farms with less than 30 dairy cows per farm due to the mountain range of the Austrian Alps. In the southeast and eastern parts of Austria (foothills and flatlands), the farms tend to be larger, with ≥30 dairy cows per farm. For details on the geographical differences, see Figure 1.

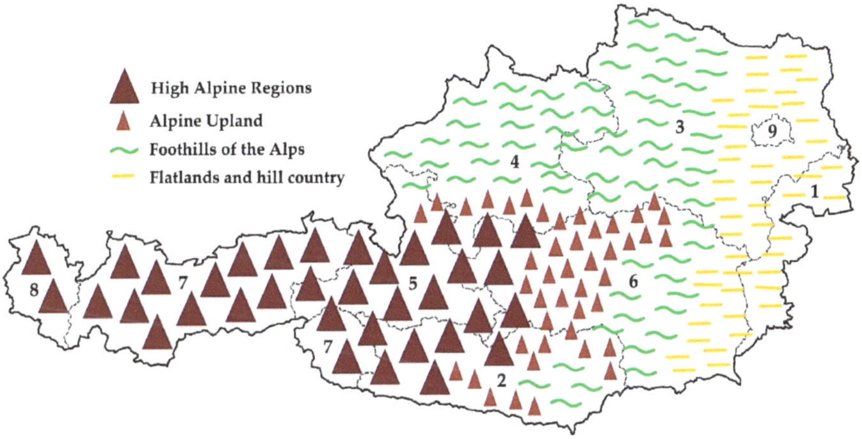

Figure 1. Schematic map of Austria's geological structure and the nine federal states. Vienna is the capital city of Austria but is also considered a federal state. (1 = Burgenland, 2 = Carinthia, 3 = Lower Austria, 4 = Upper Austria, 5 = Salzburg, 6 = Styria, 7 = Tyrol, 8 = Vorarlberg, 9 = Vienna).

The objectives of this study were to describe herd and calf management practices in Austria and describe differences in colostrum management between (1) alpine (west) and foothills/flatland (east) regions, (2) between organic and conventional production types and (3) between part-time and full-time operated farms by implementing a scoring system.

We hypothesized that there were differences in herd and calf management practices between (1) farms located in the alpine (west) or foothills/flatlands (east) regions of Austria, (2) between organic and conventional farms, and (3) between part-time and full-time operated farms.

2. Materials and Methods

2.1. Survey Distribution

An online survey was designed using Survey Monkey® (Copyright © 1999–2022 Momentive). The survey was distributed to 16,246 dairy farmers in Austria who were members of the Austrian Breeding Association (ZAR, Rinderzucht Austria) via email using their member database. The study population of Austrian breeding association members covers 68.1% of the total dairy farms in Austria. The survey was open for 8 weeks, beginning 1 February 2022 and ending 31 March 2022, with a reminder to participate in the survey sent out at the beginning of March 2022.

2.2. Survey Structure

In total, the survey included 24 questions. All 24 questions were single-choice questions, some with the possibility to provide an additional open answer. Additional supplementary questions were possible for three questions, depending on the answers given ('If yes, ...'). Overall, the questions were divided into three sections: Section (1) general farm characteristics, Section (2) information on herd-management procedures, and Section (3) information on calf-management procedures. Section one included questions on the location of the farm (federal state), Austrian Animal Health Service membership (yes/no), farm size (in terms of livestock units and number of cows), production type (organic/conventional), operation type (full-time/part-time), housing types (tie stall/free stall) and cattle breeds. Within the second Section, more specific information on herd management practices with a special focus on herd-level colostrum management was gathered, such as the availability of a calving area, udder cleaning methods, duration between parturition and colostrum milking and colostrum storage procedure. In the third Section, detailed information on calf management practices was gathered, including questions on colostrum feeding procedure (bucket feeding, nipple bottle feeding), feeding time, and colostrum quality assessment methods. The original survey (in German and English) has been provided in the Supplement Materials (Table S1).

2.3. Implementation of a Scoring System

In order to compare herd and calf management practices, a scoring system was implemented for 13 questions (Sections 2 and 3). A high score agrees with evidence-based recommendations on herd and calf management procedures. A low score indicates that the answers were not in accordance with evidence-based recommendations. Each answer was translated into a point system (minimum point = 0 points; maximum points = 4 points). In total, a maximum of 32 points could be obtained per farm: 14 points for herd management practices and 18 points for calf management practices. The point allocation was based on the current evidence-based recommendations published in peer-reviewed journals. In detail, the information from published peer-reviewed articles was used as a basis for the discussion round by the authors (NH, KL, TW). Answers according to the current evidence-based recommendations were assessed as 'correct' and received higher scores (3 and 4 points, mostly or fully meets the evidence-based recommendation) than 'incorrect' answers (1 and 2 points, does not or only partially meet evidence-based recommendations). Questions solely ending in a yes/no decision were categorized as a correct (1 point) and an incorrect answer (0 points). If superiority to a specific method/procedure over another could not be determined, an equal number of points was given. For example, Section Two included the information on the udder cleaning method, and two answers were possible (yes/no). The answer "yes—the udder was cleaned before colostrum harvest" resulted in one point, and the answer "no—the udder was not cleaned before colostrum harvest" resulted in zero points. The allocation of points was based on the publication by S. Steward et al. 2005 [33], where the authors found that a high standard of udder cleaning is essential to harvest high-quality colostrum with low bacterial counts. Our survey did not ask about the detailed udder cleaning routine. Therefore, the question about the cleaning method was excluded from the scoring system, and only the question cleaning 'yes/no' was scored. Figure 2 provides two more sample questions to illustrate the allocation of points within the scoring system.

The complete survey, including the point allocation system (including the references of the current evidence-based recommendations), has been provided in the Supplementary Materials (Table S2).

2.4. Extracting Data from Two Official Databases

Two main Austrian databases were used to gather data regarding farm structure, production type, and Animal Health Service membership. The first database which was used is owned by the AgrarMarktAustria (AMA). This institution was established in 1992 by law and is a subject of the Bundesministerium für Land- und Forstwirtschaft (Federal Ministry of Agriculture and Forestry). The AMA is authorized to report on national and international agriculture markets and advancements in agriculture. The second database used was the Verbrauchergesundheitsinformationssystem (VIS), which is a governmental database operated for the Bundesministerium für Soziales, Gesundheit, Pflege und Konsumentenschutz (Federal Ministry of Social Affairs, Health, Care and

Consumer Protection). All livestock farms and private establishments keeping animals (Zoos, equine holdings, etc.) are required by law to be registered in the VIS.

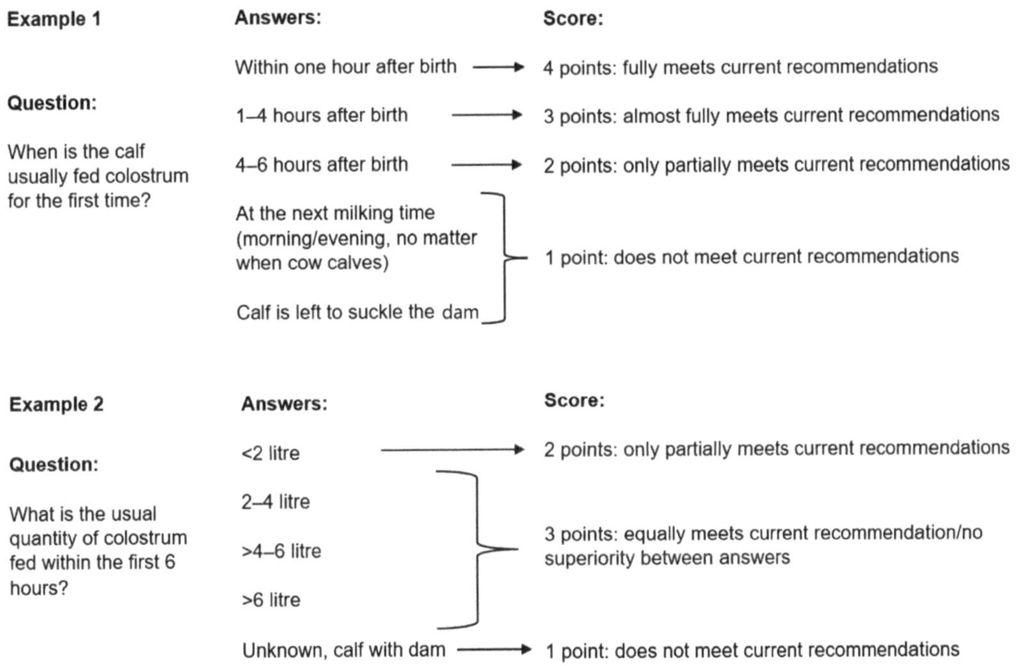

Figure 2. The figure shows the allocation of points based on two example questions.

2.5. Statistical Analysis

The answers to the internet-based survey were transferred to Microsoft Excel 2016 (Microsoft Office Professional Plus 2016 © Microsoft, Redmond, WA, USA). All answers to the survey were viewed individually for plausibility and contradiction; not plausible and contradictory answers were removed from the statistical analysis. The federal state of Vienna, which is the capital city of Austria, was excluded from the study since there is no commercial dairy farming. The survey was distributed throughout Austria via the member database of the Austrian breeding association, and the farmers participated on a voluntary basis. In order to quantify the representativity of the survey, the official Austrian databases (Section 2.4) were used to compare the survey results to the overall Austrian population using descriptive statistics (% survey response versus % overall Austria). The single-choice questions were coded, and the open answers were assessed individually and categorized if suited or not taken into statistical account. The data were described using descriptive statistics expressed as median, 10, 25, 75, and 90 percentiles, minimum and maximum values. The number of dairy cows on the farm (Ncow) and the average 305-day milk yield were categorized. The number of dairy cows on the farm was categorized as follows: ≤10 dairy cows, 11 to 20 dairy cows, 21 to 30 dairy cows, 31 to 40 dairy cows, and ≥41 dairy cows. The average 305-day milk yield was categorized as follows: no or implausible information available; 2000 to 6500 L; 6501 to 7500 L; 7501 to 8700, and 8701 to 14,000 L. If there was no information (missing values) on the location of the farm (federal state), Animal Health Service membership, production type (organic/conventional farming), and/or operation type (full-time/part-time), the questionnaire was excluded from further statistical analysis.

The final data file was transferred to SPSS® statistics software Version 28 (IBM®, New York, NY, USA) for further investigations. The herd management score and the calf management score were tested for normality using the Kolmogorov–Smirnov test, including the Lilliefors correction. The data were not normally distributed ($p < 0.05$). Therefore, the non-parametric Mann–Whitney U test (two independent variables) and the Kruskal–Wallis test (>2 independent variables including the Bonferroni correction) was applied to test if there are statistically significant differences between herd management and calf management practices between organic and conventional farms, part-time and full-time farms and between the different federal states of Austria. According to similar geological structures, the federal states of Austria were divided into alpine regions (high alpine and alpine uplands) in the west, including Vorarlberg, Tyrol, Salzburg, and Carinthia and foothills and flatland/hill country regions in the east, including Upper and Lower Austria, Styria and Burgenland. The level of significance was set at $p < 0.05$.

3. Results

3.1. Response Rate

A response rate of 14.3% (16,246 surveys sent/2328 answers returned) was calculated. In total, 72/2328 (3.1%) answers had to be removed due to missing values for the questions of the federal state, Animal Health Service, production type, and operating type. All the answers given by cow-calf operations (46/2328; 2.0%) were removed since the number was too low for a meaningful analysis. After removing these 118 answers, a total of 2210 answers (94.9% of all answered surveys; response rate: 13.6%) were used for the final statistical analysis. A detailed overview of the overall number of dairy farms in Austria, the production type (organic/conventional), operation type (part-time/full-time), and the Animal Health Service Membership (Tiergesundheitsdienst, TGD), including the survey results are shown in Table 1.

Table 1. The table gives an overview of the response rate (n = 2210) per federal state and the data from the official institutions/databases AgrarMarkt Austria (AMA) and VIS (Verbrauchergesundheitsinformationssystem). [1] Data extracted from AMA. [2] Data extracted from VIS. * Results from the underlying survey (n = 2210). BGL = Burgenland, CAR = Carinthia, LOAT = Lower Austria, UPAT = Upper Austria, SBG = Salzburg, STY = Styria, T = Tyrol, VBG = Vorarlberg.

Farm Structure	Federal State								Total Austria
	BGL	CAR	LOAT	UPAT	SBG	STY	T	VBG	
n [1] dairy cattle farmers per federal state (%)	76 (0.3)	1651 (6.9)	4012 (16.8)	6084 (25.5)	3270 (13.7)	3753 (15.7)	3879 (16.3)	1143 (4.8)	23,868 (100.0)
n * dairy cattle farmers responses (%)	8 (0.4)	151 (6.8)	404 (18.3)	451 (20.4)	296 (13.4)	341 (15.4)	439 (19.9)	120 (5.4)	2210 (100.0)
n [2] organic dairy cattle farmers per federal state (%)	7 (9.2)	373 (22.6)	839 (20.9)	1081 (17.8)	2017 (61.7)	993 (26.5)	1086 (28.0)	163 (14.3)	6559 (27.5)
n * organic dairy cattle farmers responses (%)	0 (0.0)	51 (33.8)	93 (23.0)	91 (20.2)	196 (66.2)	100 (29.3)	126 (28.7)	21 (17.5)	678 (30.7)
n [2] TGD membership per federal state (%)	69 (90.8)	1565 (94.8)	3848 (95.9)	5503 (90.5)	2575 (78.8)	3409 (90.8)	3738 (96.4)	1146 (100.0)	21,853 (91.5)
n * TGD members (%)	8 (100.0)	148 (98.0)	398 (98.5)	436 (96.7)	278 (93.9)	325 (95.3)	425 (96.8)	120 (100.0)	2138 (96.7)
n * part-time operated dairy cattle farmers (%)	0 (0.0)	50 (33.1)	88 (21.8)	117 (25.9)	147 (49.7)	101 (29.6)	271 (61.7)	60 (50.0)	835 (37.7)
n * full-time operated dairy cattle farmers answers (%)	8 (100.0)	101 (66.9)	316 (78.2)	334 (74.1)	149 (50.3)	240 (70.4)	168 (38.3)	60 (50.0)	1376 (62.3)

3.2. General Farm Characteristics

The results of this survey section on general farm characteristics were split. Regarding the farm size, Tyrol had most farms in the category of less than 10 dairy cows per farm (43.6% of all answers from Tyrol). Vorarlberg (32.5%), Salzburg (36.4%), and Carinthia (41.1%) showed most of the farms within the category of 11 to 20 dairy cows per farm. In

Styria (29.6%) and Lower Austria (28.5%), the majority of farms were within the category of 21 to 30 dairy cows per farm. The largest farms were located in Upper Austria (31.4%) and Burgenland (75.0%), within the category of more or equal to 41 dairy cows per farm.

Simmental is the major breed in all federal states (n = 1636, 74.3%), Brown Swiss/Original Brown (n = 287, 13.0%) primarily in Tyrol (24.9%) and Vorarlberg (67.5%), Holstein-Fresian 7.1% (n = 156), Pinzgauer (n = 65, 3.0%) primarily in Salzburg (14.3%) and Tyrolean Grey (n = 37, 1.7%) primarily in Tyrol (7.6%). All other breeds (Jersey, Tuxer, Belgian Blue, etc.) and crossbreeds were summarized as one category (n = 22, 1.0%).

Regarding the average milk yield in Salzburg (35.8%) and Tyrol (33.5%), most respondents were within the category of producing 2000 to 6000 L per cow per year. In Vorarlberg, most respondents were in the category of 7501 to 8700 L (30.8%). In Burgenland, Styria, Upper and Lower Austria, most respondents were in the category of 8701 to 14,000 L (62.5%, 32.3%, 31.0%, 35.1%, respectively). In Carinthia, there were equal answers (24.5%) in categories 6501 to 7500 and 8701 to 14,000 L per cow per year.

In Burgenland (50.0%), Carinthia (49.7%), Salzburg (49.1%), and Styria (47.2%), most farms house their lactating and dry dairy cows in a free-stall barn with an outdoor loafing area (OLA) and/or a pasture. In Upper (42.7%) and Lower Austria (44.1%), a free-stall barn without an OLA and/or pasture is the most common housing type. In Tyrol (68.2%) and Vorarlberg (49.2%), the cows were primarily housed in tie stalls with OLA and/or pasture.

The livestock units were not evaluated due to many missing values (383, 17.33%) and many contradictory answers (219, 9.9%). Therefore, the number of dairy cows on the farm (Ncows) was used to compare farm sizes. The question on the membership of the official national milk performance recording organization produced confusing answers. Obviously, the respondents did not know the official name of the milk performance organization or they state their membership in a breeding association. Consequently, due to multiple contradictory answers, this open question was excluded.

An overview of the results from the survey section one on general farm characteristics is provided in Table 2. The survey results for all federal states of Austria are shown in detail in the Supplementary Materials Table S3.

Table 2. Survey results on general farm characteristics of the 2210 included dairy farms. The alpine region (West) shows the summary of answers from Vorarlberg, Tyrol, Salzburg and Carinthia. The foothills/flatlands region (East) shows the summary of answers from Burgenland, Styria, Upper and Lower Austria. Additionally, the production types (ORG = organic; CON = conventional) and the operating types (Part = part-time farming; Full = Full-time farming) are shown. N.A. = no answer, OLA = outdoor loafing area. * The breed category "brown Swiss" also included the "original brown Swiss" (Original Braunvieh).

Question	Answer Category	Region		Production Type		Operation Type		n (%) Total
		n (%) West	n (%) East	n (%) ORG	n (%) CON	n (%) Part	n (%) Full	
Farm size	≤10	314 (31.4)	69 (5.7)	136 (20.1)	247 (16.2)	325 (39.2)	58 (4.2)	383 (17.4)
	11–20	348 (34.8)	284 (23.6)	261 (38.6)	371 (24.3)	320 (38.6)	312 (22.7)	632 (28.7)
	21–30	172 (17.2)	335 (27.9)	155 (22.9)	352 (23.1)	133 (16.0)	374 (27.2)	507 (23.0)
	31–40	84 (8.4)	206 (17.1)	74 (10.9)	216 (14.2)	35 (4.2)	255 (18.6)	290 (13.2)
	≥41	83 (8.3)	308 (25.6)	51 (7.5)	340 (22.3)	17 (2.0)	374 (27.2)	391 (17.7)
	Total	1001	1.202	677	1.526	830	1373	2203
Breed	Simmental	592 (59.1)	1044 (86.9)	523 (77.5)	1113 (72.8)	554 (66.7)	1082 (78.8)	1636 (74.3)
	Holstein-Friesian	90 (9.0)	66 (5.5)	33 (4.9)	123 (8.0)	37 (4.5)	119 (8.7)	156 (7.1)

Table 2. *Cont.*

Question	Answer Category	Region		Production Type		Operation Type		n (%) Total
		n (%) West	n (%) East	n (%) ORG	n (%) CON	n (%) Part	n (%) Full	
Breed	Brown Swiss *	205 (20.5)	82 (6.8)	62 (9.2)	225 (14.7)	152 (18.3)	135 (9.8)	287 (13.0)
	Pinzgauer	61 (6.1)	4 (0.3)	37 (5.5)	28 (1.8)	40 (4.8)	25 (1.8)	65 (3.0)
	Tyrolean Grey	34 (3.4)	3 (0.2)	10 (1.5)	27 (1.8)	34 (4.1)	3 (0.2)	37 (1.7)
	Others and Crossbreeds	19 (1.9)	3 (0.2)	10 (1.5)	12 (0.8)	13 (1.6)	9 (0.7)	22 (1.0)
	Total	1001	1202	675	1.528	830	1373	2203
Average milk yield per cow per year in liter	N.A.	61 (6.1)	53 (4.4)	29 (4.3)	85 (5.5)	57 (6.8)	57 (4.1)	114 (5.2)
	2000–6500	309 (30.7)	185 (15.4)	273 (40.3)	221 (14.4)	271 (32.5)	223 (16.2)	494 (22.4)
	6501–7500	267 (26.5)	264 (21.9)	234 (34.5)	297 (19.4)	218 (26.1)	313 (22.7)	531 (24.0)
	7501–8700	218 (21.7)	305 (25.3)	108 (15.9)	415 (27.1)	173 (20.7)	350 (25.4)	523 (23.7)
	8701–14,000	151 (15.0)	397 (33.0)	34 (5.0)	514 (33.6)	115 (13.8)	433 (31.5)	548 (24.8)
	Total	1006	1204	678	1532	834	1376	2210
Housing type for lactating and dry cows	Freestall barn with OLA/pasture	386 (38.7)	507 (42.1)	427 (63.2)	466 (30.6)	258 (31.2)	635 (46.2)	893 (40.6)
	Freestall barn without OLA/pasture	52 (5.2)	451 (37.5)	9 (1.3)	494 (32.4)	88 (10.6)	415 (30.2)	503 (22.9)
	Tie stalls with OLA/pasture	541 (54.2)	202 (16.8)	238 (35.2)	505 (33.1)	446 (53.9)	297 (21.6)	743 (33.8)
	Tie stalls without OLA/pasture	19 (1.9)	43 (3.6)	2 (0.3)	60 (3.9)	36 (4.3)	26 (1.9)	62 (2.8)
	Total	998	1203	676	1.525	828	1373	2201

3.3. Herd Management Practices

The second Section contained questions on herd management practices on dairy farms with a special focus on herd-level colostrum management practices. The results for alpine (west) and foothills/flatland (east) regions, organic/conventional production types, and part-time/full-time operated farms are provided in Tables 3 and 4. The total data set on herd management practices in the eight federal states of Austria is provided in the Supplementary Materials (Tables S4 and S5).

Table 3. Survey results on herd-management practices part 1 with a special focus on herd-level colostrum management practices. The alpine region (West) shows the summary of answers from Vorarlberg, Tyrol, Salzburg and Carinthia. The foothills/flatlands region (East) shows the summary of answers from Burgenland, Styria, Upper and Lower Austria. Additionally, the production types (ORG = organic; CON = conventional) and the operating types (Part = part-time farming; Full = Full-time farming) are shown. [1] nested question to the previous question.

Question	Answer Category	Region		Production Type		Operation Type		n (%) Total
		n (%) West	n (%) East	n (%) ORG	n (%) CON	n (%) Part	n (%) Full	
Availability of a separate calving area	Yes	429 (42.8)	841 (70.0)	394 (58.3)	876 (57.3)	332 (40.0)	938 (68.3)	1270 (57.6)
	No	574 (57.2)	360 (30.0)	282 (41.7)	652 (42.7)	499 (60.0)	435 (31.7)	934 (42.4)
	Total	1003	1201	676	1.528	831	1.373	2.204
[1] Cows actually calving in the separate calving area in %	All (100%)	110 (25.5)	226 (26.9)	78 (19.7)	258 (29.5)	84 (25.1)	252 (26.9)	336 (26.4)

Table 3. *Cont.*

Question	Answer Category	Region		Production Type		Operation Type		n (%) Total
		n (%) West	n (%) East	n (%) ORG	n (%) CON	n (%) Part	n (%) Full	
[1] Cows actually calving in the separate calving area in %	Almost all (90%)	176 (40.8)	376 (44.7)	162 (40.9)	390 (44.5)	136 (40.7)	416 (44.3)	552 (43.4)
	The majority (75%)	70 (16.2)	122 (14.5)	75 (18.9)	117 (13.4)	51 (15.3)	141 (15.0)	192 (15.1)
	Half (50%)	32 (7.4)	64 (7.6)	33 (8.3)	63 (7.2)	28 (8.4)	68 (7.2)	96 (7.5)
	Less than half (<50%)	43 (10.0)	53 (6.3)	48 (12.1)	48 (5.5)	35 (10.5)	61 (6.5)	96 (7.5)
	Total	431	841	396	876	334	938	1.272
Colostrum harvesting method	Milking machine	600 (59.9)	725 (60.6)	399 (59.3)	926 (60.7)	531 (63.9)	794 (58.1)	1325 (60.3)
	By hand	365 (36.5)	437 (36.5)	235 (34.9)	567 (37.2)	277 (33.3)	525 (38.4)	802 (36.5)
	Calf stays with dam	36 (3.6)	35 (2.9)	39 (5.8)	32 (2.1)	23 (2.8)	48 (3.5)	71 (3.2)
	Total	1001	1197	673	1.525	831	1367	2198
Availability of frozen colostrum stocks	Yes	763 (76.3)	1053 (87.9)	550 (81.2)	1266 (83.2)	637 (76.7)	1179 (86.2)	1816 (82.6)
	No	237 (23.7)	145 (12.1)	127 (18.8)	255 (16.8)	193 (23.3)	189 (13.8)	382 (17.4)
	Total	1000	1198	677	1521	830	1368	2198

Table 4. Survey results on herd-management practices part 2 with a special focus on herd-level colostrum management practices. The alpine region (West) shows the summary of answers from Vorarlberg, Tyrol, Salzburg and Carinthia. The foothills/flatlands region (East) shows the summary of answers from Burgenland, Styria, Upper and Lower Austria. Additionally, the production types (ORG = organic; CON = conventional) and the operating types (Part = part-time farming; Full = Full-time farming) are shown. [1] nested question to the previous question.

Question	Answer Category	Region		Production Type		Operation Type		n (%) Total
		n (%) West	n (%) East	n (%) ORG	n (%) CON	n (%) Part	n (%) Full	
Time to calf/dam separation	Not at all (<20 min)	542 (54.3)	724 (69.3)	318 (47.1)	948 (62.2)	516 (62.2)	750 (54.7)	1266 (57.5)
	Up to 1 h	199 (19.9)	192 (16.0)	126 (18.7)	265 (17.4)	168 (20.2)	223 (16.3)	391 (17.8)
	1–4 h	101 (10.1)	147 (12.2)	79 (11.7)	169 (11.1)	67 (8.1)	181 (13.2)	248 (11.3)
	>4 h up to 1 Day	92 (9.2)	96 (8.0)	83 (12.3)	105 (6.9)	58 (7.0)	130 (9.5)	188 (8.5)
	>1 Day	51 (5.1)	27 (2.2)	43 (6.4)	35 (2.3)	15 (1.8)	63 (4.6)	78 (3.5)
	Nurse cow-calf rearing	2 (0.2)	4 (0.3)	6 (0.9)	0 (0.0)	1 (0.1)	5 (0.4)	6 (0.3)
	Dam-bound calf rearing	12 (1.2)	11 (0.9)	20 (3.0)	3 (0.2)	5 (0.6)	18 (1.3)	23 (1.1)
	Total	999	1.201	675	1.525	830	1370	2200
Colostrum harvest after parturition	Within 1 h	549 (55.1)	667 (55.4)	327 (48.4)	889 (58.3)	447 (53.9)	769 (56.1)	1216 (55.3)
	1–6 h	346 (34.7)	375 (31.2)	234 (34.6)	487 (32.0)	307 (37.0)	414 (30.2)	721 (32.8)
	Next milking time	84 (8.4)	144 (12.0)	96 (14.2)	132 (8.7)	67 (8.1)	161 (11.7)	228 (10.4)
	Calf stays with dam	18 (1.8)	17 (1.4)	19 (2.8)	16 (1.0)	8 (1.0)	27 (2.0)	35 (1.6)
	Total	997	1.203	676	1.524	829	1.371	2200

Table 4. *Cont.*

Question	Answer Category	Region		Production Type		Operation Type		n (%) Total
		n (%) West	n (%) East	n (%) ORG	n (%) CON	n (%) Part	n (%) Full	
Udder cleaning before colostrum milking	Yes	831 (83.4)	1069 (88.9)	583 (86.2)	1317 (86.4)	717 (86.6)	1183 (86.2)	1900 (86.4)
	No	166 (16.6)	134 (11.1)	93 (13.8)	207 (13.6)	111 (13.4)	189 (13.8)	300 (13.6)
	Total	997	1203	676	1524	828	1.372	2.200
[1] Udder cleaning methods	Wood wool	163 (19.7)	219 (20.6)	169 (29.1)	213 (16.3)	137 (19.2)	245 (20.8)	382 (20.2)
	Udder cloth wet	281 (34.0)	337 (31.7)	164 (28.2)	454 (34.7)	250 (35.1)	368 (31.3)	618 (32.7)
	Udder cloth dry	370 (44.7)	488 (46.0)	243 (41.8)	615 (47.0)	322 (45.2)	536 (45.6)	858 (45.4)
	Automatic (robotic system)	13 (1.6)	18 (1.7)	5 (0.9)	26 (2.0)	4 (0.6)	27 (2.3)	31 (1.6)
	Total	827	1062	581	1308	713	1176	1889

3.4. Calf Management Practices

The third Section contained questions on calf management practices on dairy farms. The results for alpine (west) and foothills/flatland (east) regions, organic/conventional production types, and part-time/full-time operated farms are provided in Tables 5 and 6. The total data set on calf management practices in the eight federal states of Austria is provided in Supplementary Materials Tables S6 and S7.

Table 5. Survey results on calf management practices part 1 with a special focus on calf-level colostrum management practices (source of colostrum and colostrum testing methods). The alpine region (West) shows the summary of answers from Vorarlberg, Tyrol, Salzburg and Carinthia. The foothills/flatlands region (East) shows the summary of answers from Burgenland, Styria, Upper and Lower Austria. Additionally, the production types (ORG = organic; CON = conventional) and the operating types (Part = part-time farming; Full = Full-time farming) are shown. [1] nested question to the previous question.

Question	Answer Category	Region		Production Type		Operation Type		n (%) Total
		n (%) West	n (%) East	n (%) ORG	n (%) CON	n (%) Part	n (%) Full	
Colostrum from mother	Yes, always	721 (72.0)	739 (61.5)	458 (67.8)	1002 (65.6)	602 (72.4)	858 (62.5)	1460 (66.3)
	Mostly	8 (0.8)	7 (0.6)	8 (1.2)	7 (0.5)	4 (0.5)	11 (0.8)	15 (0.7)
	Yes, if dam has good colostrum quality	223 (22.3)	394 (32.8)	150 (22.2)	467 (30.6)	196 (23.6)	421 (30.7)	617 (28.0)
	Calf stays with dam	44 (4.4)	42 (3.5)	55 (8.1)	31 (2.0)	25 (3.0)	61 (4.4)	86 (3.9)
	No, not always	5 (0.5)	20 (1.7)	5 (0.7)	20 (1.3)	4 (0.5)	21 (1.5)	25 (1.1)
	Total	1001	1202	676	1527	831	1372	2203
[1] Colostrum source if not from mother	Frozen colostrum	23 (95.8)	36 (97.3)	17 (100.0)	42 (95.5)	14 (100.0)	45 (95.7)	59 (96.7)
	Colostrum replacer	1 (4.2)	1 (2.7)	0 (0.0)	2 (4.5)	0 (0.0)	2 (4.3)	2 (3.3)
	Total	24	37	17	42	14	47	61
Assessment of colostrum quality	Yes	272 (27.0)	318 (26.4)	162 (23.9)	428 (27.9)	219 (26.3)	371 (27.0)	590 (26.7)
	No	734 (73.0)	886 (73.6)	516 (76.1)	1104 (72.1)	615 (73.7)	1005 (73.0)	1620 (73.3)
	Total	1006	1204	678	1532	834	1376	2210

Table 5. *Cont.*

Question	Answer Category	Region		Production Type		Operation Type		n (%) Total
		n (%) West	n (%) East	n (%) ORG	n (%) CON	n (%) Part	n (%) Full	
Colostrum quality assessment method	Colostrometer	23 (8.6)	28 (9.0)	12 (7.5)	39 (9.4)	15 (6.9)	36 (10.0)	51 (8.8)
	Refractometer	35 (13.1)	71 (22.8)	25 (75.5)	81 (19.4)	25 (11.5)	81 (22.4)	106 (18.3)
	Visually	190 (71.2)	178 (57.2)	112 (69.6)	256 (61.4)	162 (74.7)	206 (57.1)	368 (63.7)
	ColostroCheck	19 (7.1)	34 (10.9)	12 (7.5)	41 (9.8)	15 (6.9)	38 (10.5)	53 (9.2)
	Total	267	311	161	417	217	361	578

Table 6. Survey results on calf management practice part 2 with a special focus on calf-level colostrum management practices (colostrum feeding time, quantity of colostrum delivered to the newborn, and feeding methods). The alpine region (West) shows the summary of answers from Vorarlberg, Tyrol, Salzburg, and Carinthia. The foothills/flatlands region (East) shows the summary of answers from Burgenland, Styria, Upper and Lower Austria. Additionally, the production types (ORG = organic; CON = conventional) and the operating types (Part = part-time farming; Full = Full-time farming) are shown. (p.n. = post natum).

Question	Answer Category	Federal State		Production Type		Operation Type		n (%) Total
		n (%) West	n (%) East	n (%) ORG	n (%) CON	n (%) Part	n (%) Full	
Time from parturition to colostrum feeding	Within 1 h p.n.	600 (60.2)	692 (57.8)	375 (55.6)	917 (60.3)	480 (58.0)	812 (59.4)	1292 (58.9)
	1–4 h p.n.	296 (29.7)	366 (30.6)	201 (29.8)	461 (30.3)	265 (32.0)	397 (29.0)	662 (30.2)
	4–6 h p.n.	27 (2.7)	23 (1.9)	16 (2.4)	34 (2.2)	24 (2.9)	26 (1.9)	50 (2.3)
	Next standard milking time	43 (4.3)	77 (6.4)	38 (5.6)	82 (5.4)	40 (4.8)	80 (5.9)	120 (5.5)
	Calf suckles the dam	31 (3.1)	40 (3.3)	44 (6.5)	27 (1.8)	19 (2.3)	52 (3.8)	71 (3.2)
	Total	997	1198	674	1521	828	1367	2195
Quantity of colostrum fed within the first 6 h after birth	<2 L	182 (18.2)	230 (19.1)	94 (13.9)	318 (20.8)	177 (21.3)	235 (17.1)	412 (18.7)
	2–4 L	680 (67.9)	868 (72.2)	486 (72.0)	1062 (69.5)	556 (66.8)	992 (72.3)	1548 (70.2)
	>4–6 L	101 (10.1)	66 (5.5)	51 (7.6)	116 (7.6)	77 (9.3)	90 (6.6)	167 (7.6)
	>6 L	11 (1.1)	5 (0.4)	4 (0.6)	12 (0.8)	9 (1.1)	7 (0.5)	16 (0.7)
	Unknown, calf with dam	27 (2.7)	34 (2.8)	40 (5.9)	21 (1.4)	13 (1.6)	48 (3.5)	61 (2.8)
	Total	1001	1203	675	1529	832	1372	2204
Colostrum feeding equipment	Bucket	524 (52.4)	366 (30.6)	304 (45.0)	586 (38.5)	353 (42.5)	537 (39.3)	890 (40.5)
	Nipple bottle	412 (41.1)	788 (65.8)	307 (45.4)	893 (58.7)	434 (52.3)	766 (56.0)	1200 (54.6)
	Esophageal tube	14 (1.4)	1 (0.1)	5 (0.7)	10 (0.7)	14 (1.7)	1 (0.1)	15 (0.7)
	Calf stays with dam	50 (5.0)	42 (3.5)	60 (8.9)	32 (2.1)	29 (3.5)	63 (4.6)	92 (4.2)
	Total	1000	1197	676	1521	830	1367	2197
Calves not drinking well receive colostrum	Immediately esophageal tube	80 (8.0)	153 (12.8)	44 (6.5)	189 (12.5)	58 (7.0)	175 (12.8)	233 (10.6)
	Esophageal tube within 2–6 h	186 (18.7)	277 (23.1)	127 (18.8)	336 (22.2)	155 (18.8)	308 (22.5)	463 (21.1)
	By esophageal tube, in general	18 (1.8)	2 (0.2)	6 (0.9)	14 (0.9)	16 (1.9)	4 (0.3)	20 (0.9)

Table 6. *Cont.*

Question	Answer Category	Federal State		Production Type		Operation Type		n (%) Total
		n (%) West	n (%) East	n (%) ORG	n (%) CON	n (%) Part	n (%) Full	
Calves not drinking well receive colostrum	Colostrum offered multiple times, in no case esophageal tube	694 (69.8)	742 (61.9)	491 (72.5)	945 (62.4)	583 (70.7)	853 (62.4)	1436 (65.5)
	Colostrum offered multiple times, later esophageal tube	15 (1.5)	22 (1.8)	8 (1.2)	29 (1.9)	11 (1.3)	26 (1.9)	37 (1.7)
	Others (vet, supplements, etc.)	1 (0.1)	2 (0.2)	1 (0.1)	2 (0.1)	2 (0.2)	1 (0.1)	3 (0.1)
	Total	994	1198	677	1515	825	1367	2192

3.4.1. Colostrum Quantity

The usual quantity of colostrum within the first six hours is 2–4 L in Burgenland, Carinthia, Lower Austria, Upper Austria, Salzburg, Styria, Tyrol, and Vorarlberg (37.3%, 65.6%, 72.3%, 73.3%, 73.8%, 71.2%, 65.8%, 64.2%). Less than two liters are fed in 25.0% (Burgenland), 22.5% (Carinthia), 19.3% (Lower Austria), 19.3% (Upper Austria), 9.2% (Salzburg), 18.5% (Styria), 21.3% (Tyrol) and 23.3% (Vorarlberg). For detailed results, see Supplementary Materials Table S7.

3.4.2. Colostrum Feeding Methods

The nipple bottle is the most reported feeding method in Burgenland, Carinthia, Lower Austria, Upper Austria, and Styria (50.0%, 64.9%, 64.8%, 67.0%, and 65.8%). In Salzburg, Tyrol, and Vorarlberg, it is the bucket (61.3%, 54.6%, and 53.8%). If the calf does not drink colostrum willingly, colostrum is offered multiple times, and in no case, an esophageal tube be used in 42.9% (Burgenland), 62.7% (Carinthia), 66.2% (Lower Austria), 56.6% (Upper Austria), 68.4% (Salzburg), 64.4% (Styria), 71.0% (Tyrol) and 78.2% (Vorarlberg). On the other hand, an esophageal tube is used within 2–6 h in 28.6% (Burgenland), 22.7% (Carinthia), 22.9% (Lower Austria), 26.3% (Upper Austria), 21.0% (Salzburg), 19.1% (Styria), 15.9% (Tyrol) and 18.5% (Vorarlberg). For detailed results, see Supplementary Materials Table S7.

3.5. Assessing Herd- and Calf Management Practices by the Use of a Scoring System

The overall median for Austria in herd management practices was 9 (range 1–14), and in calf management practices, 12 (range 3–18). Comparisons of herd and calf management scores for alpine (west) and foothills/flatland (east) regions, production types (organic versus conventional), and operating types (full-time versus part-time) are shown in Figures 3 and 4.

3.5.1. Differences between Alpine and Foothills/Flatland Regions

In the alpine regions (west), there was a statistically significant difference in herd management between Tyrol and Salzburg ($p < 0.01$) and Tyrol and Carinthia ($p < 0.01$); for details see Supplementary Materials Figure S1. In foothills/flatland regions (east), there was no statistically significant difference between each other for herd management. In all of Austria there were statistically significant differences for herd management between Lower Austria and Salzburg ($p < 0.01$), Vorarlberg ($p < 0.01$), and Tyrol ($p < 0.01$); between Upper Austria and Vorarlberg ($p < 0.01$), Salzburg ($p < 0.01$) and Tyrol ($p < 0.01$); between Styria and Vorarlberg ($p < 0.01$), Salzburg ($p < 0.025$) and Tyrol ($p < 0.01$). In calf management, there were only statistically significant differences between Upper Austria and Vorarlberg ($p < 0.01$), Tyrol ($p < 0.01$), and Styria ($p < 0.01$). The details are shown in the Supplementary Materials Figure S1.

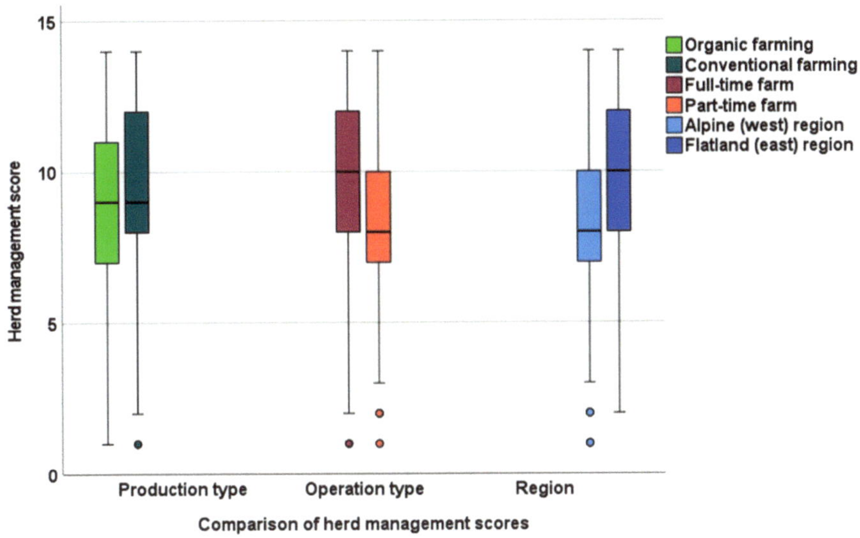

Figure 3. Comparisons on herd management scores in the alpine (west) and foothills/flatland (east) region of Austria, organic and conventional production types, and part-time and full-time operated farms.

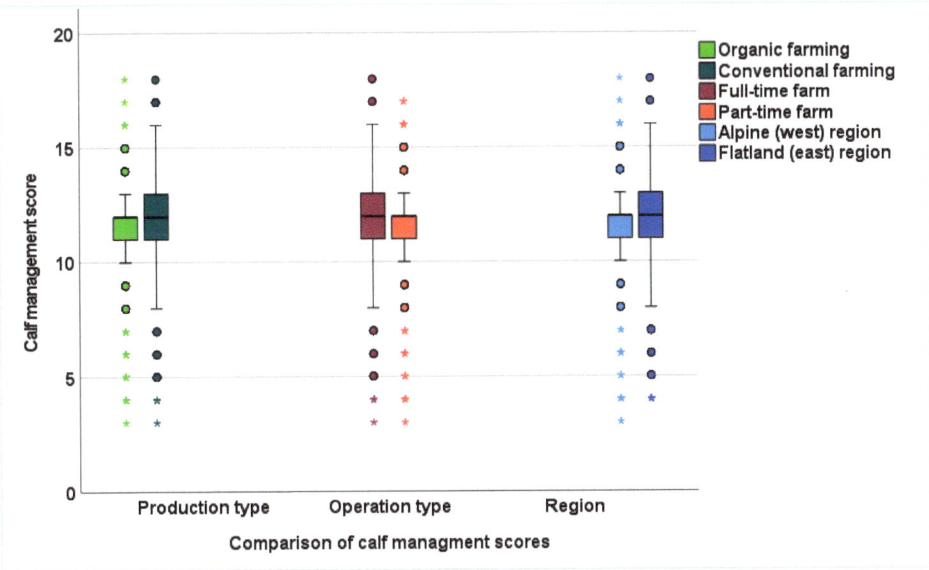

Figure 4. Comparisons on calf management scores in the alpine (west) and foothills/flatland (east) region of Austria, organic and conventional production types, and part-time and full-time operated farms.

3.5.2. Differences between Organic and Conventional Farms

Herd Management Score

Organic and conventional farms had a median score of 9 in herd management practices. Percentiles 10, 25, 75, and 90 for organic-producing farms were 6, 7, 11, and 12, respectively ($p < 0.01$). Percentiles 10, 25, 75, and 90 for conventional farms were 6, 8, 12, and 13, respectively ($p < 0.01$). Organic farms had a larger statistical spread than conventional farms.

Calf Management Score

The median for organic and conventional farms was 12. The percentiles 10, 25, 75, and 90 were 9, 11, 12, 13 for organic farms and 10, 11, 13, 14 for conventional farms ($p < 0.001$).

3.5.3. Differences between Part-Time and Full-Time Farms

Herd Management Score

Full-time operated farms had a median score of 10 in herd management practices, whereas part-time operated farms had a median score of 8. The percentiles 10, 25, 75, and 90 for full-time operated farms were 6, 8, 12, 13, and for part-time operated farms, 6, 7, 10, 12 ($p < 0.01$).

Calf Management Score

For full-time and part-time operated farms, the median was 12 with a 10, 25, 75, and 90 percentile for full-time: 9, 11, 13, 14 and part-time: 10, 11, 12, 13, respectively.

4. Discussion

4.1. Survey Design and Response Rate

The response rate was reasonable, with 14.3% in the present study. Other similar internet-based, voluntary surveys conducted in Austria showed response rates of 12.2% (1287 respondents) [26] and 11.3% (1018 respondents) [34]. Another similar email-based survey was conducted in Germany on the evaluation of fresh cow management with a response rate of 12.0% (429 respondents) [35]. Due to the fact that all studies carried out convenience sampling, the external validity is limited. The study was not designed as a representative survey. Still, the survey provides a good overview of the herd and calf-level colostrum management practices in Austria since the number of farms per region (federal states) was comparable to the relative number of farms per federal state, as provided in Table 1. The responses of organic-producing farms were, in total, overrepresented as only 27.5% of dairy farms in Austria are organically producing, and in the present study, 30.7% were organic-producing farms. For details, see Table 1. The number of farms with an Animal Health Service Membership was higher in the present study (96.7%) than it actually is in Austria (91.5%, see Table 1). It can be hypothesized that more educated and motivated farmers participated in the survey. The Animal Health Service operates in all eight federal states of Austria (except Vienna, which is the capital city of Austria but also a federal state) and is an association that farmers can join on a voluntary basis. The Animal Health Service membership includes regular farm visits from veterinarians, including herd health checks, discussing management procedures, and biosecurity measurements. Members also have to fulfill a certain number of continuing education courses throughout the year. In the federal state of Vorarlberg, the Animal Health Service was mandatory for all cattle farmers in 2022.

4.2. Scoring System

The scoring system, including the allocation of points, was implemented in order to have a tool to objectively compare the different herd management and calf management practices. The authors are aware of the fact that scientific opinion is on the lowest level on the pyramid of evidence [36]. The authors based their opinion on articles published in peer-reviewed journals. When the farmers receive the results from the scoring system, they have the possibility to focus specifically on the weaknesses in colostrum management on their farms. In the future, the scoring system might be used to predict the likelihood of FTPI in calves on farms, as similar scoring systems have been implemented in human medicine for decades to predict disease (e.g., prediction of coronary heart diseases [37]). Further studies will be needed. Currently, it is not possible to see if calves are at risk of facing

FTPI since we do not have data from the calves. At this stage, the scoring system should rather be used to educate farmers regarding current evidence-based recommendations. Three questions were not included in the scoring system due to the fact that there was no conclusive research available when the study was conducted. Therefore, no clear distinction between the methods was feasible (see Table S2). Additionally, our scientific panel consisted of three panelists. In future studies, it is recommended to set up a panel with more scientists discussing the point allocation and/or collect actual data on FTPI in calves on each farm.

Our results show higher scores for the herd ($p < 0.01$) and calf management ($p = 0.02$) practices in foothills/flatlands regions. The possible reasons for this might be more full-time operated farms, the availability of land to build new facilities due to the low population density, and/or more conventional producing farms. Calf management practices seem similar in all federal states of Austria and do not differ much in the different production and operating types, although all findings were statistically significant ($p < 0.01$). Although the median score was the same (12) for all regions, production, and operating types, alpine regions, organic producing, and part-time operated farms have a bigger statistical spread of answers. It seems that the knowledge of the importance of fast colostrum administration and high colostrum quantities is widely known in Austria and does not depend on the region, production, or operating type. Our study showed that there are small farms with a high knowledge of good colostrum management but also large farms with an insufficient knowledge of good colostrum management in Austria. In the future, the scoring system might be used by farms to find points of improvement for their colostrum management practices and to implement a guideline to standard operating procedures (SOP) for small-scale dairy farms in Austria as has been carried out in Germany [38].

4.3. General Farm Characteristics

Farm size in Austria (19 dairy cows [31]) is in accordance with the present study (28.7% 11–20 dairy cows per farm). In Austria, there is a large difference between farms in alpine and foothills/flatland regions. In Tyrol, an average of 21.6 cattle are kept per farm, whereas in Burgenland, on average, 48.9 cattle are kept on farms [39]. This might lead to more part-time operated farms in alpine regions. Additionally, agricultural tourism plays a significant role in the alpine regions of Austria. The distribution of major cattle breeds in Austria is Simmental 74.7%, Holstein-Fresian 7.3%, Brown Swiss 5.7%, Pinzgauer 2.0%, Tyrolean Grey 0.9%, which is very similar to the present study (see Table 2). Most Brown Swiss and Tyrolean Grey are on farms in Tyrol and Vorarlberg [40], also similar to our study.

The average milk yield was lower in alpine (most respondents within the category of producing 2000 to 6000 L per cow per year) than in foothills/flatland regions (category 8701 to 14,000 L per cow per year). This is probably because there are more organic producing and part-time operated farms in alpine regions. Those farm types seem, according to our study, to be more extensive, producing farms with lower milk yields and fewer dairy cows.

Barn types differ in alpine and foothills/flatland regions. Alpine regions might run tie-stall barns more often due to the farm size of $\leq$30 dairy cows. Additionally, in the alpine regions, there are plenty of part-time farms where the farmers gain additional income besides farming.

4.4. Herd Management Practices

The results indicate that it is more common for organic farms (18.7%) to leave the calf with the dam for more than 4 h than it is for conventional farms (9.2%). If the calf is left with the dam, there is no way of knowing in what time frame and how much quantity of colostrum the calf consumed. Furthermore, it can be assumed that those farmers also do not test colostrum quality. This practice might be due to the philosophy of organic producing farms ('close to nature'). However, it is indicated by studies that calves suffer less likely from FTPI when separated from the dam within 3 h [41,42]. Nevertheless, the apparent efficiency of colostrum absorption might be better if the dam is present. Therefore, further studies are needed to verify this effect on the dam and the calf. The results of the present study (Table 3) showed that in the survey, 57.6% of Austrian farms have a separate calving area. In a previous survey conducted in Austria, 47.0% of farms answered that they do have a separate calving area, and 51.1% did not [26]. This shows an improvement in Austrian management practices. The availability of a separate calving area is recommended to ensure high hygiene at birth for the calf and to minimize stress for the cow [41,43]. The present study showed that alpine regions and part-time operated farms frequently do not have a separate calving area. On organic and part-time operated farms, results showed that even if there is a separate calving area available at the farm, in 20.4% (organic farms) and 18.9% (part-time farms), half or even less than half of the cows calve in this area. Our survey did not ask for the reasons why cows do not calve in a separate area if one is available. The reasons might be that in Austria, oftentimes, the separate calving area is also used for diseased animals and, therefore, not available for periparturient cows and not disinfected regularly, which represents a risk factor for spreading diseases [44]. Other reasons might be the time management on part-time operated farms. Another possible explanation might be that in tie-stall barns, cows usually calve in their stall as there is no separate calving area available. A prolonged time period (>6 h) until first milking the dam is associated with a low colostrum quality [9]. Milking the dam at the next standard milking time as a usual procedure may conclude in a longer time period than six hours and, therefore, may result in FTPI in calves. Therefore, this practice as a standard procedure is not recommended. That knowledge seems to be widely known amongst Austrian farmers, although, according to the underlying results, farms from foothills/flatland regions (12.0%) rely on that practice more often than farms in alpine regions (8.4%).

Storing frozen colostrum is recommended by many authors [45,46]. Results from our study show that in foothills/flatland regions, the information on storing frozen colostrum is more well-known than in alpine regions, see Table 3. That is probably because, in alpine regions, there are more part-time operated farms. Additionally, the results showed that part-time-operated farms were less likely to have frozen colostrum storage.

In other studies, critical control points for bacterial contamination of colostrum have been established, one of them being the harvesting process [33]. We assumed that when colostrum is harvested by hand, the udder is not sufficiently cleaned beforehand, but only 13.6% explicitly stated that they do not clean the udder before milking. So, it can be presumed that some of those farms that harvest colostrum by hand also clean the udder beforehand. We also did not ask if they use gloves when milking colostrum by hand. No usage of gloves might again lead to bacterial contamination of the human skin [47]. In the future, this question needs to be altered in order to gain more information on the hygiene practices when harvesting colostrum by hand. Another study comparing total bacterial counts (TBC) in colostrum samples collected directly from the teat and from feeding equipment showed that TBC was higher in samples collected from feeding equipment, suggesting that the focus should lie on the hygiene of colostrum harvest and feeding equipment [48]. It is shown in different studies that dry teat cleaning lowers the bacterial count, and wet cloths should only be used if the teat is dried before milking [49]. Concerning the methods used for udder cleaning, our survey did not ask about the detailed udder cleaning routine. This should be improved in future studies using a scoring system.

4.5. Calf and Colostrum Management Practices

High-quality colostrum from the respective dam should always be preferred to pooled colostrum [3,50]. In our study, 66.3% always use colostrum from their own dam. In Austria, it is not common practice to pool colostrum. On small farms, there is often only one cow in parturition at any one time. This particular question may have produced some confusion because 29.1% checked that they do 'not always' (1.1%) or only 'if the dam has good colostrum quality' (28.0%) use colostrum from their own dam. The supplementary question 'what they use instead' was only answered by 2.8% of participants. The reason for this can only be hypothesized; it might be a possibility that farmers know they should keep a storage of frozen colostrum (82.6% stated they have one) and therefore answered that question accordingly when, in fact, they do not have a storage of frozen colostrum on their farm. The use of colostrum replacers is not very common in Austria, as shown by our results, probably because we have many organic producing farms for whom there is no permitted product available or because colostrum replacers, in general, are very expensive.

It is essential to test colostrum before feeding it to the calf, so dependent on the outcome of the test, the calf can be provided with a sufficient amount of immunoglobulins [51]. Our results show that only a quarter of farms (26.7%) in Austria test colostrum quality before feeding it to the calf. In a previous Austrian study, only 20.8% checked colostrum quality [26], so our results at least show an improvement. For colostrum testing, visual appraisal is wholly inadequate in predicting high (>50 g/L IgG) or low colostrum quality [52]. Other methods such as a colostrometer [53], brix-refractometer [54,55], or ColostroCheck® (a cone-shaped device to rate the flow velocity of colostrum, $\geq$24 s = colostrum quality > 50g IgG/L) [56] should be used. Although in the present study, more farms test colostrum quality, and fewer farms use visual appraisal for testing than in a previous study (20.8%, visual appraisal: 86.1% [26]), there is still room for improvement.

The time and quantity of first feeding colostrum to a calf is crucial. Calves fed with >2.5 L colostrum within the first three hours are less likely to suffer from FTPI [5]. Our study showed that many Austrian farmers do seem to be aware of this aspect of colostrum management. However, these two aspects of colostrum management alone are not sufficient to prevent FTPI in calves, as was shown in another study in Austria (Hartsleben et al., under review).

Besser et al. (1991) described different feeding methods and their likelihood for calves to develop FTPI. In the aforementioned study, 61.4% of calves nursed by the dam, 19.3% of calves fed by nipple bottle, and 10.8% of calves fed by esophageal tube were diagnosed with FTPI [57]. The use of an esophageal tube in general on every newborn calf is no standard practice in Austria. That might be because of the small farm structures and mostly family-owned farms where it is easy to care for each newborn calf individually. This fact could also be the reason why the most common answer was 'colostrum is offered multiple times' (65.5%). Our study showed that in foothills/flatland regions with larger farm structures, farms do use an esophageal tube within 2–6 h (23.1%) if the calf has not been drinking colostrum willingly by that time.

To the best of our knowledge, this was the first attempt to implement a scoring system to quantify herd and calf management practices. The scoring system might be used by farmers to assess the herd and calf management practices on their farms and to become aware of the areas that need to be improved. Additionally, standard operating procedures might be established, especially for small Austrian dairy farms. The findings of our study might be used by veterinarians in farm consultancies or other specialists and organizations providing advisory services to farms in order to reduce the risk of FTPI in calves.

5. Conclusions

The results of this study provide data on herd and calf management practices on dairy farms with a special focus on colostrum management practices in Austria. Furthermore, significant differences could be determined between alpine (west) and foothills/flatland (east) regions, organic and conventional farms, and full-time or part-time operated farms. In summary, eastern regions (foothills/flatland region: Burgenland, Styria, Upper and Lower Austria), conventional producing and full-time operated farms received higher scores than western regions (alpine regions: Vorarlberg, Tyrol, Carinthia, and Salzburg) than organic producing and part-time operated farms.

Supplementary Materials: The following supporting information can be downloaded at: https://www.mdpi.com/article/10.3390/ani13172758/s1, Figure S1: Overview of differences between herd and calf management score for each federal state; Table S1: Original survey on herd-, calf- and colostrum management practices on 2210 Austrian farms (in German); Table S2: Allocation of points for the scoring system; Table S3: Overview of answers given in survey section one on general farm characteristics for each federal state; Tables S4 and S5: Overview of answers given in survey section two on herd-management practices for each federal state; Tables S6 and S7: Overview of answers given in survey section three on calf management practices for each federal state.

Author Contributions: Conceptualization, N.H., K.L., D.K.-J. and T.W.; methodology, N.H., K.L., D.K.-J., A.T. and T.W.; software, N.H., K.L. and A.T.; validation, N.H., K.L., A.T. and T.W.; formal analysis, N.H., K.L. and T.W.; investigation, N.H. and D.K.-J.; resources, N.H. and T.W.; data curation, N.H., K.L. and A.T.; writing—original draft preparation, N.H.; writing—review and editing, N.H., K.L., D.K.-J. and T.W.; visualization, N.H.; supervision, K.L. and T.W.; project administration, N.H., K.L., D.K.-J. and T.W.; funding acquisition, N.H., K.L. and T.W. All authors have read and agreed to the published version of the manuscript.

Funding: The work of the veterinary doctorate student (N.H.) was financially supported by the Austrian Association for Buiatrics (ÖBG) and the Animal Health Service Salzburg (Tiergesundheitsdienst Salzburg).

Institutional Review Board Statement: Ethical review and approval were waived for this study, due to the fact that it was an online questionnaire and no animals participated in this study.

Informed Consent Statement: Informed consent was obtained from all participants involved in the study.

Data Availability Statement: Data are available within the article or in its Supplementary Materials.

Acknowledgments: Open Access Funding by the University of Veterinary Medicine Vienna.

Conflicts of Interest: The authors declare no conflict of interest.

References

1. Schnorr, B.; Kressin, M. *Embryologie der Haustiere: Ein Kurzlehrbuch*, 4th ed.; Enke Verlag Stuttgart: Erlangen, Germany, 2001.
2. Arthur, G.H. (Ed.) *Veterinary Reproduction and Obstetrics*, 7th ed.; Saunders: Philadelphia, PA, USA, 1996.
3. Weaver, D.M.; Tyler, J.W.; VanMetre, D.C.; Hostetler, D.E.; Barrington, G.M. Passive Transfer of Colostral Immunoglobulins in Calves. *J. Vet. Intern. Med.* **2000**, *14*, 569–577. [CrossRef]
4. Barrington, G.M.; Parish, S.M. Bovine Neonatal Immunology. *Vet. Clin. N. Am. Food Anim. Pract.* **2001**, *17*, 463–476. [CrossRef]
5. Morin, M.P.; Dubuc, J.; Freycon, P.; Buczinski, S. A calf-level study on colostrum management practices associated with adequate transfer of passive immunity in Québec dairy herds. *J. Dairy Sci.* **2021**, *104*, 4904–4913. [CrossRef]
6. Fischer, A.J.; Malmuthuge, N.; Le Guan, L.; Steele, M.A. Short communication: The effect of heat treatment of bovine colostrum on the concentration of oligosaccharides in colostrum and in the intestine of neonatal male Holstein calves. *J. Dairy Sci.* **2018**, *101*, 401–407. [CrossRef]
7. Li, M.; Penner, G.B.; Hernandez-Sanabria, E.; Oba, M.; Guan, L.L. Effects of sampling location and time, and host animal on assessment of bacterial diversity and fermentation parameters in the bovine rumen. *J. Appl. Microbiol.* **2009**, *107*, 1924–1934. [CrossRef]
8. Smeaton, T.C.; Simpson-Morgan, M.W. Epithelial cell renewal and antibody transfer in the intestine of the foetal and neonatal lamb. *Aust. J. Exp. Biol. Med. Sci.* **1985**, *63 Pt 1*, 41–51. [CrossRef]
9. Reschke, C.; Schelling, E.; Michel, A.; Remy-Wohlfender, F.; Meylan, M. Factors Associated with Colostrum Quality and Effects on Serum Gamma Globulin Concentrations of Calves in Swiss Dairy Herds. *J. Vet. Intern. Med.* **2017**, *31*, 1563–1571. [CrossRef]
10. McGuirk, S.M.; Collins, M. Managing the production, storage, and delivery of colostrum. *Vet. Clin. N. Am. Food Anim. Pract.* **2004**, *20*, 593–603. [CrossRef]
11. Staley, T.E.; Bush, L.J. Receptor Mechanisms of the Neonatal Intestine and Their Relationship to Immunoglobulin Absorption and Disease. *J. Dairy Sci.* **1985**, *68*, 184–205. [CrossRef]
12. Johnson, J.L.; Godden, S.M.; Molitor, T.; Ames, T.; Hagman, D. Effects of feeding heat-treated colostrum on passive transfer of immune and nutritional parameters in neonatal dairy calves. *J. Dairy Sci.* **2007**, *90*, 5189–5198. [CrossRef]
13. Raboisson, D.; Trillat, P.; Cahuzac, C. Failure of Passive Immune Transfer in Calves: A Meta-Analysis on the Consequences and Assessment of the Economic Impact. *PLoS ONE* **2016**, *11*, e0150452. [CrossRef]
14. Cordero-Solorzano, J.; de Koning, D.-J.; Tråvén, M.; de Haan, T.; Jouffroy, M.; Larsson, A.; Myrthe, A.; Arts, J.A.J.; Parmentier, H.K.; Bovenhuis, H.; et al. Gentic parameters of colostrum and calf serum antibodies in Swedisch dairy cattle. *Genet. Sel. Evol.* **2022**, *54*, 68. [CrossRef]
15. Andrée O'Hara, E.; Båge, R.; Emanuelson, U.; Holtenius, K. Effects of dry period length on metabolic status, fertility, udder health and colostrum production in 2 cow breeds. *J. Dairy Sci.* **2019**, *102*, 595–606. [CrossRef]
16. Zentrich, E.; Iwersen, M.; Wiedrich, M.-C.; Drillich, M.; Klein-Jöbstl, D. Short communication: Effect of barn climate and management-related factors on bovine colostrum quality. *J. Dairy Sci.* **2019**, *102*, 7453–7458. [CrossRef]

17. Immler, M.; Büttner, K.; Gärtner, T.; Wehrend, A.; Donat, K. Maternal Impact on Serum Immunoglobulin and Total Protein Concentration in Dairy Calves. *Animals* **2022**, *12*, 755. [CrossRef]
18. Puppel, K.; Gołębiewski, M.; Grodkowski, G.; Solarczyk, P.; Kostusiak, P.; Klopčič, M.; Sakowski, T. Use of somatic cell count as an indicator of colostrum quality. *PLoS ONE* **2020**, *15*, e0237615. [CrossRef]
19. Faber, S.N.; Faber, N.E.; Mccauley, T.C.; Ax, R.L. Case Study: Effects of Colostrum Ingestion on Lactational Performance1. *Prof. Anim. Sci.* **2005**, *21*, 420–425. [CrossRef]
20. DeNise, S.K.; Robison, J.D.; Stott, G.H.; Armstrong, D.V. Effects of Passive Immunity on Subsequent Production in Dairy Heifers. *J. Dairy Sci.* **1989**, *72*, 552–554. [CrossRef]
21. Heinrichs, A.J.; Heinrichs, B.S. A prospective study of calf factors affecting first-lactation and lifetime milk production and age of cows when removed from the herd. *J. Dairy Sci.* **2011**, *94*, 336–341. [CrossRef]
22. Godden, S. Colostrum management for dairy calves. *Vet. Clin. N. Am. Food Anim. Pract.* **2008**, *24*, 19–39. [CrossRef]
23. Tautenhahn, A.; Merle, R.; Müller, K.E. Factors associated with calf mortality and poor growth of dairy heifer calves in northeast Germany. *Prev. Vet. Med.* **2020**, *184*, 105154. [CrossRef] [PubMed]
24. Lichtmannsperger, K.; Hartsleben, C.; Spöcker, M.; Hechenberger, N.; Tichy, A.; Wittek, T. Factors Associated with Colostrum Quality, the Failure of Transfer of Passive Immunity, and the Impact on Calf Health in the First Three Weeks of Life. *Animals* **2023**, *13*, 1740. [CrossRef]
25. Østerås, O.; Gjestvang, M.S.; Vatn, S.; Sølverød, L. Perinatal death in production animals in the Nordic countries—Incidence and costs. *Acta Vet. Scand.* **2007**, *49* (Suppl. S1), S14. [CrossRef]
26. Klein-Jöbstl, D.; Arnholdt, T.; Sturmlechner, F.; Iwersen, M.; Drillich, M. Results of an online questionnaire to survey calf management practices on dairy cattle breeding farms in Austria and to estimate differences in disease incidences depending on farm structure and management practices. *Acta Vet. Scand.* **2015**, *57*, 44. [CrossRef] [PubMed]
27. Robbers, L.; Bijkerk, H.J.C.; Koets, A.P.; Benedictus, L.; Nielen, M.; Jorritsma, R. Survey on Colostrum Management by Dairy Farmers in the Netherlands. *Front. Vet. Sci.* **2021**, *8*, 656391. [CrossRef]
28. Santos, G.D.; Bittar, C.M.M. A survey of dairy calf management practices in some producing regions in Brazil. *R. Bras. Zootec.* **2015**, *44*, 361–370. [CrossRef]
29. Vasseur, E.; Borderas, F.; Cue, R.I.; Lefebvre, D.; Pellerin, D.; Rushen, J.; Wade, K.M.; de Passillé, A.M. A survey of dairy calf management practices in Canada that affect animal welfare. *J. Dairy Sci.* **2010**, *93*, 1307–1315. [CrossRef] [PubMed]
30. Kehoe, S.I.; Jayarao, B.M.; Heinrichs, A.J. A survey of bovine colostrum composition and colostrum management practices on Pennsylvania dairy farms. *J. Dairy Sci.* **2007**, *90*, 4108–4116. [CrossRef] [PubMed]
31. Nachhaltige Tierhaltung Österreich—NTÖ. Geschäftsbericht Arge Rind. 2020. Available online: https://www.nutztier.at/daten/ (accessed on 26 September 2022).
32. Bundesministerium für Land- und Forstwirtschaft; Regionen und Wasserwirtschaft. Grüner Bericht. 2022. Available online: https://gruenerbericht.at/cm4/jdownload/category/2-gr-bericht-terreich (accessed on 18 May 2023).
33. Stewart, S.; Godden, S.; Bey, R.; Rapnicki, P.; Fetrow, J.; Farnsworth, R.; Scanlon, M.; Arnold, Y.; Clow, L.; Mueller, K.; et al. Preventing Bacterial Contamination and Proliferation During the Harvest, Storage, and Feeding of Fresh Bovine Colostrum. *J. Dairy Sci.* **2005**, *88*, 2571–2578. [CrossRef]
34. Pothmann, H.; Nechanitzky, K.; Sturmlechner, F.; Drillich, M. Consultancy to dairy farmers relating to animal health and herd health management on small- and medium-sized farms. *J. Dairy Sci.* **2014**, *97*, 851–860. [CrossRef]
35. Heuwieser, W.; Iwersen, M.; Gossellin, J.; Drillich, M. Short communication: Survey of fresh cow management practices of dairy cattle on small and large commercial farms. *J. Dairy Sci.* **2010**, *93*, 1065–1068. [CrossRef]
36. Kapoor, M.C. Types of studies and research design. *Indian J. Anaesth.* **2016**, *60*, 626–630. [CrossRef]
37. Wilson, P.W.; D'Agostino, R.B.; Levy, D.; Belanger, A.M.; Silbershatz, H.; Kannel, W.B. Prediction of coronary heart disease using risk factor categories. *Circulation* **1998**, *97*, 1837–1847. [CrossRef] [PubMed]
38. Hesse, A.; Bertulat, S.; Heuwieser, W. Survey of work processes on German dairy farms. *J. Dairy Sci.* **2017**, *100*, 6583–6591. [CrossRef] [PubMed]
39. ZAR—Jahresbericht 2020. 2020-zar-Jahresbericht. Available online: https://www.rinderzucht.at/downloads/jahresberichte.html?file=files/rinderzucht-austria/01-rinderzucht-austria/downloads/jahresberichte-rza/2020-zar-jahresbericht.pdf (accessed on 14 December 2022).
40. Rinderzucht Austria. 2021-rza-Jahresbericht. Available online: https://www.rinderzucht.at/downloads/jahresberichte.html?file=files/rinderzucht-austria/01-rinderzucht-austria/downloads/jahresberichte-rza/2021-rza-jahresbericht.pdf (accessed on 4 March 2023).
41. Mee, J.F. Newborn dairy calf management. *Vet. Clin. N. Am. Food Anim. Pract.* **2008**, *24*, 1–17. [CrossRef] [PubMed]
42. Trotz-Williams, L.A.; Leslie, K.E.; Peregrine, A.S. Passive immunity in Ontario dairy calves and investigation of its association with calf management practices. *J. Dairy Sci.* **2008**, *91*, 3840–3849. [CrossRef]
43. Svensson, C.; Lundborg, K.; Emanuelson, U.; Olsson, S.-O. Morbidity in Swedish dairy calves from birth to 90 days of age and individual calf-level risk factors for infectious diseases. *Prev. Vet. Med.* **2003**, *58*, 179–197. [CrossRef]
44. Klein-Jöbstl, D.; Iwersen, M.; Drillich, M. Farm characteristics and calf management practices on dairy farms with and without diarrhea: A case-control study to investigate risk factors for calf diarrhea. *J. Dairy Sci.* **2014**, *97*, 5110–5119. [CrossRef]

45. Foley, J.A.; Otterby, D.E. Availability, Storage, Treatment, Composition, and Feeding Value of Surplus Colostrum: A Review1,2. *J. Dairy Sci.* **1978**, *61*, 1033–1060. [CrossRef]
46. Allen, N.N. The use of stored colostrum to replace marketable milk for calf feeding. *J. Dairy Sci.* **1944**, *27*, 652–653.
47. Lam, T.J.G.M. Mastitis control: From science to practice. In Proceedings of the International Conference, Hague, The Netherlands, 30 September–2 October 2008; Academic Publishers: Cambridge, MA, USA, 2008. [CrossRef]
48. Hyde, R.M.; Green, M.J.; Hudson, C.; Down, P.M. Quantitative Analysis of Colostrum Bacteriology on British Dairy Farms. *Front. Vet. Sci.* **2020**, *7*, 601227. [CrossRef]
49. Galton, D.M.; Petersson, L.G.; Merrill, W.G.; Bandler, D.K.; Shuster, D.E. Effects of premilking udder preparation on bacterial population, sediment, and iodine residue in milk. *J. Dairy Sci.* **1984**, *67*, 2580–2589. [CrossRef]
50. Barry, J.; Bokkers, E.A.M.; Sayers, R.; Murphy, J.P.; de Boer, I.J.M.; Kennedy, E. Effect of feeding single-dam or pooled colostrum on maternally derived immunity in dairy calves. *J. Dairy Sci.* **2022**, *105*, 560–571. [CrossRef] [PubMed]
51. Godden, S.M.; Lombard, J.E.; Woolums, A.R. Colostrum Management for Dairy Calves. *Vet. Clin. N. Am. Food Anim. Pract.* **2019**, *35*, 535–556. [CrossRef]
52. Maunsell, F.P.; Morin, D.E.; Constable, P.D.; Hurley, W.L.; McCoy, G.C. Use of mammary gland and colostral characteristics for prediction of colostral IgG1 concentration and intramammary infection in Holstein cows. *J. Am. Vet. Med. Assoc.* **1999**, *214*, 1817–1823. [PubMed]
53. Pritchett, L.C.; Gay, C.C.; Hancock, D.D.; Besser, T.E. Evaluation of the hydrometer for testing immunoglobulin G1 concentrations in Holstein colostrum. *J. Dairy Sci.* **1994**, *77*, 1761–1767. [CrossRef]
54. Chigerwe, M.; Hagey, J.V. Refractometer assessment of colostral and serum IgG and milk total solids concentrations in dairy cattle. *BMC Vet. Res.* **2014**, *10*, 178. [CrossRef] [PubMed]
55. Quigley, J.D.; Lago, A.; Chapman, C.; Erickson, P.; Polo, J. Evaluation of the Brix refractometer to estimate immunoglobulin G concentration in bovine colostrum. *J. Dairy Sci.* **2013**, *96*, 1148–1155. [CrossRef] [PubMed]
56. Kritzinger, F. Die Qualitätseinstufung von Kolostrum Mit Einem Einfachen Präzisionstrichter. Ph.D. Thesis, Ludwig Maximilian University of Munich, Munich, Germany, 2017.
57. Besser, T.E.; Gay, C.C.; Pritchett, L. Comparison of three methods of feeding colostrum to dairy calves. *J. Am. Vet. Med. Assoc.* **1991**, *198*, 419–422. [PubMed]

animals

MDPI

Review

A Comprehensive Review of Bovine Colostrum Components and Selected Aspects Regarding Their Impact on Neonatal Calf Physiology

Flávio G. Silva [1,2,3,*], Severiano R. Silva [1], Alfredo M. F. Pereira [2], Joaquim Lima Cerqueira [1,3] and Cristina Conceição [2]

[1] Veterinary and Animal Research Centre (CECAV), Associate Laboratory of Animal and Veterinary Science (AL4AnimalS), University of Trás-os-Montes e Alto Douro, Quinta de Prados, 5000-801 Vila Real, Portugal; cerqueira@esa.ipvc.pt (J.L.C.)

[2] MED—Mediterranean Institute for Agriculture, Environment and Development & CHANGE—Global Change and Sustainability Institute, Department of Zootechnics, School of Science and Technology, University of Évora, Pólo da Mitra Apartado 94, 7006-554 Évora, Portugal; apereira@uevora.pt (A.M.F.P.); cristinaconceicao@uevora.pt (C.C.)

[3] Center for Research and Development in Agrifood Systems and Sustainability, Polytechnic Institute of Viana do Castelo, Agrarian School of Ponte de Lima, Rua D. Mendo Afonso, 147 Refóios do Lima, 4990-706 Ponte de Lima, Portugal

* Correspondence: fsilva@uevora.pt

Simple Summary: Colostrum is the "starter kit" for every calf, containing nutrients, bioactive components, cells, and microorganisms that provide energy and protection against pathogens and contribute to the maturation and development of the gastrointestinal tract. Components such as fatty acids, whey proteins, oligosaccharides, immune cells, and microorganisms are present in colostrum and are essential for the health and well-being of the newborn calf, so its importance should not be underestimated. In this review, the variability of these components, as well as their impact on selected aspects of the newborn calf's physiology, are discussed.

Abstract: Colostrum contains macro- and micronutrients necessary to meet the nutritional and energy requirements of the neonatal calf, bioactive components that intervene in several physiological aspects, and cells and microorganisms that modulate the calf's immune system and gut microbiome. Colostrum is sometimes mistaken as transition milk, which, although more nutritive than whole milk, has a distinct biochemical composition. Furthermore, most research about colostrum quality and colostrum management focuses on the transfer of maternal IgG to the newborn calf. The remaining components of colostrum and transition milk have not received the same attention, despite their importance to the newborn animal. In this narrative review, a large body of literature on the components of bovine colostrum was reviewed. The variability of these components was summarized, emphasizing specific components that warrant deeper exploration. In addition, the effects of each component present in colostrum and transition milk on several key physiological aspects of the newborn calf are discussed.

Keywords: bioactive components; cell; gut maturation; nutrition; microbiome; passive immune transfer; thermoregulation

1. Introduction

Colostrum is the first lacteal secretion of mammals after parturition. It is highly energy-dense and contains many bioactive compounds essential for the newborn's survival and development [1,2]. Colostrum is the 'starter kit' for any mammal, providing molecules for all the biological processes the newborn needs to survive now that it is no longer protected by the womb and nourished by the placenta [3,4]. The importance of colostrum is also species-dependent. Due to its physiological characteristics, the female ruminant's placenta does not allow for the passage of macromolecules to the fetus, which prevents it from acquiring immunoglobulins during gestation, unlike other species such as humans [5,6]. This peculiarity makes it essential for ruminants to consume colostrum immediately after birth, since colostrum has a high concentration in immunoglobulins. In addition to hypogammaglobulinemia, the newborn ruminant also has other physiological constrains, like hypoglycaemia, hypoxia, and hypercapnia, and possible metabolic and respiratory acidosis, the severity of which may depend on the calving difficulty. Fortunately, in addition to immunoglobulins, ruminant colostrum contains macronutrients like proteins, lipids, and carbohydrates, and micronutrients, such as minerals and vitamins, which provide energy and elements for basic cell functions, thermoregulation, and growth [3,7]. Additionally, it contains: (i) many immune factors, antimicrobial peptides, and immune cells that actively protect against specific pathogens and help regulate and "teach" the newborn's immature immune system [8–10]; (ii) endocrine factors that are crucial for the maturation and development of the gastrointestinal tract (GIT) [11]; (iii) signaling molecules that modulate and program gene expression and other cell-related regulatory functions [12,13]; and (iv) microorganisms that help establish the newborn's microbiota [14]. All these components are required because newborn calves undergo major morphophysiological changes in their first days of life as they adapts to extrauterine life [15], and during this time, they cannot initiate a proper immune response against pathogens [16].

Immunoglobulins, particularly IgG, are the most well-studied bioactive compounds, as colostrum quality is usually defined by its concentration in IgG [17–20]. However, calves that were fed maternal colostrum instead of colostrum replacer (which has a defined concentration of IgG) showed improved immunocompetence and higher body gains [21], which demonstrates the importance of the whole colostrum matrix for calf wellbeing. The nutritional value and other bioactive components of BC have received less attention from the academic community. Nevertheless, it has been noticed that colostrum contains a wide variety of properties other than the transfer of IgG, which may be just as important. Growth factors are a good example and perhaps the second most studied component of BC. Immune cells have been studied for a long time, and their importance as protectors and immune regulators is a given fact, but it seems that freezing deactivates these cells, which is very problematic as freezing colostrum for later use is a common practice in dairy farms. Some other specific examples of bioactive components that are now receiving attention are the milk fat globule membrane (MFGM), oligosaccharides, lactoferrin, miRNA and nucleic acids, and the BC microbiome. It is known that the most studied components (mainly IgG) vary with season, maternal nutrition, breed, and age, but the variability of other less studied components are not known. These compounds have different roles in protecting and developing the newborn calf, with possible long-term effects. However, there are still questions to be answered, and further research is needed. Colostrum is a unique feed containing hundreds to thousands of different bioactive components [2], which means that there is still so much to learn about colostrum, and with new properties being discovered all the time and with the increasing need for production efficiency, reduced use of antimicrobials, and increased demands for animal welfare, research into colostrum is necessary.

Therefore, in this review, we aimed to provide a comprehensive list of the most studied nutritional and bioactive components in BC. We summarized the range of concentrations for most of the components discussed, distinguishing them from transition and whole milk, and the source of variation of these components was discussed. A characterization of the cellular and microbial populations in colostrum and their interactions with the newborn calf is presented. The role of colostrum components, cells, and microorganisms on selected aspects of neonatal calf physiology was discussed in each section. We also aim to identify compounds that warrant further research, either because they are present in higher concentrations in colostrum than in milk or because there is currently limited information available.

2. Bovine Colostrum, Transition Milk, and Inter-Species Variability

Colostrum is a lacteal secretion secreted by mammary epithelial cells prior to parturition, which is physically and chemically distinct from mature milk [22,23]. In ruminants, transition milk is secreted after birth during the first few days of lactation and gradually changes to mature milk in both its physical and chemical proprieties [22]. The definition of the number of days during which transition milk is produced varies between studies, mainly because the transition from colostrum to mature milk varies between cows and the number of milkings and may also be dependent on the starting point, i.e., transitional milk from a cow that produced colostrum with a higher solids content may take more days to become mature milk, compared to a cow that produces colostrum with lower solids. However, it is important to note that lactogenesis II (copious milk production) begins immediately after parturition [23], which means that transition milk secretion theoretically also starts immediately after birth, so the time to first milking may determine whether the fluid milked is more similar to colostrum or transition milk, which may explain some of the variations seen in some colostrum constituents between studies. Nevertheless, in the literature, transition milk can be considered as the lacteal secretion produced during the first and second days after parturition [24], from the second to the fourth [25,26], from the second to the fifth [27], or from the second to the sixth milking [28]. Alternatives to BC are colostrum replacers and colostrum supplements. These products contain around 20% and 50% of fat and protein, respectively, and are highly concentrated in immunoglobulins derived from plasma, serum, or dried colostrum, available in doses containing a concentration of 60 to 200 but usually between 100 and 130 g of IgG, depending as well on if they are used as a replacer or a supplement [29]. However, colostrum replacers are lacking in some bioactive components [30]. Therefore, if good quality colostrum is available, it should not be replaced by colostrum replacers, but it can be provided or added to maternal colostrum if this is not the case or if cows are positive for transmissible diseases such as paratuberculosis (John's disease). Additionally, it can possibly be used as a nutraceutical, as it has been shown to reduce the likelihood of bovine respiratory disease outbreaks in calves aged 14 to 50 days [31].

BC is the most studied prepartum lacteal secretion in domestic animals, while Human colostrum is perhaps the most studied. Colostrum from the human species, like many other mammals, is characterized by having a higher concentration in IgA, while colostrum from ungulates is more concentrated in IgG. The mechanisms that lead to the transfer of components to colostrum are species-specific [2]. Nevertheless, in all mammal species, colostrum is important for systemic and enteric immunization, especially in species where placental passive immunization does not occur, as in the case of ungulates. Within ruminant farm species, macronutrients and IgG concentrations vary considerably. The concentration of fat (4.0–8.0%), protein (10.5–17.1%) and lactose (3.2–4.4%) in goat colostrum has been shown to vary between breeds [32]. According to these results, the variation in fat and protein seems to be similar between goats and dairy cows, but lactose may be higher in goat colostrum (macronutrient concentrations in BC are discussed in the following chapter). However, studies with Majorera goats showed different results, with colostrum higher in fat, lower in protein, and similar in lactose concentration to that of dairy cows [33–35]. Ewe colostrum from different breeds was higher in fat (4.0–13.7%) and protein (13.8–22.5%) concentrations and similar in lactose (2.4–3.8%) concentration to dairy cow colostrum [32,36]. Colostrum from Santa Inês sheep had concentrations of fat and lactose normally higher than colostrum from dairy cows but lower in protein concentration [37]. There is a marked variation in IgG concentration in colostrum [2], and the average values from these three species are similar but they fall in different ranges across studies [32,34]. With regard to other components, there is not yet enough data to make a proper comparison, but there is evidence of differences in minor components [34]. Therefore, there is no specific rule for colostrum components between these three species, although some tendencies can be assumed. However, it is reasonable to assume that sheep colostrum has the highest solids content compared to cow and goat colostrum [34,38] and that colostrum components may vary between breeds [32].

Moreover, the importance of colostrum, mainly BC, for farm animals is gaining popularity as a food ingredient [39] and as a nutraceutical for humans [40,41]. It is also being explored for potential applications in anti-tumor therapeutics for specific cancers [42]. Due to its nutritional value and low availability, BC has a high market value compared to other by-products of the sector [43], and its demand is projected to increase [44]. This heightened market value accentuates the significance of conducting research to enhance both the availability and quality of bovine colostrum (BC), both on the farm and throughout the processing stages.

3. Macro- and Micronutrients in Bovine Colostrum

Colostrum contains water, proteins, fats, carbohydrates (macronutrients), minerals and vitamins (micronutrients), as well as bioactive components such as immunoglobulins, hormones, enzyme inhibitors, growth factors, and nucleic acids, albeit in smaller quantities. Some of these components can be described as having both nutritional and bioactive roles; however, despite their nutritional value, the importance of specific molecules goes beyond straightforward caloric content, which is why they will be considered as bioactive molecules in this review.

All nutrient and non-nutrient factors except water make up the total solids (TS) of colostrum, which is officially measured using the air-oven loss-on-drying technique (Method 925.23 of the AOAC). TS in colostrum reflect its overall quality and can be measured on the farm level with a with a Brix % refractometer, which also correlates well with protein concentration [45,46]. Another instrument is the colostrometer, which measures specific gravity, relating to total protein and immunoglobulin content in colostrum [47]. Dry matter, protein, fat, and lactose analysis have official methods. Still, in most recent studies spanning approximately 25 years, these components have been analyzed using an automated mid-infrared method, employing Milkoscan equipment. Several peripartum dam-related factors significantly influence the yield of nutrient and non-nutrient components in colostrum. However, more knowledge is still needed to understand how these factors affect the production and transfer of these components other than IgG [48]. Factors such as farm size, prepartum feed, breed, parity, season, calving month, udder health, time interval from calving to milking, colostrum weight, and method of analysis can contribute to IgG variation [49–52]. Some of these factors also affect the nutrient component of colostrum, but there are still very few studies regarding this subject [53]. Table 1 and Supplementary Figure S1 show the variability of the main components in BC. The coefficient of variation was calculated to highlight the variability of each component across the literature. The authors tried to provide a broad range of studies across different regions, but the majority are concentrated in the subcontinent of North America. Specific information on each component is provided in the respective subchapters.

Table 1. Variability of the main components of bovine colostrum in relation to breed, region, and method of analysis.

Component	Mean	S.D.	Min	Max	CV	*n*	Breed	Region	Method of Analysis	Reference
Total Solids (%)	23.9	3.41			14.3	10	H	Kansas, USA	AOAC	[54]
	24.2					36	H	Cairo, Egypt	AOAC, 1999	[55]
	27.2	5.8	12.9	47.2	21.3	365	H	Isfahan, Iran	Milkoscan	[52]
	27.6	5.84	18.3	43.3	21.2	55	H	Pennsylvania, USA	AOAC, 1975	[56]
	25.8	4.68			18.1	1074	H	Northern Greece	Brix	[53]

Table 1. *Cont.*

Component	Mean	S.D.	Min	Max	CV	*n*	Breed	Region	Method of Analysis	Reference
Total Solids (%)	24.7	0.51			2.1	72	H	Central Denmark	Milkoscan	[57]
	26.3		13.5	37		288	HF	Ontario, Canada and Minnesota, USA	Brix	[58]
	25.3	4.9	11.0	42.0	19.4	709	HF	Évora, Portugal	Brix	[59]
	24.8	0.26	18	31	1.0	73	F	Pisa, Italy	Brix	[60]
	22.6	4.7	1.7	33.1	20.8	496	H + J	USA	Milkoscan	[61]
	22.5	6.74			30.0	16	J	Kansas, USA	AOAC	[54]
	27.7	8.76			31.6	99	J	Northern India	Gravimetric	[62]
	23.6	5.56	12.8	36.6	23.6	86	J	USA	Dried overnight	[18]
	23.4	0.74			3.2	32	J	Central Denmark	Milkoscan	[57]
	21.2	4.43	10.5	28.6	20.9	58	J	Iowa, Canada	Brix	[63]
	24.3		10.4	52.6		569		Alberta, Canada	Brix	[45]
Protein (%)	14.0	2.59			18.5	8	H	Kansas, USA	Subtration	[54]
	16.6					6	H	Tuscia, Italy (TN)	Milkoscan 104	[64]
	13.9					6	H	Tuscia, Italy (HT)	Milkoscan 104	[64]
	13.5					36	H	Cairo, Egypt	AOAC, 1999	[55]
	14.9	3.32	7.10	22.6	22.3	55	H	US-PA	Kjeldahl	[56]
	18.5	4.9	4.90	29.6	26.5	365	H	Isfahan, Iran	Milkoscan	[52]
	17.8	3.97			22.3	1074	H	Northern Greece	Milkoscan	[53]
	14.7	3.51	4.55	25.22	23.9	559	H	Northen Italy	Kjeldahl	[65]
	15.4	0.42			2.7	72	H	Central Denmark	Milkoscan	[57]
	14.0	3.67			26.2	1226	H + F *	Northen Ireland	Milkoscan	[66]
	12.6	2.9	3.34	17.12	23.0	76	HF	Swiszterland	Milkoscan	[67]
	18.2	3.94	11.28	24.6	21.6	21	HF	Germany	Milkoscan	[67]
	16.1	1.64			10.2	8	F	Reading, UK	DairyLab II (NIR)	[68]
	12.7	3.3	2.60	20.5	26.0	542	H + J	USA	Milkoscan	[61]
	13.1	4.08			31.1	99	J	Northern India	Kjeldahl	[62]
	23.6	5.07	9.16	31.63	21.5	88	J	USA	Kjeldahl	[18]
	14.2	5.26			37.0	11	J	Kansas, USA	Subtration	[54]
	14.6	0.62			4.2	32	J	Central Denmark	Milkoscan	[57]

Table 1. *Cont.*

Component	Mean	S.D.	Min	Max	CV	*n*	Breed	Region	Method of Analysis	Reference
Fat (%)	6.7	2.65			39.6	29	H	Kansas, USA	Babcock	[54]
	6.0					6	H	Tuscia, Italy (TN)	Milkoscan 104	[64]
	5.9					6	H	Tuscia, Italy (HT)	Milkoscan 104	[64]
	8.0					36	H	Cairo, Egypt	Gerber	[55]
	6.7	4.16	2.0	26.5	62.1	54	H	US-PA	Babcock	[56]
	4.6	3.4	0.3	20.9	73.9	365	H	Isfahan, Iran	Milkoscan	[52]
	6.4	3.3			51.6	1074	H	Northern Greece	Milkoscan	[53]
	5.2	0.34			6.5	72	H	Central Denmark	Milkoscan	[57]
	4.6	3.04	0.12	14.95	66.1	557	H	Northen Italy	VDLUFA, 2013	[65]
	6.4	3.32			51.9	1226	H + F *	Northen Ireland	Milkoscan	[66]
	4.4	1.75	2.16	8.78	39.8	21	HF	Germany	Milkoscan	[67]
	5.5	2.8	1.32	14.21	50.9	76	HF	Swiszterland	Milkoscan	[67]
	3.55	1.82			51.3	8	F	Reading, UK	DairyLab II (NIR)	[68]
	5.6	3.2	1.0	21.1	57.1	531	H + J	USA	Milkoscan	[61]
	3.3		0.1	8.7			J	USA	Infrared	[18]
	3.4	0.51			15.0	32	J	Central Denmark	Milkoscan	[57]
	4.2	1.81			43.1	32	J	Kansas, USA	Babcock	[54]
	8.0	7.96			99.5	99	J	Northern India	Gerber	[62]
Lactose (%)	2.7	0.91			33.7	8	H	Kansas, USA	AOAC 1945	[54]
	3.2					6	H	Tuscia, Italy (TN)	Milkoscan 104	[64]
	2.6					6	H	Tuscia, Italy (HT)	Milkoscan 104	[64]
	2.5	0.65	1.2	5.2	26.0	55	H	US-PA	Colorimetric	[56]
	2.0	0.9	0.3	5.2	45.0	365	H	Isfahan, Iran	Milkoscan	[52]
	3.68	0.04			1.1		H	Central Denmark	Milkoscan	[57]
	2.15	0.73			34.0	1074	H	Northern Greece	Milkoscan	[53]
	2.36	0,51	0.74	4.06	21.6	577	H	Northen Italy	HPLC	[65]
	2.7	0.55			20.4	1226	H + F *	Northen Ireland	Milkoscan	[66]
	3.2	0.53	1.94	4.6	16.6	76	HF	Swiszterland	Milkoscan	[67]
	2.9	0.59	1.82	3.81	20.3	21	HF	Germany	Milkoscan	[67]
	2.7	0.46			17.0	8	F	Reading, UK	DairyLab II (NIR)	[68]
	2.9	0.5	1.2	4.6	17.2	538	H + J	USA	Milkoscan	[61]

Table 1. *Cont.*

Component	Mean	S.D.	Min	Max	CV	*n*	Breed	Region	Method of Analysis	Reference
Lactose (%)	3.73	0.06			1.6	32	J	Central Denmark	Milkoscan	[57]
	2.4	0.77			32.1	11	J	Kansas, USA	AOAC 1945	[54]
	3.0	0.20			6.7	99	J	Northern India	Lane-Eynon	[62]
Minerals (%)	1.11	0.16			14.4	10	H	Kansas, USA	Evaporation	[54]
	0.05	0.01	0.02	0.07	20.0	55	H	US-PA	AOAC, 1975	[56]
	1.9	0.17			8.9	8	F	Reading, UK	DairyLab II (NIR)	[68]
	1.02	0.40			39.2	99	J	Northern India	Incineration	[62]
	1.22	0.14			11.5	16	J	Kansas, USA	Evaporation	[54]

We only considered values from first milking after birth. When not provided, SD (standard deviation) was calculated from SE (standard error) with the following formula: SD = SE × Sqrt(n). CV—Coefficient of variation was calculated based on the mean and SD reported in each reference. Total solids, dry matter, and brix were considered the same component. Missing values were not reported. TS, Total Solids; H, Holstein breed; F, Friesian breed; HF, Holstein–Friesian breed; J, Jersey breed. * Mostly Holstein and Friesian. H, F, and HF differentiation: The Holstein–Friesian breed started as Friesian in Northern Europe in the second half of the XVIII century; after the XIX century, Friesian cows were exported to Canada and the United States of America, where they were renamed Holstein–Friesian and Holstein, respectively. Different genetic traits were selected in the continents of Europe and America; in Europe, animals were selected for mixed traits (milk and meat production), while in America, animals were selected purely for milk capacity [69]. All three names, "Friesian", "Holstein", and "Holstein–Friesian" can be found in the literature. In the analysis presented in this table, Friesian cows were only used in European countries, while the others were used in different continents. The identification has been done according to the information provided by the authors in the methodology section of the paper.

3.1. Macronutrients

3.1.1. Total Solids

During the first few weeks of a calf's life, colostrum and milk are the primary sources of water. Voluntary intake of free water is low during this period and gradually increases with age, weight, and dry matter intake [70,71]. The quality of colostrum is determined by the proportion of its non-water constituents, commonly referred to as TS. Most review articles on colostrum composition report the same mean value of 23.9% for the TS content of BC [4,22,72], all of which (indirectly) trace back to a paper published in 1950 with Holstein cows [54]. However, more recent research should be considered when considering an overall value for BC. More recent papers also report a percentage of total solids similar to Parrish et al. [54]; for example, a study with Jersey cows in the United States (n = 96; 23.6%) [18], another with dairy cows from Switzerland (n = 24; 24%) [73], and another with Egyptian Holstein cows (n = 12; 24.2%) [55]. A more extensive study (n = 496) showed a higher variation, including differences between different regions of the United States [61]. The mean values ranged from 20.6 to 24.1%, the minimum reported value was as low as 1.7%, and the maximum reached 33.1%. In a study in Pennsylvania (n = 55), the reported mean TS (27.6 ± 5.8%) was higher than in the aforementioned studies [56]. Two other studies reported similar values in Iran with 365 Holstein cows [52] and in India with Jersey cows [62]. Jersey cows are known for their ability to produce milk with a higher concentration of solids than Holstein cows [74]. However, Morrill et al. [61] found no differences in the total solids of colostrum between Holstein (n = 494) and Jersey cows (n = 87). In contrast, a study in Brazil with Jersey cows reported higher values of TS when compared to other studies (31.3%) [75]. Local dairy breeds may have lower TS than Jersey and Holstein cows [62,76], but there are very few studies evaluating the colostrum composition of local dairy breeds to support this statement. There is still uncertainty regarding whether the reduced solids content in indigenous breeds has a similar impact on their offspring as observed in calves of Holstein or Jersey breeds. Further exploration is necessary to consider other factors, such as calf size, which directly affects their dietary needs, and any regional discrepancies across countries. Another interesting factor is related to the number of parities. In some studies, colostrum from first and second lactation cows was of lower quality [19,77], which was somewhat expected, as older cows tend to have more antibodies than younger cows. However, other studies have shown that colostrum from first lactation cows did not differ from that of third lactation cows and that colostrum from second lactation cows had lower TS [53,59] and IgG [18,49]. No explanation has yet been found for these results.

These studies indicate a degree of homogeneity in the mean TS concentration of colostrum within the herd, with some possible variations between regions and breeds, as seen in Table 1. From these studies, a mean concentration of total solids of 24.7% can be inferred. This is slightly higher than the previously reported value (23.9%) and above the limit for good quality colostrum (22%), which should estimate an IgG concentration of 50 mg/mL. However, the minimum values are represented by colostrum of lower quality (<22%), so it is still important to regularly assess colostrum quality [4]. This variation can compromise the calves' nutritional requirements and passive immune transfer if colostrum quality is not assessed prior to feeding or if the calf suckles directly from the dam [78].

3.1.2. Proteins

Protein is the most abundant nutrient in colostrum and can be four times higher than in milk [72]. In BC, it is found at a concentration of 15.5%, ranging from 2.6 to 31.6% (Table 1).

Breed can affect the protein percentage in colostrum [62], and dairy cows with lower milk yield tend to have higher milk solids, yet Morrill et al. [61] found no differences in colostrum protein content between Holstein and Jersey cows. The protein percentage of BC was lower in first and second parity cows than in cows with three or more parities [66]. Zarei et al. [52] and Soufleri et al. [53] found significantly higher values only in cows with four or more parities. At the same time, Morril et al. [61] found no differences in the protein percentage between parities. It would be interesting to investigate the variation in protein fractions in the colostrum from cows with different parities, as the lower protein concentrations in younger cows could be attributed to lower immunoglobulin concentration; however, as previously mentioned, this may not always be the case, at least for cows in first, second, and third parity. A higher protein percentage in colostrum was recorded during autumn and winter in Northern Greece [53] and during winter in Northern Ireland [66]. In contrast, a study conducted in Egypt found no differences in colostrum protein content between seasons [52]. Nevertheless, it was observed that protein percentage was lower under heat stress conditions compared to thermoneutral conditions [64]. To better understand these differences, controlled experiments with consistent climatic conditions could help clarify whether the observed differences are due to environmental factors, such as temperature and photoperiod, or to confounding effects. There appears to be no marked effect of dry period length on protein percentage in BC [53], but a higher protein percentage was observed in colostrum from cows with a very long dry period ($\geq$16 weeks) [66]. Cows receiving bovine rhinotracheitis vaccinations produced 11 g/L more protein in colostrum than unvaccinated cows, probably due to the increase in immunoglobulins, since it was significantly higher in vaccinated compared to unvaccinated cows [66]. Colostrum sampled from cows vaccinated against neonatal calf diarrhea also showed increased Brix values [79].

Colostrum has a higher concentration of caseins (insoluble fraction) and whey proteins (soluble fraction) than milk [22]. Casein is present in the form of colloidal casein micelles (CMs); these CMs are larger in colostrum than in milk [80]. CM size is important in milk processing, but the role it plays for the newborn calf is unknown. Compared to milk, BC has a higher percentage of κ-casein, a lower percentage of α_s-casein, and a similar percentage of β-casein [81]. Casein is involved in the transport of minerals and trace elements [72,82], and it can also reduce proteolytic degradation, acting as an enzyme-inhibiting protein, preserving intestinal integrity and function (protecting epidermal growth factor from digestion), and aiding in the activity of other biologically active peptides [83,84].

The whey fraction, in addition to its nutritional value, is composed of a large number of different proteins that play an important role in several biological processes [85], of which the most studied in milk are: α-lactalbumin, β-lactoglobulin, bovine serum albumin (BSA), immunoglobulins (IgG, IgM, IgA, IgE, and IgD), bovine lactoferrin, and lactoperoxidase [86]. According to our research, their concentration in bovine colostrum varies widely (Table 2 and Supplementary Figures S2 and S3). Some proteins such as casein, α-lactalbumin, and β-lactoglobulin are synthesized in the mammary gland, while others such as immunoglobulins, BSA, and lactoferrin are actively transported from the blood into the lumen of the mammary gland [17]. The importance of these proteins for the human species is well-established (Madureira et al., 2007), but their importance for newborn ruminants is not as well known, except for immunoglobulins. Colostrum proteins are a source of amino acids that the newborn calf can use for protein synthesis. However, the newborn gastrointestinal tract (GIT) exhibits low proteolytic activity, and colostrum, despite containing trypsin inhibitors [87], still allows for some exogenous protein hydrolysis alongside intracellular protein degradation, which contributes to amino acid supply [88]. Whilst some proteins, such as α-lactalbumin, can be rapidly hydrolyzed to amino acids in the abomasum, caseins form a clot, which undergoes slower degradation, thus providing a more spaced source of amino acids [3,88]. Curd formation retains caseins and lipids in the abomasum and slowly releases the whey fraction into the gut, where it can be effectively absorbed. This may explain why calves without curd formation had lower serum IgG and IgA levels [89].

Immunoglobulins and β-lactoglobulin are proteins that are highly resistant to degradation [88]. Compared to milk, colostrum has a higher β-lactoglobulin/α-lactalbumin ratio, suggesting that β-lactoglobulin may play a specific role for the newborn [90]. B-lactoglobulin is involved in the digestion of milk lipids by improving the activity of pregastric lipases [91], enhancing the digestion of colostrum fat, promptly providing energy to the newborn. β-lactoglobulin also functions as a transporter of small hydrophobic ligands, such as retinol, cholesterol, and vitamin D, to specific intestinal receptors [86]. α-Lactalbumin is present in the bovine colostrum and milk at lower concentrations than in human colostrum and milk, playing a role in the synthesis of lactose [92]. BSA helps maintain oncotic pressure and is involved in the binding and transporting of molecules such as fatty acids, bilirubin, hormones, and minerals [93]. It is transferred from blood to colostrum and is usually considered an indicator of the permeability of the blood–milk barrier [19,94,95]. It has been reported that a higher proportion of BSA in colostrum can affect the transfer of immunoglobulins from the intestinal lumen of the calf to the circulatory system due to the limited capacity of the macromolecular transport mechanism [96].

Table 2. Concentrations of the major proteins in bovine colostrum.

Protein	Concentration	Reference
Caseins (αs, β, κ) (mg/mL)	48–96	[22,82,97–99]
α-lactalbumin (mg/mL)	2.0–7.5	[94,98–100]
β-lactoglobulin (mg/mL)	4.8–24	[94,98–100]
Bovine serum albumin (mg/mL)	0.45–2.5	[19,94,101]
Immunoglobulins (IgG, IgA, IgM) (mg/mL)	47–106	[19,56,102]
Lactoferrin (μg/mL)	34–4400	[56,97,99,100,102–105]
Transferrin (μg/mL)	187–1070	[103–106]
Lysozyme (μg/mL)	0.4–1262	[100,102,107]
Lactoperoxidase (μg/mL)	22.8–22.8	[102]

Values represent the range of mean concentrations.

3.1.3. Lipids

Lipids are the second most important macronutrients of colostrum after proteins and are the main source of energy that the calf receives at birth. It is also the component with the highest variability, ranging from almost 0 to around 26%. The average fat percentage in BC is around 5.5% (Table 1). Breed can affect the fat percentage in colostrum [62], and dairy cows with lower milk production capacity tend to have higher milk solids, but Morrill et al. [61] found no differences in colostrum fat content between Holstein and Jersey cows. The fat percentage of BC is higher in primiparous cows than in multiparous cows [52,53,61,66]. This can be explained by the generally lower yields of primiparous animals; however, protein concentration is generally lower in primiparous cows. Season may have a significant effect on fat percentage, but this is likely to be region-dependent. In Northern Ireland [66] and Northern Greece [53], colostrum fat percentage was higher in spring. In Egypt, there were no differences between seasons [52]. Nevertheless, it was observed that fat percentage was lower under heat stress conditions compared to thermoneutral conditions [64], which is usually associated with lower dry matter intake. Without climatic information, comparing seasons across different studies and regions could be misleading, and other factors that are probably indirectly related, like feed availability, could be influencing these results. There is not much information on the effect of dry period length on fat percentage in BC, but it seems that a dry period length of more than 60 days may result in colostrum with a higher fat content [53,66], which could be explained by increased mobilization of the body's energy reserves associated with increased body condition during a prolonged dry period [53]. In addition, a significant ($p = 0.001$) positive association was observed between fat percentage in BC and feeding straw and grass silage 7 to 9 weeks before calving and vaccination against leptospirosis [66]. The impact of diet on colostrum fat concentration needs to be better understood, especially as the onset of colostrogenesis is not well understood. Thus, dietary differences between the dry period and the peripartum period may influence colostrum composition. Nevertheless, feeding forages increase ruminal acetic–acid formation, which influence the synthesis of lipids in colostrum and milk [108].

The main classes in the lipid fraction are triglycerides or triacylglycerols (TAG; the main class within the lipid fraction), phospholipids (the two main classes are glycerophospholipids and sphingolipids), free fatty acids (FA; short- and long-chain, saturated and unsaturated), eicosanoids, and sterols (the most abundant of which is cholesterol) [7,109]. Lipids are an energy source, serve as structural components of membranes, are precursors for other molecules, and act as actuators in various biological processes [110]. The lipids present in colostrum and milk are almost entirely in the form of milk fat globules (MFGs), a lipid droplet containing mostly TAG that is formed in the endoplasmic reticulum of the alveolar epithelial cells of the mammary gland [111]. MFGs are then secreted by fusion with the plasma membrane of the alveolar epithelial cell, acquiring a peripheral bilayer called the milk fat globule membrane (MFGM), which contains polar lipids, cholesterol, glycoproteins, gangliosides, and enzymes in its structure [112,113]. The functions and mechanisms of action of MFG and MFGM are not fully understood, but they may contribute to antimicrobial effects, gut maturation, structural development, and the establishment of early neonatal gut microbiota [112,114].

Lipids also act as a source of energy, with FA oxidation providing energy to sustain gluconeogenesis [3]. This source of energy is particularly important in neonates, as lactose intake from colostrum is insufficient to meet glucose requirements [30]. In fact, a high postanal capacity to oxidize FA has been reported in several species [115]. FA oxidation capacity increases during the first 24 h after birth, especially medium-chain FA [15]. Most of the energy comes from TAG, since they are the major constituents in milk fat, as around 95–98% of the bovine milk fat are TAG [116]. Gastric and pancreatic lipases hydrolyze TAG to FA and monoglycerides, which are absorbed in the intestine. These lipids then undergo several processes of re-esterification and hydrolysis until they reach the endothelial cells, where they can serve as a substrate for bioactive molecules, be oxidized to provide immediate energy, or be resynthesized into TAG, where they are transported to extrahepatic tissues such as muscle and adipose tissue for storage and later use as a source of FA and energy [110,115,117].

The energy provided with FA oxidation can also be used as a heat source, as newborns are born poorly insulated and have low metabolic rates [3,4,48]. Rectal temperature drops by an average of 0.5 °C in the first 90 min, mainly due to the wet hair coat, which has a low thermal insulation, and due to the evaporation of amniotic fluid, which removes heat from the surface of the skin [118]. Other factors like wind velocity can also increase the rate of heat loss. Thus, heat production in the first hours of life is not enough to meet heat losses at 20 °C [118]. The newborn calf is born hypoglycaemic and has very limited energy reserves, as only 3% or less of its body weight is made up of lipids, most of which are structural [61]. It is speculated that the endogenous lipid content of a 40 kg newborn calf could support summit metabolism for 15 h and that glycogen reserves would be depleted in $\leq$3 h [119]. In cold environments where hypothermia may occur or in the event of dystocia, the situation may be exacerbated [118]. Nevertheless, the lower critical temperature for a newborn calf is close to 22 °C. Rectal temperature may rise around 15 h after birth due to more efficient peripheral vasoconstriction and the increased insulating capacity of the now dried hair coat [118], although it would depend on the environmental climatic conditions and housing characteristics. Approximately 1.5% of calf BW is brown adipose tissue (BAT), mainly located in the perineal, prescapular, pericardial, and abdominal regions [120]. BAT contributes to non-shivering thermogenesis and is mediated by the cold-induced secretion of norepinephrine by the sympathetic nervous system, which stimulates the expression of the UCP1 protein. UCP1 uncouples electron transfer from ATP synthesis in the mitochondria of brown adipocytes, so the energy derived from FA oxidation is dissipated as heat instead [121]. This strategy allows hypothermia to be avoided until nursing occurs. However, some gestational factors can affect the amount of BAT that the newborn has [121]. In addition, BAT is gradually replaced by white adipose tissue during the first few days of life [120]. Therefore, the high caloric content of colostrum makes it a good source of energy for heat production [122]. The energy content of BC can vary substantially, considering the variability in the nutrient composition of colostrum (Table 1). Reported average values for the metabolizable energy (ME) of BC range from 24 to 25.6 MJ/kg DM, however, when considering extreme values, can range from 14.2 to 34.8 MJ/kg DM [53,56,123,124]. As expected, it is higher than the ME of whole milk, which is around 18.1 to 22.5 MJ/kg DM [123–125]; however, colostrum on the lower end can have lower energy than whole milk. According to the NRC, a calf with a body weight of 40 kg has a maintenance ME requirement of 6.7 MJ/day [123]. Considering an average value of 24.8 MJ/kg DM in colostrum and the energy requirements of a 40 kg calf (6.7 MJ/day), this calf would need 0.27 kg of DM as colostrum per day. Considering a 24.7% of DM in colostrum, this calf would need to consume a total of 1.1 L of colostrum to meet its ME requirements. This value is way below the current recommendations of 10% of the calf's body weight, which in this case would be 4 L of colostrum (26.8 MJ of ME). This recommendation is based on immunological requirements (mostly IgG) and not on energy requirements. The rest of the energy would then be used for thermogenesis (heat production) and growth (protein turnover), both of which are demanding activities in early life. Therefore, the energy provided by colostrum is essential soon after birth to maintain normal functions, including normothermia.

3.1.4. Carbohydrates

Colostrum contains carbohydrates in the form of lactose, oligosaccharides, glycoproteins, glycolipids, and nucleotide sugars. Lactose is the major carbohydrate in colostrum, with an average concentration of 2.8%, ranging from less than 1% to 5.2% (Table 1). Nazir et al. [62] observed that local dairy cows had lower lactose in colostrum than Jersey cows, and Zarcula et al. [126] observed lower lactose levels in Romanian Black and White cows compared to Holstein–Friesian cows. Morrill et al. [61] found no differences in colostrum lactose content between Holstein and Jersey cows. The effect of parity on lactose content in colostrum is not marked when compared to protein or fat. However, there is a tendency to decrease when the number of parities increases [52,53,66], though this number varies between studies. In a study in Northern Ireland [66] and in Egypt [52], lactose percentage was higher in autumn but lower in a study in Northern Greece [53].

Nevertheless, lactose percentage was observed to be lower under heat stress conditions compared to thermoneutral conditions [64]. As colostrum yield increased, lactose percentage also increased, but protein percentage decreased [53]. Lower lactose percentage was also observed in colostrum from cows with a very long dry period (≥16 weeks) but not in cows with a shorter dry period [66]. The effect of dry period length on colostrum composition is not entirely understood. The majority of studies focus on IgG levels and in shorter or no dry periods at all, but even these results are not consistent [127–129]. It would be interesting to understand the effect of longer dry periods on colostrum composition, despite the economic constraints associated with longer periods.

The concentration of lactose is lower in colostrum than in milk [7,54], and as it has osmoregulatory functions, a high concentration of lactose would increase the movement of water from the cytoplasm of the mammary epithelium into the secretory vesicles and subsequently into the colostrum [7]. This would reduce its overall quality, since it would have fewer components per litter. Therefore, lactose concentration in colostrum increases with the time until colostrum collection, but the concentration of other components decreases [53,66]. This can explain why older cows tend to have lower lactose levels in colostrum than younger cows, since TS and protein levels usually increase with the parity number. The low levels of lactose in BC make lipids the main source of energy available to the newborn calf in its first feed. However, lactose in colostrum is necessary to increase the water content of colostrum during the final phase of prepartum, otherwise the colostrum would be too dense for the calf to suckle [2].

Lactose is a disaccharide that, when hydrolyzed, provides glucose and galactose as energy sources [130]. Carbohydrates, mainly glucose and lactate, are the fetus' main form of constant energy supply, which changes after birth to a high-fat inconstant energy supply [15]. Calves are born hypoglycaemic, and lactose intake from colostrum does not fulfil the calf's glucose requirements [30]. Therefore, calves must rapidly establish endogenous glucose production (glycogenolysis and gluconeogenesis processes) [3,30]. During the first hours of life, the calf can use the hepatic glycogen stored during the last phase of gestation to maintain normal blood glucose levels [30]. It has also been shown that colostrum feeding improves glucose absorption [131]. However, prolonged absence of feeding can lead to hypoglycaemia when there is no exogenous glucose intake, the glycogen stores are depleted, and gluconeogenesis still needs maturation [30,131,132]. Lactate (and to a limited extent, glycerol) is the main substrate for gluconeogenesis in the early stages of life, and as the rumen develops, propionate becomes the main substrate as the production of volatile fatty acids increases [15].

3.2. Micronutrients

3.2.1. Minerals

There is a large variation in the total mineral concentration of BC, which can vary with the time interval between parturition and milking [56,133,134], the milking strategy [26], and the number of milkings [134,135]. Unlike most colostrum constituents, mineral content does not seem to be negatively associated with colostrum yield [135]; however, further research is needed in this area, as well as with other variation factors. The most abundant minerals in colostrum are Ca, P, K, Na, and Mg, while Zn, Fe, Cu, and Mn are present in lesser amounts. Colostrum has a high concentration of Ca, P, Na, Mg, Fe, Se, Cu, and Zn compared to milk but a lower concentration of K and Mn [72,133,135,136]. Some minerals, such as Ca, P, and Mg, seem to be in higher concentration in colostrum from first or second parity cows compared to cows with three or more parities [134,135].

Each mineral has different physiological functions in the organism, and therefore, adequate levels of macro- and microminerals are essential for neonatal health and growth. However, the role of colostrum minerals in newborn calves is not fully understood, and there is some disagreement as to whether maternal mineral supplementation improves colostrum quality [137]. In a recent study, both cows and calves showed increased levels of blood IgA, IgM, and total antioxidant capacity, following prepartum supplementation with

Mn, Zn, and Cu, but results were dependent on the source of supplementation [138]; protein, fat, and lactose in colostrum did not change with treatment, and immunoglobulins were not measured in colostrum. Ca is essential for skeletal development, and absorption rates can reach 99% in young calves (10 days), decreasing with age to 22% in adult cattle [139]. Mineral supplementation over dietary requirements did not improve colostrum mineral concentrations in Brahman cattle [133] or sows [140], except for selenium. The efficiency of absorption is likely to decrease when dietary minerals, such as Ca, are supplemented above requirements [133,135], which also depends on the source of the mineral, as in the case of selenium [141]. More recently, it has been shown that pre-natal mineral and vitamin supplementation can alter the newborn calf microbiome at different sites of the body, affecting early microbial colonization [142]. Research into the mineral composition of colostrum is limited, and therefore, very few conclusions can be drawn.

3.2.2. Vitamins

Vitamins can also be included as bioactive components, but for the purposes of this review, they will be referred to as micronutrients. Colostrum is an important source of vitamins essential for the health and growth of the newborn calf, and delayed colostrum intake can impair cell growth and differentiation and increase susceptibility to infectious diseases [4,73]. Fat-soluble vitamins (vitamins A, E, and D) are present in higher concentrations in colostrum than in milk, but water-soluble vitamins (vitamins C and B complex) do not follow the same trend [7,22,56,143–148]. Compared to milk, colostrum has only slightly higher concentrations of vitamin C, and within the B complex, vitamins B1, B2, B6, B9, and B12 may be more concentrated in colostrum, B5 and B7 are usually less concentrated, and B3 seems to be as concentrated as in milk. [7,72]. Nevertheless, vitamin C is the most concentrated vitamin in BC. Table 3 shows the concentrations of vitamins in BC, as well as their physiological roles in cattle. There is a lack of recent research on the vitamin content of BC. As genetics and feeding strategies have changed in recent decades, updating these values may be relevant. Considering the available results, the variation in vitamin concentrations in colostrum appears to be more related to analytical difficulties rather than variable secretion patterns [72]. The factors influencing vitamin variation in BC are poorly understood but are known to be influenced by the time between parturition and milking and by the prepartum diet [7,149]. Heat treatment of colostrum (60 °C for 60 min) does not appear to affect levels of vitamin A, vitamin E, or β-carotene [150]. More research is required to comprehend the factors influencing the vitamin content of colostrum. Vitamin deficiencies can notably impact calf health, and since colostrum provides a substantial amount of these nutrients, further investigation is warranted.

Vitamin A is important for protection against infection, immune function, cell growth and differentiation, maintenance of epithelial surfaces, and vision [151]. However, calves are born deficient in vitamin A and β-carotene; supplementing cows with vitamin A during the dry period can increase plasma retinol concentrations in calves [152]. Vitamin E appears to be transferred into colostrum by a mechanism involving low density proteins [7,153]; it acts in the lipid phase as a radical scavenger, protecting phospholipid membranes from peroxidative damage, and increases the functional activity of neutrophils [154]. Vitamin B2 is found in higher concentrations in colostrum compared to other B vitamins. The concentration of B2 in colostrum exceeds the requirement of 100 μg/100 g of dry matter consumed by pre-weaned calves [155], suggesting a specific role for the newborn calf. To the authors' knowledge, its specific role in the newborn calf has not been studied, but a deficiency of B2 results in mucosal lesions and growth-related problems in young calves [156]. Calves can be born with vitamin D deficiency and remain at low levels throughout the pre-weaning period, which has been shown to affect the immune system [148,157]. Cows are able to synthesize vitamin C primarily in the liver [158]. However, calves do not begin to synthesize endogenous vitamin C until about 2–3 weeks of age, making them dependent on the vitamin C provided by milk [159]. Cases of scurvy (a sign of vitamin C deficiency) have been reported in calves [160]. Non-colostrum-fed calves with a plasma vitamin C concentration of less than 0.15 mg/dL showed active infections and swollen navels, whereas injections of 500 mg/dL alleviated the symptoms [159].

Table 3. Mean concentrations and physiological roles of vitamins present in the bovine colostrum.

Vitamin	Mean	Physiological Role
Fat-soluble vitamins		
Vitamin A (μg/100 mL)	233–369	Immune function, cell-growth, and vision.
Vitamin E (μg/100 g)	191–530	Antioxidant function.
Vitamin D (IU/100 g fat)	120–181	Ca and P absorption, bone health, and immune function.
Vitamin K (μg/100 mL)	>2	Blood clotting and bone health.
Water-soluble vitamins		
Thiamine (B1) (μg/100 mL)	58–90	Energy metabolism and nervous system.
Riboflavin (B2) (μg/100 mL)	455–610	Energy production and cell growth.
Niacin (B3) (μg/100 mL)	34–96	Redox reactions (synthesis of NAD), energy metabolism.
Pantothenic acid (B5) (μg/100 mL)	224	Acetyl-transfer reactions (synthesis of coenzyme A), energy metabolism.
Pyridoxal (B6) (μg/100 mL)	15.0	
Pyridoxamine (B6) (μg/100 mL)	21.0	Brain development, immune function, and production of hemoglobin.
Pyridoxine (B6) (μg/100 mL)	4.0	
Biotin (B7) (μg/100 mL)	1.0–2.7	Carboxylation reactions, glucose, amino acids, and fatty acids metabolism.
Folate (B9) (μg/100 mL)	0.75–0.8	Single-carbon-transfer reactions (nucleic acids synthesis), DNA, and methionine metabolism.
Cobalamin (B12) (μg/100 mL)	0.2–60	Red blood cell production, neurological function, and DNA synthesis.
Vitamin C (μg/100 mL)	1620–3200	Antioxidant, immune function, skin, and blood vessel integrity.

Values represent the range of mean concentrations and were obtained from refs [7,22,56,143–148].

4. Bioactive Components in Bovine Colostrum

Bioactive components (that can also be referred to as nutraceuticals) are natural essential and non-essential molecules that act in the animal organism, usually with health benefits beyond basic nutritional value [161]. Compounds such as vitamins, hormones, growth factors, certain proteins, carbohydrates, and lipids, as well as molecules such as nucleotides, polyamines, and miRNA have been identified as bioactive components in colostrum [11,13,48,55,97,162–165]. These substances can be either of blood origin or produced in the lactocytes of the mammary gland [162].

Bioactive components from colostrum stimulate the development and function of the GIT, modulate the GIT microbiota, and provide local protection. There are multiple receptors for bioactive components in the GIT that trigger multiple events, making it difficult to explain the function of a specific factor [166]. Although there are many bioactive components in colostrum [2], only a few have been the target of research. Therefore, only the currently most relevant bioactive components will be reviewed.

4.1. Bioactive Proteins

In addition to being a source of amino acids, the proteins in colostrum can perform several functions for the newborn. These proteins are found in the whey fraction, and some of them have already been mentioned above. These bioactive proteins act through a variety of mechanisms, but their role is largely related to host defence.

4.1.1. Immunoglobulins

Colostral immunoglobulins are responsible for protecting the immunological naïve newborn calf against pathogens by activating and regulating the innate immune system. Although there is still much to learn, they have been extensively studied and reviewed in other papers, so only a brief mention of their transport mechanisms and concentrations in colostrum follows.

Immunoglobulins are a family of high molecular weight proteins with similar physicochemical properties and antigenic determinants [167]. The main immunoglobulin classes present in bovine colostrum are IgG, IgA, and IgM [17], with IgD and IgE also present [4].

IgG is the predominant immunoglobulin in bovine colostrum, whereas IgA is predominant in primate colostrum [23]. IgG represents over 50% of the total protein of BC [168]. IgG is divided into three subtypes: IgG1, IgG2, and IgG3 [168]. IgG1 predominates over IgG2 in bovine colostrum by approximately 7:1, although blood concentrations are similar [169]. Nevertheless, IgG1 in blood decreases during colostrogenesis, which may be due to the passage of IgG1 from blood into colostrum [2]. IgG3 is present at even lower levels and has only recently been detected in BC [168].

IgG in colostrum is found in concentrations ranging from 0.68 to 216.70 mg/mL; IgA concentrations range from 0.13 to 22.14 mg/mL and IgM from 0.18 to 14.01 mg/mL [19]. As IgG is the predominant immunoglobulin in BC, only this class is considered in quality assessments, and a colostrum with 50 mg/mL of IgG is considered good quality colostrum [4]. However, for effective immunization, factors such as the amount of colostrum ingested, the time between birth and ingestion, microbiological contamination, and the method of administration must be taken into account [4].

IgG1 is selectively transported from the bloodstream into colostrum by specific receptors on mammary alveolar epithelial cells, a pH-dependent process called transcytosis [4,23], while IgG2 (and BSA) are recycled [170]. IgG1 and IgG2 differ in their amino acid structure, which may explain their different transport mechanisms [170]. IgA and IgM are produced in smaller quantities by plasma cells in the mammary gland [17]. While immunoglobulin transport appears to be selective at the mammary gland, the intestinal tract of the calf is non-specific for any immunoglobulin class until 24 h after birth [17]. Immunoglobulins and complement proteins are resistant to gastric acids [3,72], which can increase their bioactive functionality along the GIT or allow for absorption without compromising their structural integrity [112]. IgG1 can also be re-secreted into the GIT via FcRn receptors, providing local specific immunity [29,171].

4.1.2. Lactoferrin and Transferrin

Lactoferrin (LF), a member of the transferrin protein family, is an iron-binding glycoprotein synthesized in the mammary gland and in other exocrine glands and is therefore present in colostrum, milk, saliva, and bronchial, cervicovaginal, and gastrointestinal fluids [72,84,172,173]. LF is not present in the lacteal secretions of all mammalian species; for example, it has not been detected in dogs, rats, and rabbits, but it is one of the most abundant glycoproteins in ruminant and human milk, although human milk is much more concentrated in LF than bovine milk (100 times or more) [173,174]. In contrast, rat and rabbit milks are more concentrated in transferrin than human milk, in which it is undetectable [175]. In cattle, both LF and transferrin are found in higher concentrations in colostrum than in milk [103], and transferrin is higher in blood [106]. The mammary gland secretes a high mass of LF during the dry period and colostrogenesis, and it may influence the release of IgG1 in colostrum by increasing intracellular pH, facilitating the release of IgG1 from FcRn, since this receptor depends on a pH < 6.5 or >6.5 to either bind or release IgG1 [2]. The concentrations of LF and transferrin in bovine colostrum are shown in Table 2 and Supplementary Figure S3.

LF has been described as having many functions and may be relevant to GIT growth regulation in neonates [84]. Some functions are the ability to regulate iron absorption [176,177]; improvement in the intestinal epithelial barrier by promoting cell growth, decreasing paracellular permeability, and increasing alkaline phosphatase activity and transepithelial electrical resistance [178]; ability to be released from plasma neutrophils during infection or inflammation, possibly contributing to the activation of other immune cells, thus providing protection against pathogens [179,180]. However, it can also regulate the inflammatory process by inhibiting the progressive inflammatory cascade [72].

It appears that LF plays a greater role as a protective mechanism against infection in the dry mammary gland than in the lactating mammary gland, probably because of the low concentration of LF and the high concentration of citrate in the latter stage [180,181]. However, when LF is saturated with iron, it loses its bacteriostatic effect [182], and given that citrate is an iron chelator, it seems that the low concentration of LF in the lactating mammary gland may be the main reason for the reduced antimicrobial effect of LF at this stage. Nevertheless, LF has a 300-fold higher affinity for iron than transferrin, which may not be relevant at physiological pH but allows LF to retain its iron-binding capacity in a more acidic environment, particularly in the presence of citrate [177]. LF has received increasing attention as a multifunctional protein, but its mechanisms are not yet fully understood.

4.1.3. Proline-Rich Polypeptide

Proline-rich polypeptide (PRP), also known as colostrinin, is a complex of at least 32 peptides present in colostrum of various species, including bovine [183,184]. PRP is probably derived from the partial proteolysis of annexin and β-casein [184,185]. The therapeutic effects of PRP have been studied in laboratory animals and in humans, particularly in patients suffering from Alzheimer's disease [185,186]. PRP can help prevent oxidative damage [187], which is important for newborn calves, as they are susceptible to oxidative stress in the early stages of life [188]. It has also been shown to be effective in improving long-term memory in newborn chickens [189] and in modulating adaptive and innate immunity [190,191]. Immunocompromised rats infected with enterotoxigenic *E. coli* had reduced endotoxin levels and infected lymph nodes when treated with PRP [192]. PRP helps regulate cytokine production and has been shown to reduce allergic inflammation in murine [193]. These antiallergic properties could be helpful for calves fed milk replacers, since allergic reactions in calves to soybean flour in milk replacers have been reported [194].

4.1.4. Enzymes

Colostrum, like milk, contains many enzymes that perform functions associated with the host defence mechanism against microorganisms and oxidative damage, as well as with many essential metabolic processes such as catalysis, lipolysis, and proteolysis.

Lactoperoxidase (LPO) is one of the most frequently mentioned enzymes in the literature. It is a glycoprotein, a member of the family of haem peroxidase enzymes, secreted by the mammary gland into colostrum [84,184]. The main biological function of LPO is to defend against microorganisms by generating reactive oxygen species (ROS), which is effective against a wide range of bacteria but also has antiviral and tumoricidal activities [180]. This enzyme has been shown to be very resistant to proteolysis, highlighting its importance in the defence of the calf's GIT [180]. LPO and LF in milk lose activity during heat treatment above 70 °C for 30 min [195]. In this study, where different temperatures were tested, raw milk samples showed a lower growth rate per hour than treated milk of *Streptococcus thermophilus*, *Lactococcus lactis*, *Pseudomonas fluorescens*, and *Escherichia coli* in whey because as temperature increased, the total protein content in whey decreased significantly, which also reduced the bacteriostatic activity in milk.

Lysozyme is an enzyme present in colostrum and milk [100,102,107] with specific hydrolytic activity against the peptidoglycan in cell walls of Gram-positive and Gram-negative bacteria [196]. It is more effective against Gram-positive bacteria because their cell walls contain up to 90% peptidoglycan [197]. Compared to humans or other species, such as horses, this enzyme is present in lower concentrations in bovine milk, probably too low to contribute effectively to the overall bacteriostatic and bactericidal activity [180,197]. Reported concentrations of lysozyme in bovine colostrum vary widely and are usually lower than other enzymes with antimicrobial activity (Table 2), but it generally increases after the first milking [198].

Other enzymes present in bovine colostrum include catalase, superoxide dismutase, and glutathione peroxidase, which have antioxidant properties, and β-1,4-galactosyltransferase, lactate dehydrogenase, alkaline phosphatase, and gamma-glutamyltransferase, which catalyze important biological reactions, and esterases, lipases (such as lipoprotein lipase), proteases (such as tissue plasminogen activator), and ribonucleases (such as ribonuclease II-1), as well as enzyme inhibitors, which are present in very high concentrations but rapidly decrease with time after birth [7,11,154,166,197,199–201].

4.1.5. Cytokines

Cytokines in colostrum are divided into interleukins, interferons, and tumor necrosis factors, which are responsible for modulating the immune system [7]. Some of these pro-inflammatory cytokines, such as IL-1β, IL-6, and tumor necrosis factor-α (TNF-α), and acute phase proteins, such as serum amyloid A and haptoglobin, can influence the concentration of these molecules in the serum of calves during their first weeks of life [202]. Weaned pigs supplemented with colostrum also showed changes in cytokine mRNA expression in spleen- and gut-associated lymphoid tissues, with increased expression of IL-2, IL-4, IL-10, IL-12 and decreased expression of IFN-γ [203]. IL-1Ra, IL-1β, IL-6, TNF-α, and IFN-γ are present in higher concentrations in bovine colostrum than in mature milk; for example, IL-1Ra can be 180 times higher in colostrum than in mature milk [204], suggesting their importance as immunomodulatory factors in the newborn calf. Concentrations of these cytokines in colostrum can range from 77 to 5206 ng/mL [204]. There is still a lack of knowledge about the variation of other cytokines between the first milking and subsequent milkings.

Cytokines are essential in the immune response, but some cytokines need tight regulation; otherwise, the inflammatory process can have nefarious effects on the organism. For example, IL-6 plays important roles in inflammatory processes at the intestinal level but can also cause tissue damage, compromise the integrity of the intestinal barrier, and lead to systemic infections when overproduced [205,206]. IL-6 is mediated by the nuclear factor kB (NF-kB), which is involved in the pathogenesis of inflammatory diseases [207]. Bovine colostrum was able to reduce NF-kB activation and, consequently, IL-6 production in an in vitro model [206]. In another in vitro model, colostrum inhibited the NF-kB pathway in human colon cancer HT29 cells, protecting against intestinal epithelial cell inflammation [208]. These results are in agreement with Blais et al. [209], who observed a negative effect of bovine colostrum on the transcriptional activation of NF-kB. BC seems to have a strong capacity to increase the production of some cytokines and decrease others. It also appears to inhibit the production of cytokines that are present in higher concentrations in colostrum than in milk, such as IL-6 and INF-γ, probably as a means of regulating and establishing a balance between exogenous and endogenous cytokine levels.

4.1.6. Complement System

Colostrum and milk contain various proteins of the complement system [210–212]. They are involved in innate and specific immunity and play an important role in defence against pathogens. Complement proteins can be activated by the classical pathway, the alternative pathway, and the lectin pathway [210].

It seems that proteins related to the complement system are upregulated in colostrum compared to milk, such as clusterin (clearance of cellular debris and apoptosis in MFGM), complement factor B (related to the alternative pathway of the complement system), C3, C7, and C9 (classical pathway of the complement system) [211,212].

The newborn calf can acquire these proteins from colostrum but complement activity can also be found in the fetal serum right after birth. For example, C3 was not found in fetal serum and was only detected in calves between 1–3 days old [213]. In contrast, complement activity was not found in lambs at birth, only becoming detectable from day 1 (after colostrum consumption) and increasing until day 20 (end of the study) [214]. In this study, there were no differences in the complement activity, IgG, IgM, and chitotriosidase activity between lambs fed colostrum at 2 or 14 h post-natal. It has also been shown that the type of milk (goat milk or formula) after the colostrum feeding period can influence the activity of the complement system in goat kids [215]. It is also important to mention that temperature affects the expression of complement proteins (positively and negatively, depending on the proteins) [98,213]. For example, C1, C2, and C8 showed good heat resistance (56 °C), C7 showed moderate resistance, and C3 and C6 increased after heat treatment [213].

Classical and lectin pathways are compromised in the newborn calf, but complement-mediated cytotoxic functions normalize between days 7 and 28 after birth [16]. During this period, the incidence of neonatal enteric diseases is increased, so adequate transfer of immune factors through colostrum is essential to protect the calf while its immune system becomes competent. The combination of maternal and innate complement systems helps in the defence against microorganisms during the first days of life, but it is important to note that the effectiveness of the complement system in the GIT of the newborn calf is not clear due to the activity of proteases [210].

In this section, only those proteins in colostrum with the most scientific evidence that have important biological functions in the newborn calf have been mentioned, but there are many more that have not yet been extensively studied [212]. Colostrum has at least 253 proteins in the whey fraction, of which 36 are uniquely present in colostrum compared to milk [34,98,211]. Proteomic analysis of colostrum from different species could provide more knowledge about the factors of variation as well as the role of these proteins in the newborn [34,98,211].

4.2. Amino Acids

Amino acids (AA) can be divided into essential amino acids (EAA), non-essential amino acids (NEAA), free amino acids (FAA), and insoluble–proteome amino acids (IPAA; e.g., AA in casein) [216]. Lysine, methionine, phenylalanine, threonine, tryptophan, valine, leucine, isoleucine, histidine, and arginine are EAA for cattle. Calves receive AA via placental transfer for protein synthesis and energy utilization [15]. After birth, a continuous supply of AA is important for gluconeogenesis and for the high rate of protein turnover that is characteristic of the perinatal period [217]. Protein synthesis and nitrogen turnover are higher in the post-natal period (enteral feeding) than in the pre-natal period (parenteral feeding), which may indicate that the GIT is involved in protein metabolism [218]. Colostrum and transition milk are a source of EAA and NEAA. Calves fed colostrum and transition milk for 3 days had higher plasma EAA than calves fed colostrum and milk replacer and calves fed milk replacer alone [218].

A continuous supply of AA from milk is important, as AA intervene in regulatory and immunological activities that are essential for the growth and development of the young animal [216]. As mentioned above, several factors such as breed, age, and diet influence the protein concentration in colostrum and milk and thus the AA content. It has been shown that BC contains 0.32 g/L FAA and 4.2 g/L hydrolytic AA [219]. Total AA content decreases gradually from colostrum to mature milk, and FAA is present in lower concentrations than protein-bound AA [220]. 33 FAA and 29 IPAA in BC and 31 FAA and 30 IPAA were identified in milk [216]. In the FAA domain, 14 AA were significantly upregulated in colostrum compared to milk (histidine, threonine, isoleucine, leucine, methionine, phenylalanine, tryptophan, valine, glycine, proline, ethanolamine, hydroxyproline, 3-methylhistidine, and sarcosine) and six were significantly downregulated (glutamate, glutamine, alpha-aminoadipic acid, citrulline, beta-alanine, and 1-methylhistidine); in the IPAA domain, 6 AA were upregulated in colostrum (taurine, threonine, cysteine, arginine, serine, and alanine) and 6 AA were downregulated (lysine, sarcosine, ornithine, citrulline, alpha–aminoadipic acid, and beta–alanine) [216]. Taurine and valine were found in higher concentrations in colostrum (≈93–118 and 55 ug/mL as FAA; ≈18 and 379–398 ug/mL as protein-bound, respectively) and milk (12 and 3.8 ug/mL as FAA; 2.7 and 360 ug/mL as protein-bound, respectively) compared to other AA [216,219]. Isoleucine is also one of the most abundant FAA in colostrum (18.31 ug/mL) but not in milk (0.17 ug/mL) [216]. NEAA such as phosphorylethanolamine, ethanolamine, alanine, and glutamate also appear to be upregulated in colostrum and milk compared to other AA [216]. Some AA may only be present in colostrum or milk as FAA or bound to proteins. For example, free cysteine was detected only in colostrum and transition milk in the study by Zanker et al. [220]; however, in the study by Li et al. [216], it was not detected either in colostrum or in milk. In both studies, cysteine was present as a protein-bound AA in all samples. Glutamine was not found in colostrum but only in milk, and protein-bound glutamate was very high in colostrum in the study by Zanker et al. [220]. Li et al. [216] also found high concentrations of glutamate (mainly protein-bound) and lower concentrations of glutamine, but the authors considered colostrum from 0–7 days after birth, which may explain the differences between the studies. Conversely, Liang et al. [219] found much higher concentrations of glutamine than glutamate in colostrum. These differences could possibly be explained by breed and dietary factors, differences in proteolytic activity and microbial load of the samples collected in different studies, as well as what is considered colostrum by different authors.

Valine, isoleucine, and leucine are branched-chain AA that are essential for the regulation and activation of signaling pathways, such as the mammalian target of rapamycin (mTOR) signaling pathway, which controls protein synthesis and degradation, hence resulting in muscle tissue development [221]. Thus, increased levels of these AA in colostrum may be associated with the upregulation of the mTOR system and the high neonatal protein turnover [222]. Taurine can contribute to innate immunity in pre-weaned calves by reducing pro-inflammatory signals in PMNL and by increasing their resistance to oxidative stress; however, supplementation above 8 ug/mL may lead to toxicity [223]. Hammon and Blum [218] observed an increase in plasma glutamate and a decrease in plasma glutamine in calves after colostrum and transition milk ingestion, demonstrating the important role of glutamine as a fuel for post-natal intestinal development. Glutamine plays an important role in the metabolism of macronutrients, serves as a precursor for molecular components, and intervenes in cell regulatory processes and immune functions [220]. Glutamine is a conditionally essential AA, and supplementation does not appear to be necessary except in specific catabolic situations [224]. Supplementation of pre-weaned calves with L-glutamine has been shown to reduce gastrointestinal permeability and biomarkers of physiological stress [225]. L-glutamine supplementation during weaning was also shown to have beneficial effects on growth performance, gut morphology, and physiological stress [226].

The AA content of colostrum appears to be necessary for various biological processes, particularly those related to protein turnover. However, even if EAA are not provided after birth, the calf can mobilize them from tissue stores until colostrum is provided for at least up to 24–25 h of life, after which plasma EAA levels normalize, unlike what happens with IgG, for example [220].

4.3. Fatty Acids

Although the general trend is similar, the lipid profile differs between colostrum, transition milk, and milk [227] (Table 4), highlighting possible specific needs of newborn calves [27]. Colostrum, like milk, is particularly rich in saturated fatty acids (SFA) and monounsaturated fatty acids (MUFA), with a lower proportion of polyunsaturated fatty acids (PUFA) [27,113]. Around 65.6–74.1% of the total FA are SFA, 24.5–28.4% are MUFA, and 3.88–4.28% are PUFA [27,227,228]. Compared to milk, colostrum seems to have a lower content of SFA, branched chain FA, MUFA, and conjugated linoleic acid (CLA) but a higher content of PUFA, ω-3, ω-6, and cholesterol (see Supplementary Table S1), and this is more pronounced in the first hours after birth [27,227–229]. In transition milk, the lipid profile generally gradually changes to a profile more similar to that found in milk. However, some lipids, such as high carbon number TAG, can maintain their initial concentration beyond the first milking [228]. Nevertheless, there is some variation within each group of FA, phospholipids, and TAG, so a general statement may not be entirely appropriate. There is disagreement in the literature regarding the lipid profile between colostrum and milk, so more research is needed to understand what factors may be involved in the differences observed among studies, such as supplementation during pregnancy, which appears to have a considerable effect [230,231]. There is a general agreement that palmitic and oleic acid are the most abundant FA in colostrum, but myristic and stearic acid are also present in higher proportions than the rest of the FA, although stearic acid is present in lower proportions in colostrum compared to milk [27,113,227]. This is also true for other species, with the exception of myristic acid [232]. The FA or FA groups present in higher concentrations in colostrum than in transition milk or milk are shown in Table 4.

Table 4. Predominant trends of fatty acids in colostrum compared to transitional milk or milk.

Fatty Acid	Predominant Trend
C4:0 Butyric Acid	↓
C6:0 Caproic Acid	↓
C8:0 Octanoic Acid	?
C12:0 Lauric Acid	?
C14:0 Myristic acid	↑
C14:1 ω-5 Myristoleic acid	?
C15:0 pentadecanoic acid	?
C16:0 Palmitic acid	↑
C16:1 ω-7 Palmitoleic	?
C17:0 Heptadecanoic acid	?
C18:0 Stearic acid	?
C18:1 ω-9 Oleic acid	?
C18:2 ω-6 Linoleic acid (LA)	↑
C18:3 ω-3 α-Linolenic acid (ALA)	?
C21:0 Behenic acid	?
C20:3 ω-6 Dihomo-γ-linolenic acid	?
C23:0 Tricosanoic acid	?
SFA	?
Branched-chain FA	?
MUFA	?
Trans-MUFA	↓
Conjugated linoleic acid (CLA)	↓
PUFA	↑
ω-3	↑
ω-6	↑

↓ Indicates that the fatty acid or fatty acid group is present at a lower concentration in colostrum than in transitional milk or milk; ? indicates that the results are inconclusive; only one author has presented results or there is an inconsistency between references; ↑ indicates that the fatty acid or fatty acid group is present at a higher concentration in colostrum than in transitional milk or milk. Based on refs [27,227,228].

Linoleic acid (LA) and alpha–linolenic acid (ALA) are essential fatty acids (EFA) [233], classified as n-3 and n-6 FA, respectively They cannot be synthesized by mammals and must therefore be ingested, and they are precursors to n-3 and n-6 PUFA, which are essential for metabolic regulation, cell membrane function, and gene regulation [231]. It is not clear whether LA and ALA are more concentrated in colostrum than in milk, but it appears that n-3 and n-6 fatty acids are more concentrated in colostrum (Table 5). However, further research is needed to clarify this due to the variability found between studies. Factors that can influence this variability in colostrum are not yet understood.

Table 5. Fatty acids present in higher concentration (g/100 g total fatty acid) in colostrum compared to milk (at 5th day of lactation).

Fatty Acid	Colostrum	Milk (5 d)
C14:0 Myristic acid	12.8–13.7	8.3–11.2
C16:0 Palmitic acid	32.5–40.4	27.2–29.7
C18:2 ω-6 Linoleic acid (LA)	1.95–2.79	1.53–2.23
PUFA	3.88–4.28	2.97–3.62
ω-3	0.56–1.18	0.33–0.70
ω-6	2.57–3.72	2.64–3.00

Values represent the range of mean concentrations and were obtained from refs [27,227,228].

It appears that the placenta has different permeability to different FA; for example, supplementation of cows with docosahexaenoic acid or ALA during late gestation affected plasma levels of docosahexaenoic acid but not ALA in calves [234]. EFA and CLA supplementation altered the FA composition of skeletal muscle and adipose tissue in calves [235] and influenced their metabolism [236], growth, and feed efficiency [237] via placental, colostrum, or dietary transfer.

Garcia et al. [238] found that pre-weaned calves supplemented with LA had lower plasma concentrations of acid-soluble protein and platelets, higher plasma n-3 FA, glucose, and IGF I, haematocrit and blood lymphocyte concentrations, increased IFN-γ production by peripheral blood mononuclear cells, and higher feed efficiency. Calves from cows supplemented with stearic acid (SFA) before parturition had higher dry matter intake and average daily gain [238] and higher plasma IgG and AEA [230]. Supplementation of cows with PUFA (LA or eicosapentaenoic and docosahexaenoic acid) resulted in higher colostrum IgG concentrations [239], and calves fed colostrum from supplemented cows also had higher plasma IgG levels and higher AEA [240]. The mechanisms underlying this increased IgG uptake are not fully understood. Nevertheless, the authors postulate that it could be related to an effect of the FA in the membrane of enterocytes [230], probably at the microvillar membrane, a specialized part in the luminal surface of the membrane, which is a phospholipid bilayer aggregated with specific proteins. However, it is important to note that immature cells found in newborns have different functionalities [241]. On the other hand, Hiltz and Laarman [242] found that calves fed colostrum replacer supplemented with 2.5% (*w*/*v*) butyrate decreased serum IgG concentrations and AEA, which may be related to the ability of butyrate to promote enterocyte differentiation and proliferation, leading to faster maturation and consequent loss of intestinal macromolecules absorption capacity, ultimately reducing IgG uptake [242,243].

A certain degree of oxidative stress is normal to occur during the fetal-to-neonatal transition; however, unresolved inflammation and insufficient antioxidant defence mechanisms can lead to functional and structural damage to cell components [244,245]. While generally, n-6 FA have pro-inflammatory proprieties, n-3 FA have anti-inflammatory capacity by either inhibiting the formation of eicosanoids, such as PGE2, PGF2α, and LTB4, or by forming anti-inflammatory lipid mediators, such as resolvins and protectins [246,247]. n-3 eicosapentaenoic acid can also be metabolized through cyclooxygenase and lipoxygenase pathways into prostaglandins, thromboxanes, and leukotrienes with lesser pro-inflammatory capacity [246]. Additionally, they can compete with AA for the cyclooxygenase and lipoxygenase enzymes, which are also necessary for highly pro-inflammatory prostaglandin formation (e.g., PGE2, PGF2, and PGD2), derived from the metabolization of n-6 arachidonic acid [246]. These mechanisms can reduce the transcriptional activation of NF-kB and the production of pro-inflammatory cytokines and enzymes [246]. Calves receiving colostrum supplemented with n-3 FA had increased plasma concentrations of n-3 FA and decreased oxidative stress but no change in health and growth parameters [245,248]. The authors suggest that continuous supplementation rather than one-time supplementation may have altered the results. However, Masmeijer et al. [249] found no benefit of continuous supplementation with n-3 FA on the health and growth of pre-weaned calves; furthermore, neutrophiles' ROS production was increased in calves supplemented with microalgae compared to the control group. Further studies are needed to clarify the role of n-3 and n-6 FA in the neonate. The growth and immunomodulatory effects in calves vary depending on the type of n-3 FA, the dose, and the diet in which they are supplied, and may even worsen the outcome [249,250]. Overall, it seems to be beneficial to supplement the dam or calf with FA, but inclusion limits and the lipid source should be considered. However, it is clear that lipids are crucial for the newborn, highlighting the importance of the fat content of colostrum, which is sometimes overlooked in favor of the protein fraction.

4.4. Oligosaccharides

Oligosaccharides (OS) are carbohydrates containing three to ten monosaccharides linked by glycosidic bonds and can be divided into two classes: neutral and acidic. While the neutral OS do not contain charged carbohydrate residues, acidic OS contain one or more negatively charged residues of sialic acid (the most prominent in BC is 5-Nacetylneuraminic) [251]. There have been 52 OS identified in BC and milk [252–254], but recently, another OS has been found in BC [255]. Sialylated oligosaccharides are the major OS in colostrum and milk, representing more than 70% and 50% of the total fraction, respectively [253]. 3′-Sialyllactose (3′SL) and 6′-Sialyllactose (6′SL), 6′-Sialyllactosamine (6′SLN), and Disialyllactose (DSL) are the most abundant in colostrum (Supplementary Figure S4) and are also more concentrated in colostrum than in mature milk, especially 3′SL [256–258] (Table 6).

Table 6. Most abundant sialylated oligosaccharides in bovine colostrum and milk with corresponding concentrations (mg/L).

Oligosaccharide	Raw Colostrum	Mature Milk
3′-Sialyllactose (3′SL)	341–867	42–114
6′-Sialyllactose (6′SL)	103–243	17–89
6′-Sialyllactosamine (6′SLN)	117–239	11–170
Disialyllactose (DSL)	84–520	4–38

Values represent the range of mean concentrations and were obtained from refs [24,256–260].

There are few studies quantifying OS in bovine colostrum and milk, mainly due to the difficulty of the analytical process [253]. Nevertheless, concentrations of the most abundant OS have been reported [24,256–260]. OS in bovine milk or colostrum can vary with breed [254,260], parity [24,258], days in lactation [253,258], hours postpartum [256], and heat treatment [259]. As seen in Table 6, 3′SL is the most abundant OS in colostrum, with considerable variation among studies. This may be related to the time of sampling after birth and individual genetic variation [252,258]. 3′SL was highly affected by heat treatment in comparison to other OS [259] and represents a major source of variation (see Supplementary Table S2 and Supplementary Figure S4).

Recent research has shown that OS in bovine milk have beneficial effects on neonates in several species. Milk OS may promote gut health by acting as a prebiotic for beneficial bacteria; increasing beneficial bacterial colonization of the surface of epithelial tissues; help defend against infection by acting as a decoy for pathogens, thus inhibiting pathogen adhesion to host target cells; by being able to modify epithelial glycan receptor expression; and also by competitive binding with the host cell surface receptor [261].

In an in vitro model, OS were shown to restore intestinal barrier function by promoting the formation of a mucus layer that reduces bacterial adherence, thereby increasing epithelial cell protection, and by reducing damage to the intercellular junction of intestinal epithelial cells [262]. OS ameliorated microbiota dysbiosis and intestinal barrier function in obese mice by increasing the abundance of *Lactobacillus* and reducing intestinal inflammation, as shown by the decreased expression of colonic TNF-α [263]. They can promote the growth of beneficial bacteria, which compete with pathogenic bacteria and produce metabolites such as bacteriocins and disrupt the acid–base balance in the gut, inhibiting the growth of pathogens [264]. It has also been shown that OS can help establish early colonization of beneficial bacteria, such as *Bifidobacterium*, in the newborn calf gut [259]. Bovine OS may also have an impact at the level of neurological tissues, as it has been shown to improve spatial cognition in premature pigs, with hippocampal upregulation of genes related to sialic acid metabolism, myelination, and ganglioside biosynthesis [265].

Compared to bovine, human colostrum has less protein and fat but more lactose [266]. This may be due to the different brain development between the two species, resulting in human offspring requiring more glucose available for neurological tissue development and functionality. Interestingly, the OS content of colostrum is similar in both species, but mature human milk can contain almost twice as much OS as mature bovine milk [267]. This similarity could be interesting for infant formulas. Although mammals are able to synthesize sialic acids for incorporation into neurological tissues, the rapid development of the infant may exceed the synthesis capacity; hence, the higher concentration in human milk [268]. Elephant milk has similar concentrations of lactose to bovine milk, which are lower than human milk, but the content of OS is higher than in bovine milk and three times higher than in human milk, with a very different profile [269]. OS in primate milk is more complex and diverse than in non-primate animals [268]. To date, only chimpanzees, bonobos, and Asian elephants have specific combinations of OS characteristics of the human species [270]. This information highlights the importance of OS in more neurologically advanced species, as well as their importance for neurodevelopment and cognition in neonates [267], but more research is needed to understand the functions of OS in neonates and how different OS interact in different species.

4.5. Endocrine Factors

4.5.1. Hormones

Although there has been some considerable research into the hormones present in milk in the decades of the 70s and 90s [271], research into the hormone concentrations in colostrum is limited. Colostrum appears to be more concentrated than milk in androstenedione, estrone, oestradiol, cortisol, cortisone, GnrH, GH, prolactin, TRH, insulin, glucagon, leptin, adiponectin, and motilin than milk, and the opposite seems to be true for parathyroid hormone-related peptide (PTHrP), testosterone, progesterone, and rT_3 (Table 7). There are other hormones for which this difference between colostrum and milk is not yet clear, such as corticosterone, GHIH, oxytocin, bombesine-like peptide, gastrin-releasing peptide, neurotensin, vasoactive intestinal peptide, calcitonin, melatonin, and erythropoietin. Some hormone concentrations in milk are influenced by lactation phase, pregnancy status, season, diurnal patterns, physiological state of the animal, and also differences between analysis methods [272–276], making direct comparisons with colostrum difficult.

The exact role of these hormones in the neonatal ruminant is not fully understood, but it is known that they can be absorbed into the circulation and that they may contribute to the maturation of the gastrointestinal, endocrine, and immune systems [277]. This is particularly relevant due to the immaturity of the newborn GIT, which allows the passage of these hormones into the circulation and, thus, results in a systemic effect [84]. They can be found in higher concentrations in colostrum or milk than in maternal blood plasma, highlighting their importance for the offspring [271,276]. However, although leptin concentrations are higher in colostrum than in milk (Table 7), there is no evidence of leptin absorption by the neonatal calf, as occurs in other species such as rats and humans, where this hormone has systemic effects related to food intake, insulin-dependent glucose metabolism, intestinal maturation, and thermogenesis [30].

Interestingly, some steroid hormones are present in higher concentrations in human colostrum than in BC, which could be related to the evolution of the mammalian brain, as they regulate cellular mechanisms such as synapse formation, dendritic arborization, and cell turnover, and generally contribute to physiological, behavioral, and cognitive functions [278,279]. However, steroid concentrations in foods of animal origin can be of concern, making it important to assess hormone concentrations in BC, as it is one of the foods with higher levels of these hormones [280].

These peptides exert many different essential functions in the newborn, such as regulating tissue growth and differentiation (promoting: GH, GH-releasing factor (GRF), IGFs, EGF, TGFα; inhibiting: somatostatin (growth hormone-inhibiting hormone—GHIH), and TGF-β), metabolism and thermogenesis (TRH, TSH, T_3, and T_4), energy status and glucose metabolism (insulin, glucagon, adiponectin, glucocorticoids, catecholamines), and blood serum calcium levels (parathormone (PTH) and parathyroid hormone-related protein (PHrP)) [15,277,281–283].

Neohormones, like oxytocin and relaxin, are present in milk and are involved in the regulation of vital mammalian traits, such as internal fertilization, pregnancy, and lactation [284]. Milk-derived relaxin (in colostrum) has been shown to influence the development of the neonatal reproductive system in female pigs [285]. The transfer of these bioactive factors, which act as signaling molecules in neonatal tissues, through nursing is known as lactocrine signaling [285,286]. Although relaxin is present in the milk of different species, it appears that the gene encoding ovarian relaxin-2 is deleted in bovine and ovine species [287]. Calcitonin is a hormone produced by the thyroid gland that lowers blood calcium by inhibiting the reabsorption of bone calcium and increasing urinary calcium loss [288]. An inhibitory effect of prolactin release by calcitonin has also been reported [277]. Although calcitonin has been mentioned in many publications as present in bovine colostrum [38,84,165,184,289] and milk [271], the authors could not find any research showing its concentrations in cow colostrum or milk, as opposed to human and rat milk [276,277,290].

Table 7. Reported hormone concentrations in bovine colostrum and milk.

Hormone	Colostrum	Milk	References
Gonadal Hormones			
Androstenedione (ng/mL)	0.18–8.36	0.1–3.5	[274,279,280,291]
Estrone (E1) (ρg/mL)	1300–31,070	0.6–159	[273,274,279,280,292]
17α-Estradiol (ng/mL)	8.6	0.03	[274,280]
17β-Estradiol (E2) (ρg/mL)	300–7010	0.3–14.0	[273,274,279,280,292]
Estriol (E3) (ρg /mL)	<3000	9.0–31.0	[280,292]
Testosterone (ng/mL)	0.1–≈1.6	20–120	[279,280,291]
Progesterone (ng/mL)	2.62–6.46	2.13–15.49	[274,279,280]
Adrenal gland hormones			
Corticosterone (ng/mL)	?	2.92 ± 0.26	[293]
Cortisol (ng/mL)	1.71–4.4	0.35–1.28	[279,293,294]
Cortisone (ng/mL)	2.16 ± 1.71	0.11–0.51	[274,279]
Hypothalamus-Hypophyseal Hormones			
Gonadotropin-releasing hormone (GnRH) [a] (ng/mL)	11.78 ± 0.72	0.5–3.0	[295]
Growth hormone (GH) [b] (ng/mL)	0.17–1.4	<0.03–<1.0	[11,296–298]
Growth hormone-inhibiting hormone (GHIH) [c] (ρM)	?	19.0 ± 6.0	[299]
Oxytocin (ρg/mL)	?	8.0–10.0	[300]
Prolactin (ng/mL)	280–800	3.7–57.0	[11,301–303]
Thyrotropin-releasing hormone (TRH) (ng/mL)	0.16 ± 0.03	0.05	[295]
Brain-Gut Hormones			
Bombesin-like (related to gastrin releasing peptide) (ng/mL)	?	1.17 ± 0.89	[304]
Gastrin releasing peptide (nM)	?	1.4 ± 1.0	[299]
Neurotensin	?	?	
Vasoactive intestinal peptide (ρM)	?	16 ± 9.0	[299]
Pancreatic hormones			
Insulin (ng/mL)	35.4–65	1.0	[11,305]
Glucagon (ng/mL)	0.16	0.01	[11]
Thyroid Gland Hormones			
Calcitonin	?	?	
Triiodothyronine (T_3) (ng/mL)	<0.31–2.02	0.21–0.41	[296,306–308]
Reverse Triiodothyronine (rT_3) (ng/mL)	0.57 ± 0.06	3.48–91.1	[307,308]
Thyroxin T_4 (ng/mL)	0.12–1.9	0–0.67	[296,306–309]
Other hormones			
Parathyroid hormone-related protein (PTHrP) (ng/mL)	26.0–56.0	59.0–168.0	[310,311]
Glucagon (ng/mL)	≤0.16	0.01	[11,297]

Table 7. *Cont.*

Hormone	Colostrum	Milk	References
Relaxin	-	-	
Melatonin (ρg/mL)	?	4.71–41.94	[275,312]
Erythropoietin	?	?	
Leptin (ng/mL)	13.9–30	4.4–6.1	[272,313]
Adiponectin (μg/mL)	56.1 to 75.9	0.61	[314,315]
Motilin (ng/mL)	0.23 ± 0.06	0.03 ± 0.02	[316]

Values refer to mean values reported in research papers. When a single value was found, mean and standard deviation was reported instead. [a] Also referred to as Luteinizing hormone-releasing hormone (LH-RH). [b] Also referred to as Somatotropin. [c] Also referred to as Somatostatin.

4.5.2. Growth Factors

After immunoglobulins, growth factors are probably the most studied bioactive components in BC [2,73,124,298,317–320]. There are about 50 different polypeptides that can modulate the growth, maturation, and function of the GIT [11,44,321]. Insulin-like growth factor (IGF) 1 (IGF-I) and 2 (IGF-II), transforming growth factor-β (TGF-β), epidermal growth factor (EGF), β-cellulin (BTC), fibroblast growth factor 1 and 2 (FGF1 and FGF2), platelet-derived growth factor (PDGF), and vascular endothelial growth factor (VEGF) are probably the most important growth factors in colostrum [7,321–323]. See Baumrucker and Macrina [303] for cow colostrum and milk concentrations. Growth factors have a certain degree of thermal tolerance, since they can withstand temperatures up to 60 °C for 30 min [298] or 60 min [324]. However, the need for caution when heat-treating colostrum is reinforced, as to the authors' knowledge, there is no information beyond these ranges. These factors also need to survive digestion, retain biological activity, have receptors in the GIT, or be absorbed to influence the neonate. The GIT has site-specific receptors in different regions, so it is expected that bioactive components in colostrum will have different effects in different regions of the GIT [166]. The epithelial cells have a high turnover rate, and IGFs are key regulators of cell growth and differentiation [325].

Compared to human colostrum, BC is particularly high in IGFs and low in components of the EGF family [166]. EGF binds to the epidermal growth factor receptor (EGFR), which increases intracellular tyrosine kinase activity, activating signaling cascades that ultimately promote cell proliferation and angiogenesis and reduce apoptosis [326]. Other EGFR-related ligands are TGF-β and BTC [166]. TGF- β is associated with immunomodulatory activities, inflammatory responses, oncogenesis, and proliferation of intestinal cells [44,323,327]. IGF concentration is also higher in colostrum than in blood [166]. The IGF system is composed by IGFs and IGF-binding proteins (IGFBPs) [317]. IGFs are synthesized in the liver in response to the growth hormone (GH) [328], and some IGFBPs are synthesized in the lactocytes of the mammary gland [329]. IGFs and IGFBPs are highly concentrated in prepartum secretions and colostrum and decline rapidly in subsequent milkings [11]. While EGF may be more important in the human and porcine species, it can be suggested that IGFs are more important for the ruminant neonate, as they are present in higher concentrations in ruminant colostrum [317]. Whether the growth factors in colostrum and milk are exclusively directed to the offspring or to mammary gland development (e.g., cell proliferation and differentiation) is not entirely understood. Heifers have higher circulatory concentrations of IGFs than older cows, which vary throughout the pregnancy–lactation cycle [330]. They are higher in the dry period, decrease at the onset of lactation, and then gradually increases until late lactation [331]. They are present in much higher concentrations in colostrum than in milk (IGF-I around 3000 and 5–50 ug/L, respectively) [331]. IGF-I in colostrum is also mainly in a free form, which leads to higher bioavailability of IGF-I [331]. So, it seems that these growth factors in colostrum are mainly directed to the neonate but not exclusively. Nevertheless, the neonate calf is also capable of synthesizing IGFs in different tissues [166]. The components of the IGF system have different regulatory properties in GIT epithelial cells. IGF-I appears to be more involved in mitogenesis, which promotes enterocyte proliferation and regeneration, and IGF-II may be more related to enterocyte differentiation promoting functions [325].

Delaying colostrum feeding can negatively affect some metabolic and endocrine traits of the newborn. Calves fed colostrum on day 1 had increased circulating glucose, albumin, insulin, and IGF-I concentrations compared with calves fed on day 2 of life [332]. Fischer-Tlustos et al. [333] showed that calves fed colostrum right after birth (0 h) had increased GIT development in terms of some histomorphological parameters like villi height, crypt depth, and surface area index, compared to calves fed 12 h after birth at 51 h of age. Smaller differences were observed when compared to calves fed 6 h after birth. However, there were no differences in the serum IGF-I concentrations of calves with respect to feeding time but it was 29% higher at 48 h than at 0 h after birth (312.8 ± 14.85 vs. 241.9 ± 14.06 ng/mL) [333]. On the other hand, extending colostrum feeding can promote GIT maturation, but similar results can also be obtained with a mixture of colostrum and whole milk (1:1) [334], which has a composition similar to transition milk. Even after gut closure and with theoretically increased proteolytic activity, calves fed colostrum or a mixture of colostrum and whole milk had a greater surface area of the GIT and increased villi height in the proximal and distal jejunum than calves fed whole milk, and ileal crypts tended to proliferate more with a mixture of colostrum and whole milk than with whole milk alone, but crypt depth did not differ [334]. However, calves fed colostrum had only minimally increased plasma GLP-2 and serum IGF-1 compared to calves fed whole milk, which the authors believe may be related to inadequate nutrient intake [334]. In an in vitro study, it was shown that amino acid deficiency reduced intestinal epithelial cell reconstitution through a decrease in TGF-β production [335]. The maternal diet during gestation can also affect the GIT development of the newborn, as lambs from ewes supplemented with folic acid during pregnancy were shown to have increased IGF-I expression, small intestine weight/live body weight ratios, and intestinal muscularis thickness [336]. In fact, growth factors present in colostrum may be able to influence the neonate's growth performance until later in life. A study by Buranakarl et al. [337] showed that kids fed colostrum with higher concentrations of IGF-I (>518.3 ng/mL) had increased body weight gain at the end of the first month than kids fed colostrum with lower concentrations of IGF-I ($\leq$518.3 ng/mL). Unfortunately, milk composition was not analyzed, since it could have affected the kids' development. It is important to note that IGFs are not the only molecules intervening in intestinal maturation, as it is a combination of several bioactive components [317].

4.6. Nucleic Acids

4.6.1. MicroRNA

A review on microRNA (miRNA) in bovine colostrum has recently been published [13], and this is a very new area of study within the colostrum field; therefore, we will only briefly describe these molecules in colostrum and milk and discuss some recent studies on their effect on the newborn calf. MiRNAs are a group of small (18–25 nucleotides in length), endogenously expressed, non-coding RNA molecules that act as regulators of gene expression and other cell-related regulatory functions, like survival, proliferation, apoptosis, tumor growth, and metastasis [338,339]. It has also been shown that miRNAs can act as regulators in many immunological pathways [12,340], in wound healing, and in infection processes [341,342]. Other non-coding RNA molecules include transfer RNAs (tRNAs) and ribosomal RNAs (rRNAs) [343].

It is suggested that miRNAs are produced in the mammary gland and transferred to milk and colostrum [163,344,345], protected within extracellular vesicles (EV), such as exosomes. Colostrum seems to have a higher concentration and expression of different miRNA molecules than transition milk and milk [163,346,347], and some novel miRNAs are still being discovered [344,348]. EVs are cell-secreted membrane-encased vesicles that contain proteins, miRNAs, mRNAs, and lipids, depending on their source, and are associated with various physiological and pathophysiological processes [342,349]. This allows miRNAs to be resistant to acidic conditions (pH 2 at 37 °C for 1 h) and to RNases [346], making them resistant to the GIT environment and allowing absorption in the small intestine [13]. This finding contributes to the theory that miRNAs in colostrum and milk act as signaling molecules between the mother and the offspring [13]. However, according to Hue et al. [345], the miRNAs present in the newborn calf's bloodstream are not colostrum-derived but rather from an endogenous source. The study by Kirchner et al. [349] also did not confirm that the miRNAs present in the EVs of colostrum either entered the circulation of the newborn calf, remained in the intestinal epithelium, or were rapidly transported to other tissues and were therefore not present in the bloodstream at the time of analysis. These results do not indicate that miRNAs from colostrum are absorbed by the newborn calf.

Depending on the phase of lactation, different miRNAs may be expressed [13]. It seems that during colostrogenesis, there are predominantly expressed miRNAs associated with immune pathways, while during the rest of the lactation, the miRNA expressed are more related to milk synthesis [344,350]. Colostrum is also more concentrated in miRNAs than in transition and mature milk [347]. The same applies to breed differences; compared to beef heifers, the mammary glands of dairy heifers had a more pronounced downregulation of miRNAs involved in the inhibition of genes related to the maintenance and activity of mammary stem cells, presumably required for intensive regenerative processes during puberty and pregnancy [351]. While there are miRNAs that are commonly expressed in colostrum and milk of different species, like miR-30a-5p, miR-22-3p, and miR-26a, which are related to immune functions, there is also significant variation between species [352]. It was found that miRNA abundance was not affected by heat treatment at 60 °C for 60 min and that miRNAs were higher in frozen samples than in raw samples [347].

Some miRNAs have a higher expression level in BC than in milk, such as miR-142-5p, miR-150, miR-155, miR-181a, and miR-223 [163,345]. Ma et al. [353] found that colostrum with very different IgG levels (62.8 ± 3.6 and 256.5 ± 5.7 mg/mL) had a similar expression profile of miRNAs present in small EVs, which contributes to the statement that colostrum with a higher concentration of IgG does not necessarily indicate higher concentrations of the many other BC reviewed in the present article. Nevertheless, studies correlating IgG and other BC are still lacking. Only one miRNA was more abundant in high-IgG colostrum—miR-27a-3p—which is associated with osteogenic differentiation in pre-osteoblasts [353] and with glycaemic and lipid status in women with gestational diabetes mellitus [354].

The importance of miRNAs to newborn health is clear, and it is also clear that these molecules are highly expressed in colostrum; however, it is not clear if their action is dependent on uptake into the bloodstream or if they may act in the GIT. In any case, it seems that they can act as regulators of the mucosal immune system [355], which reinforces their importance as immune modulators of GIT.

4.6.2. Nucleotides and Nucleosides

Nucleotides are monomers that constitute the basic building blocks of nucleic acids (RNA and DNA). They consist of a nitrogenous base, a pentose sugar, and a phosphate group. Nucleosides are essentially nucleotides without the phosphate group. Nucleobases are the nitrogenous bases that are the basic units of the genetic code (adenine (A), cytosine (C), guanine (G), thymine (T), and uracil (U)). They are essential for cellular function, acting as mediators of chemical energy transfer, signal transduction, and growth regulators [356,357].

Nucleo(s/t)ides are present in milk and colostrum in the sub-milligram per liter range and belong to the non-protein nitrogen (NPN) fraction [358]. The concentration of these compounds shows little variability during lactation, except during the first few days [358,359]. Immediately after birth, the concentration of nucleotides in colostrum is low but rises to a maximum 24–48 h later and then gradually decreases [360,361]. Nucleosides are generally found at lower concentrations in milk than nucleotides. They are found in higher concentrations in colostrum without a clear maximum and generally decrease to a constant level by 3 weeks after parturition [359]. Uridine and uridine 5′-monophosphate (5′UMP) are the most abundant nucleo(s/t)ides in colostrum, with concentrations ranging from 50.6–102.4 and 13.33–143.7 µmol/dL, respectively [360,361]. Concentrations and variations during lactation are species-specific, e.g., uridine levels in ewe milk can be four times higher than in cow and goat milk at about 24 h postpartum [360]. Orotic acid is present in much higher concentrations in cow milk compared to ewe and goat milk and appears to increase during the first 2 months of lactation [360]. Gill et al. [357] found a difference in the total base concentration (nucleotides and nucleosides) in colostrum collected in summer and winter (62.1 $\pm$ 6.2 and 258.7 $\pm$ 6.8 µmol/dL, respectively), with a notable difference between the concentration of 5′UMP in summer (1.2 $\pm$ 0.0 µmol/dL) and winter (143.7 $\pm$ 8.5 µmol/dL). There is still very little information on the variation of nucleo(s/t)ides in ruminants' colostrum, and there seems to be a significant impact of species, breed, season, and analytical methods on the differences obtained so far [357,359–361].

Nucleotides, especially those with an uracil nucleobase, contribute to the calf's biological functions. Mashiko et al. [362] showed that supplementation with 2 g/d of 5′UMP had a positive impact on the immune status of calves aged 4–42 days. In Katoh et al.'s study [363], an effect of 2 g/d of 5′UMP on the endocrine status of calves 4–14 days old was observed. Oral administration of 5 g of a nucleotide supplement (5′CMP, 5′UMP, 5′AMP, and 5′GMP) improved growth performance, intestinal morphology, and oxidative status in pre-weaned calves [364]. Supplementation of sows with uridine and cytidine (Ur:Cy = 1:1) reduced birth mortality, increased piglet birth weight, and modulated gene and protein expression of enzymes involved in lipid metabolism in the liver of neonatal piglets [365]. Maternal supplementation with yeast-based nucleotides containing 5′UMP increased litter weaning weight and decreased diarrhea rate, promoted ileal villus development, increased secretory IgA in the ileum and jejunal, and increased jejunal and ileal expression of interleukin IL-17, IL-8, IL-1β, IL-10, and TNF-α in piglets [366]. However, in another study where sows were supplemented with a yeast-based nucleotide without 5′UMP, there was no effect on sow or piglet performance [367]. Including 10 and 20% of a nucleotide mixture (NuPro, Alltech Inc., Nicholasville, KY, USA) in the milk replacer formula of male calves worsened growth performance and did not improve health status [368]. Feeding calves a yeast-based nucleotide mixture (NuPro, Alltech Inc., Nicholasville, KY, USA) resulted in improved intestinal function and morphology, but feeding purified nucleotides (80 µmol/L of AMP, 64 µmol/L of CMP, and 374 µmol/L of UMP) did not and resulted in lower fecal beneficial bacteria, higher harmful bacteria, fecal water loss, and calf dehydration [369]. A reduction in *Lactobacillus* spp. was also observed in calves supplemented with nucleotides (yeast-derived; 500 g/t of milk replacer; Ascogen, Chemoforma, Switzerland) [370]. Most studies on nucleotide supplementation have tested pre-weaned calves, but it appears that post-weaning nucleotide supplementation may result in better growth performance [371,372]. It may be of interest to study the effect of nucleotide supplementation, especially uracil-based, in the very first days of life, given the good results of maternal supplementation and the natural variation of nucleo(s/t)ides in BC.

5. Cells and Microorganisms in Bovine Colostrum

5.1. Cells

The cell fraction of colostrum can be divided into epithelial cells (lactocytes), erythrocytes, and leukocytes [9,162]. The mammary gland's epithelial cells contribute to host defence by expressing pathogen recognition receptors (e.g., toll-like receptors) and by synthesizing proteins with antimicrobial properties such as lactoferrin, β-defensin, and lipopolysaccharide-binding protein [9]. Epithelial cells are present in the bovine colostrum at 2 to 15% of the total cell count [373]. In contrast, epithelial cells make up more than 20% of the total cell count in sow colostrum [8]. The leukocyte fraction consists mainly of lymphocytes, macrophages, and neutrophils, and its concentration depends on diet, age, breed, and physiological and individual conditions [8,373,374]. For example, it has been shown that heifers have lower activity of colostral immune cells than cows with three or more lactations [375]. On the other hand, Kampen et al. [376] found no significant effect of cow health status on colostral lymphocyte concentration. In Park et al.'s study [377], colostrum collected 48 h before and after parturition differed in the percentage of leukocytes, with an increase in all subpopulations except macrophages, which increased, suggesting a decrease in leukocyte count with the onset of lactation, similar to other components in BC; however, only two cows were sampled in the 48 h prepartum period. In contrast, in pigs, the percentage of T lymphocytes, B lymphocytes, and macrophage/monocyte subsets in colostrum did not differ between 0 and 8 h postpartum [378].

Colostrum contains approximately 10^6 leukocytes/mL, and macrophages and neutrophils are the major fractions of colostral cells, followed by lymphocytes (Table 8) [347,373,377,379]. In the lymphocyte fraction, T lymphocytes represent a higher proportion of the total cell count than B lymphocytes, at 16 and 10.7%, respectively [377,379,380]. Although these proportions appear consistent across studies, there is still considerable variation in the number of cells. The total cell count, usually referred to as the somatic cell count (SCC) of colostrum, is around 2.3×10^5 and 5×10^5 cells/mL [374,379] but can range from 1.2×10^5 to 1.9×10^6 cells/mL [107,381]. The specific influence of factors on these variations has yet to be entirely understood.

Table 8. Mean variation in the differential leukocyte count (%) in bovine colostrum and milk.

Differential Leukocyte Count	Colostrum [a]	Milk [b,*]
Lymphocytes (total)	2–27	18–58
Lymphocytes T	16	47
Lymphocytes B	11	20
Macrophages	31–69	10–29
Neutrophils	30–65	28–49

Values represent the range of mean concentrations. [a] Values were obtained from refs [347,373,377,379]; [b] Values were obtained from refs [382–384]; * Includes results from dairy cows in various stages of lactation.

Significant effects of freezing and heat treatment on colostral cell viability have been reported. The rapid freezing of colostrum in liquid nitrogen, followed by a slow thawing, resulted in lysis of the cells, which led to an inability to detect these cells under the microscope [385]. Rapid freezing of colostrum on steel plates pre-cooled to −80 °C followed by heat treatment at 50 °C also resulted in cell lysis [386]. Similarly, no percentage of viable cells was found in pooled colostrum frozen at −20 °C for a period between 24 h and 3 months and then thawed (37 °C) prior to administration, in contrast to fresh colostrum, which had 24 ± 8% viable cells [387]. Similar results were found in Chandler et al.'s study [347], with no viable cells after freezing at −20 °C overnight or heat treatment (60 °C for 60 min), but there were more than 80% viable cells in fresh colostrum cooled on ice and stored at 4 °C overnight. In contrast, frozen (−20 °C) porcine colostrum had reduced numbers of lymphocytes (CD79a+) and conventional B cells (SWC7+CD5-) but no macrophages, granulocytes, and NK cells [378]. Similarly, feeding fresh or frozen colostrum did not significantly affect the neutrophil and monocyte activation in newborn dairy calves [388]. Heat treatment (60 °C for 60 min) reduced the SCC of bovine colostrum by 36% [320]. Martínez et al. [389] studied the effect of different treatments on the SCC of ewe milk analyzed by the Fossomatic method and concluded that freezing milk at −20 °C only reduced the SCC when azidiol was used as a preservative and when the milk was heated to 60 °C, whereas unpreserved milk or milk preserved with bronopol or potassium dichromate and analyzed at 40 °C did not affect the SCC. Significant differences in SCC between fresh and frozen BC were also not found in another study [61]. It can be assumed that freezing may cause cell lysis, but they can still be counted in the Fossomatic method, as according to Chandler et al. [347], intact (but not viable) cells and cell components are still visible after freezing. Freezing and heating may, therefore, affect cell viability in colostrum.

Colostrum management practices, such as giving priority to fresh colostrum, which can be refrigerated for about 2 days (or even longer if a preservative such as potassium sorbate [4,22] is added), rather than freezing it immediately if more cows are likely to calve in the next few days, could help preserve maternal cell function in colostrum. Another possibility could be the fermentation of colostrum as a preservation method, but there is currently no information on the cell viability of fermented colostrum.

Leukocytes are absorbed by the intestinal epithelium of the newborn, allowing them to migrate to Peyer's patches and mesenteric lymph nodes or to enter the systemic circulation and eventually reach other organs associated with the immune system, such as the liver and spleen, normally within 24 h after colostrum ingestion [373,390,391]. To be absorbed, leukocytes must undergo phenotypic changes provided by the colostrum environment in the mammary gland [391].In the swine species, there is a maternal–neonatal recognition that does not allow the passage of colostral cells or peripheral blood leukocytes from other sows through the piglet's intestinal wall [8], but it does occur in primates [392] and sheep [393] and probably in cows, although this is not clear as in most studies, calves were fed colostrum from their respective dams. It is important to identify this issue, as many farmers feed calves with colostrum from a colostrum bank rather than from the dam. Nonetheless, in one study where calves were fed pooled colostrum or colostrum from another dam, passive immune transfer, morbidity, and weaning weight did not differ from calves fed maternal colostrum [394].

At birth, the newborn calf has a naive innate immune system; a deficient and immunosuppressive state is evidenced by increased expression of TGF-β1 and TGF-β2, together with reduced functions of phagocytosis and platelet aggregation [16]. Thus, immunocompetent colostral cells (as well as immunoglobulins and other immune factors present in colostrum) are necessary to induce cell-mediated responses and enhance both innate and adaptive responses.

It is likely that maternal leukocytes migrate to different tissues depending on their ability to express the CD62L receptor. Leukocytes with little or no expression of CD26L are likely to migrate to peripheral non-lymphoid tissues where they act as memory cells; leukocytes expressing higher levels of CD26L expression are more likely to migrate to secondary lymphoid tissues where they can act as regulators of the newborn's specific immune responses [391]. In fact, memory-activated T-cell phenotypes are the most common in colostrum [8,391].

Maternal colostral cells allow for more rapid development of adaptive immunity by enhancing the antigen-presenting capacity of monocytes and lymphocytes, as indicated by increased expression of MHC class I molecules [395,396]. Maternal colostral cells also have a long-term effect on the development of the calf's immune system, improving responses to vaccines [397]. Maternal vaccination may improve neonatal defences against certain pathogens, as transfer of antigen-specific lymphocytes through colostrum can occur with antigens to which the dam has been exposed [373,386], but there are still few studies in bovine species on maternal–neonatal transfer of antigen-specific immunity.

Colostrum affects the B-cell lineage of the newborn calf [398]. Exposure to maternal commensal microbes in utero or spontaneous production of low-affinity natural antibodies causes a calf that has not received colostrum to have IgG-positive cells in the lymphoid tissues, whereas calves fed colostrum do not [398]. This occurs because the colostral immune system can suppress or eliminate IgG-positive B cells but does not affect IgM- and IgA-positive cells in the mesenteric lymph nodes, which may play an important role in mucosal immunity in the early life of the calf [398]. BC also has a much higher concentration of IgG than IgM and IgA, which may also contribute to this suppression of neonatal IgG production but not IgM and IgA. This phenomenon may be necessary for transferring maternal components (which would otherwise be foreign to the neonate), which also results in prolonged tolerance to maternal immunoglobulins, increasing their longevity during the development of the innate and acquired immune systems [398].

5.2. Microorganisms

Feeding colostrum increases gut bacterial colonization in the newborn calf [399]. Bacteria can reach the cow's mammary gland exogenously, specifically from the maternal skin, or endogenously with an immune-related translocation of intestinal bacteria to the mammary gland, referred to as the entero–mammary pathway [400]. The former can explain the presence of aerobic microorganisms, and the latter can explain the presence of anaerobic microorganisms in colostrum or milk. It is hypothesized that the aim of this process is to 'train' or 'educate' the newborn's immune system to recognize commensal microorganisms and develop an appropriate response [400].

The microbial quality of colostrum is usually assessed using total plate count (TPC) and total coliform count (TCC), which provide an overall assessment of hygiene conditions from milking to feeding. It is difficult to establish a mean value for the TPC of BC because there is a large variation between published studies. Mean TPC in raw colostrum can range from 250,000 [401] to 1062 ufc/mL [402] ufc/mL,. Mean TCC can range from 12 ufc/mL [61] to 63,000 ufc/mL [387]. Elevated TPC and TCC are associated with colostrum contamination. Current targets are a TPC below 100,000 cfu/mL and a TCC below 10,000 cfu/mL [403]. While coliforms are essentially fecal bacteria that can include *E. coli*, TPC includes a wide variety of bacteria, including beneficial bacteria. Therefore, a higher TPC does not necessarily indicate poorer microbial quality. Other recommendations from the same reference include streptococci and staphylococci below 50,000 cfu/mL [403].

Heat treatment of colostrum is aimed at reducing pathogenic bacteria. It was found that heating colostrum at 60 °C for 60 or 120 min reduced TPC and pathogenic bacteria without affecting IgG concentration [7,404]. Other combinations of time and temperature above 60 °C resulted in a significant loss of IgG concentrations [7,404]. Heat treatment can effectively reduce TPC and TCC, but staphylococci appear to be more resistant to heat treatment [320].

Colostrum also contains beneficial bacteria, like lactic acid bacteria (LAB), but current recommendations are based only on pathogenic bacteria. While heat treatment significantly reduces TPC and TCC [320,401], there is still a gap in the knowledge of the interaction between beneficial and pathogenic bacteria in raw, frozen, and heat-treated colostrum. Using quantitative real-time PCR analysis, detectable bacteria were only present in untreated colostrum samples, from which *E. coli* represented 70.6% of the total bacteria, whereas no bacteria were detected in heat-treated (60 °C/60 min) samples, but it was shown that calves fed heat-treated colostrum 6 h after birth had higher intestinal colonisation with *Bifidobacterium* [399]. In another study, a significant reduction in LAB (lactococci and lactobacilli) counts was found when heat treatment (treatment 1–56 °C/60 min or treatment 2–63 °C/30 min) was applied to raw caprine colostrum [405]. Beneficial bacteria can compete with pathogens and improve the structural integrity of intestinal epithelial cells by increasing transepithelial resistance, promoting tight junction stability and mucin expression [406]. LAB isolated from colostrum showed good in vitro antimicrobial activities and antioxidant power [407]. LAB supplementation improved body weight gain, feed conversion, and fecal condition in calves [408,409]. While pathogenic bacteria can reduce IgG absorption in the neonatal gut by binding to bacteria or by competing for epithelial receptors, the role of beneficial bacteria in this process is not well understood [410].

Culture-independent microbiome studies have shown that the BC microbiota is dominated by four phyla: Firmicutes, Bacteroidetes, Proteobacteria, and Actinobacteria [14,410–412]. The top 10 genera in BC are listed in Table 9. Of these 10 genera, *Acinetobacter*, *Pseudomonas*, and *Staphylococcus* are the most frequent in BC. Factors such as season, breed [410], parity [14], antibiotic administration [413], and mastitis [14] can affect the colostrum microbiota. At the phylum level, colostrum core microbiome is relatively stable, at the genus level, there is some variation that is not fully understood, and at the species level, there is considerable variation [411].

Table 9. Core microbiome of colostrum (genus level), according to the frequency that a certain genus has been considered part of the colostrum core microbiome in the literature.

Genus	Phyla	Frequency
Acinetobacter	Proteobacteria	+++++
Pseudomonas	Proteobacteria	++++
Staphylococcus	Firmicutes	++++
Bacteroides	Bacteroidetes	+++
Corynebacterium	Actinobacteria	+++
Streptococcus	Firmicutes	+++
Bacillus	Firmicutes	++
Chryseobacterium	Bacteroidetes	++
Flavobacterium	Bacteroidetes	++
Lactococcus	Firmicutes	++

In terms of inclusion criteria, the genera have to be designated as part of the colostrum core or top genera in at least two studies. The symbol "+" indicates the number of studies. Values were obtained from refs [14,410,411,413,414].

The genus *Acinetobacter* comprises more than 50 Gram-negative coccobacilli species, from which the majority are non-pathogenic environmental organisms [415]. *A. baumannii* is the best known species of this genus and has clinical significance in both human and veterinary medicine, as it can be associated with different types of infections in different species and due to its ability to accumulate antimicrobial resistance [416]. Patangia et al. [413] have confirmed that *Acinetobacter* had a high abundance in the colostrum of the cows that received antibiotics as part of the dry therapy. They are common in the environment and are therefore likely to be present in the cow's GIT, allowing them to migrate into colostrum, but it is not yet understood whether they have a specific role in the mammary gland or in the GIT of the neonate, as they are commonly abundant in different studies.

Pseudomonas and *Staphylococcus* can also be pathogenic and appear to be positively correlated in colostrum, as well as *Chryseobacterium* [411]. In contrast, they were negatively correlated with the *Bacteroidales-S24-7-group*, which are beneficial bacteria [411]. There also appears to be a relationship between the concentration of IgG and the prevalence of certain microorganisms in colostrum. For example, the genera *Lactococcus* and *Carnobacterium* (beneficial bacteria) were more abundant in colostrum with an IgG concentration above 100 g/L than in colostrum with an IgG concentration below 50 g/L [410]. In contrast, a species of the *Enterobacteriaceae* family (coliforms) was 97 more abundant in colostrum, with an IgG concentration of less than 50 g/L compared to colostrum with a higher IgG concentration (>50 g/L) [410]. It is not yet understood whether IgG is lower or higher depending on the microbiome or the opposite, but it seems plausible that a combination of both may occur. Lima et al. [14] found that bacteria with a pathogenic tendency were more abundant in the colostrum of multiparous cows compared to primiparous cows and that primiparous cows had a richer microbiota, but primiparous cows with lower diversity might be more susceptible to future disease events. These findings highlight the impact that the microbiome in colostrum and milk may have on mammary gland defence and consequently on the calf's mucosal defence. Calves fed milk replacers instead of milk (from healthy cows) do not benefit from this training and defence provided by the colostrum microbiome, so the inclusion of probiotics may help alleviate this deficit.

Like the colostrum microbiota, at birth, the calf's GIT is dominated by Proteobacteria, Firmicutes, Actinobacteria, and Bacteroidetes, which is similar to the in-utero microbiota, but it can change rapidly. Within 24 h, the calf's GIT can be colonized by pathogenic bacteria such as *Escherichia, Shigella, Clostridium* spp., and *Enterococcus* spp. [417]. This is particularly likely if colostrum administration is delayed [418]. From the results obtained by Malmuthuge et al. [399], calves are born with low bacterial density but with a high proportion of beneficial bacteria, such as *Lactobacillus* and *Bifidobacterium*, compared to potentially pathogenic bacteria, such as *E. coli*. However, colostrum can modulate this microbiome, and in this study, providing colostrum reduced the intestinal *Lactobacillus* content. However, when colostrum was heat-treated (60 °C/60 min), the prevalence of *Bifidobacterium* was higher at 6 h post-natal compared to untreated colostrum or no colostrum groups; however, at 12 h post-natal, there were no differences in *Bifidobacterium* prevalence between the groups of calves that received colostrum. In addition, calves that did not receive colostrum had higher levels of *E. coli* at 6 h and 12 h after birth than calves fed colostrum, especially when compared to calves fed heat-treated colostrum [399]. This confirms that environmental bacteria can colonize the calf's gut very quickly and highlights the importance of providing good quality (chemically and microbiologically) colostrum soon after birth. During the first few days, the microbiota of calves continues to change and is related to the calf's health status [414]. At 14 days of age, there were differences in the fecal microbiome between healthy and diarrheal calves, with a high prevalence of *Faecalibacterium* and *Butyricicoccus* species in healthy calves [414]. Colostrum and milk microbiome, in combination with the environment, modulate the GIT microbiome of calves; however, this area of study is relatively new and warrants further investigation.

6. Conclusions

Different considerations of colostrum are found across studies, which makes it difficult to make adequate comparisons between studies, especially in areas where information is limited. Colostrum, transition milk, and mature milk have different physical and chemical proprieties, so it is important to properly define what colostrum is in research activities. Each species has its own unique differences in colostrum composition. The value of bovine colostrum for animal (other than calves) and human consumption has increased due to its nutritional and nutraceutical value but also due to its availability on farms when compared to colostrum from other species. This demand could affect the availability of good quality colostrum for calves, making it important to better understand colostrogenesis and how to improve colostrum quality and quantity, hopefully. From a nutritional and thermoregulatory point of view, high-density colostrum should be fed to newborn calves as soon as possible, as their energy (mainly in the form of fat) and protein requirements are very high. Minerals and vitamins are micronutrients present in colostrum that are essential for normal biological functions. However, there have not been many recent studies, and many of the reported concentrations would benefit from more recent studies using updated techniques. Colostrum also has many bioactive constituents that play multiple roles in the newborn calf. They may act as mediators in the passive transfer of immunity, defend against pathogens, contribute to cellular differentiation and growth, and possibly act as maternal–offspring signaling molecules. However, only a few have been studied. The most studied to date are immunoglobulins (especially IgG), growth factors (mainly IGF-I and IGF-II), and fatty acids (mainly n-3 and n-6 fatty acids). More recently, other bioactive components such as lactoferrin, oligosaccharides, and microRNAs have received more attention. However, there is still a lack of research on some components, such as many hormones whose concentrations in colostrum are not known, as well as the role of nucleic acids, enzymes, cytokines, hormones, and amino acids in newborns. Most, if not all, of these components appear to be important for the newborn calf, as some of them have significantly higher concentrations in colostrum or in colostrum and transition milk compared to whole milk. Cells in colostrum are important modulators of the calf's immune system but have not been properly studied; they appear to lose viability during the freezing process, which is nefarious as, for practical reasons, many calves receive thawed colostrum from a colostrum bank. Finally, research into the colostrum microbiota has increased recently and its role in colonizing the neonatal gut is crucial for a better understanding of the digestive and absorptive processes of colostrum components, as well as its role in neonatal enteric disease. As morbidity and mortality rates in dairy calves remain high during the first days of life, research into the components and properties of colostrum could help improve calf health and welfare. Future areas of study could focus on colostrum proteomics and metabolomics, oligosaccharides, fatty acids, miRNAs, immune cells, and the microbiome. This opens research opportunities, as bovine colostrum can be used not only for the newborn calf but also as a nutraceutical and therapeutic agent for animals and humans.

Supplementary Materials: The following supporting information can be downloaded at: https://www.mdpi.com/article/10.3390/ani14071130/s1, Table S1: Qualitative comparison of lipid content between colostrum and either transitional or mature milk. Table S2: Most abundant sialylated oligosaccharides in bovine colostrum and corresponding concentrations (mg/L). Mean values from each reference are shown, as well as a calculated mean, standard deviation (SD), and coefficient of variation (CV). Figure S1: Box plot of the range of macronutrient concentrations in colostrum. The graphic was constructed with the means reported in Table 1. Figure S2: Dot-plot of the range of major protein concentrations in colostrum. Dots from the same study have the same colour. Figure S3: Dot-plot of the range in the minor protein concentrations in colostrum. Dots from the same study have the same colour. Figure S4: Dot-plot of the range in the major oligosaccharide's concentrations in colostrum.

Author Contributions: Conceptualization, F.G.S., J.L.C., C.C. and S.R.S.; writing—original draft preparation, F.G.S., A.M.F.P., J.L.C., C.C. and S.R.S.; writing—review and editing, F.G.S., A.M.F.P., J.L.C., C.C. and S.R.S.; visualization, supervision, project administration, funding acquisition, J.L.C., C.C. and S.R.S. All authors have read and agreed to the published version of the manuscript.

Funding: This work is funded by National Funds through FCT—Foundation for Science and Technology under the Projects CECAV UIDB/00772/2020 (https://doi.org/10.54499/UIDB/00772/2020), MED UIDB/05183/2020 (https://doi.org/10.54499/UIDB/05183/2020; https://doi.org/10.54499/UIDP/05183/2020) and CHANGE (https://doi.org/10.54499/LA/P/0121/2020) and F.G.S. PhD grant UI/BD/150834/2021.

Institutional Review Board Statement: Not applicable.

Informed Consent Statement: Not applicable.

Data Availability Statement: The original contributions presented in the study are included in the article/supplementary material, further inquiries can be directed to the corresponding author.

Acknowledgments: The author F.G.S., would like to thank Maria José Marques Gomes, Department of Zootechnics—UTAD, for the encouragement in writing this review.

Conflicts of Interest: The authors declare no conflicts of interest.

References

1. Lopez, A.J.; Heinrichs, A.J. Invited review: The importance of colostrum in the newborn dairy calf. *J. Dairy Sci.* **2022**, *105*, 2733–2749. [CrossRef] [PubMed]
2. Baumrucker, C.R.; Gross, J.J.; Bruckmaier, R.M. The importance of colostrum in maternal care and its formation in mammalian species. *Anim. Front.* **2023**, *13*, 37–43. [CrossRef] [PubMed]
3. Quigley, J.D.; Drewry, J.J. Nutrient and Immunity Transfer from Cow to Calf Pre- and Postcalving. *J. Dairy Sci.* **1998**, *81*, 2779–2790. [CrossRef] [PubMed]
4. Godden, S.M.; Lombard, J.E.; Woolums, A.R. Colostrum Management for Dairy Calves. *Vet. Clin. North Am. Food Anim. Pract.* **2019**, *35*, 535–556. [CrossRef] [PubMed]
5. Chucri, T.M.; Monteiro, J.M.; Lima, A.R.; Salvadori, M.L.B.; Kfoury, J.R., Jr.; Miglino, M.A. A review of immune transfer by the placenta. *J. Reprod. Immunol.* **2010**, *87*, 14–20. [CrossRef] [PubMed]
6. Bigler, N.A.; Bruckmaier, R.M.; Gross, J.J. Implications of placentation type on species-specific colostrum properties in mammals. *J. Anim. Sci.* **2022**, *100*, skac287. [CrossRef]
7. McGrath, B.A.; Fox, P.F.; McSweeney, P.L.H.; Kelly, A.L. Composition and properties of bovine colostrum: A review. *Dairy Sci. Technol.* **2016**, *96*, 133–158. [CrossRef]
8. Le Jan, C. Cellular components of mammary secretions and neonatal immunity: A review. *Vet. Res.* **1996**, *27*, 403–417. [PubMed]
9. Stelwagen, K.; Carpenter, E.; Haigh, B.; Hodgkinson, A.; Wheeler, T.T. Immune components of bovine colostrum and milk. *J. Anim. Sci.* **2009**, *87*, 3–9. [CrossRef]
10. Korhonen, H.J. Bioactive milk proteins, peptides and lipids and other functional components derived from milk and bovine colostrum. In *Functional Foods*; Elsevier: Amsterdam, The Netherlands, 2011; pp. 471–511. ISBN 9781845696900.
11. Blum, J.W.; Hammon, H. Colostrum effects on the gastrointestinal tract, and on nutritional, endocrine and metabolic parameters in neonatal calves. *Livest. Prod. Sci.* **2000**, *66*, 151–159. [CrossRef]
12. Sonkoly, E.; Ståhle, M.; Pivarcsi, A. MicroRNAs and immunity: Novel players in the regulation of normal immune function and inflammation. *Semin. Cancer Biol.* **2008**, *18*, 131–140. [CrossRef] [PubMed]
13. Van Hese, I.; Goossens, K.; Vandaele, L.; Opsomer, G. Invited review: MicroRNAs in bovine colostrum—Focus on their origin and potential health benefits for the calf. *J. Dairy Sci.* **2020**, *103*, 1–15. [CrossRef] [PubMed]

14. Lima, S.F.; Teixeira, A.G.V.; Lima, F.S.; Ganda, E.K.; Higgins, C.H.; Oikonomou, G.; Bicalho, R.C. The bovine colostrum microbiome and its association with clinical mastitis. *J. Dairy Sci.* **2017**, *100*, 3031–3042. [CrossRef]
15. Hammon, H.M.; Steinhoff-Wagner, J.; Schönhusen, U.; Metges, C.C.; Blum, J.W. Energy metabolism in the newborn farm animal with emphasis on the calf: Endocrine changes and responses to milk-born and systemic hormones. *Domest. Anim. Endocrinol.* **2012**, *43*, 171–185. [CrossRef]
16. Kaushik, A.K.; Kandavel, H.; Nalpathamkalam, T.; Pasman, Y. Bovine neonate is deficient in innate immunity at birth. *Mol. Immunol.* **2021**, *133*, 101–109. [CrossRef] [PubMed]
17. Larson, B.L.; Heary, H.L.; Devery, J.E. Immunoglobulin Production and Transport by the Mammary Gland. *J. Dairy Sci.* **1980**, *63*, 665–671. [CrossRef] [PubMed]
18. Quigley, J.D.; Martin, K.R.; Dowlen, H.H.; Wallis, L.B.; Lamar, K. Immunoglobulin Concentration, Specific Gravity, and Nitrogen Fractions of Colostrum from Jersey Cattle. *J. Dairy Sci.* **1994**, *77*, 264–269. [CrossRef]
19. Costa, A.; Goi, A.; Penasa, M.; Nardino, G.; Posenato, L.; De Marchi, M. Variation of immunoglobulins G, A, and M and bovine serum albumin concentration in Holstein cow colostrum. *Animal* **2021**, *15*, 100299. [CrossRef]
20. Altvater-Hughes, T.E.; Hodgins, D.C.; Wagter-Lesperance, L.; Beard, S.C.; Cartwright, S.L.; Mallard, B.A. Concentration and heritability of immunoglobulin G and natural antibody immunoglobulin M in dairy and beef colostrum along with serum total protein in their calves. *J. Anim. Sci.* **2022**, *100*, skac006. [CrossRef]
21. Urakawa, M.; Baakhtari, M.; Ramah, A.; Imatake, S.; Ahmadi, P.; Deguchi, Y.; Uematsu, M.; Nakama, Y.; Imabeppu, K.; Nomura, Y.; et al. Comparative Analysis of Maternal Colostrum and Colostrum Replacer Effects on Immunity, Growth, and Health of Japanese Black Calves. *Animals* **2024**, *14*, 346. [CrossRef]
22. Foley, J.A.; Otterby, D.E. Availability, Storage, Treatment, Composition, and Feeding Value of Surplus Colostrum: A Review. *J. Dairy Sci.* **1978**, *61*, 1033–1060. [CrossRef]
23. Baumrucker, C.R.; Bruckmaier, R.M. Colostrogenesis: IgG1 transcytosis mechanisms. *J. Mammary Gland Biol. Neoplasia* **2014**, *19*, 103–117. [CrossRef] [PubMed]
24. Quinn, E.M.; O'Callaghan, T.F.; Tobin, J.T.; Murphy, J.P.; Sugrue, K.; Slattery, H.; O'Donovan, M.; Hickey, R.M. Changes to the Oligosaccharide Profile of Bovine Milk at the Onset of Lactation. *Dairy* **2020**, *1*, 284–296. [CrossRef]
25. Van Soest, B.; Cullens, F.; VandeHaar, M.J.; Nielsen, M.W. Short communication: Effects of transition milk and milk replacer supplemented with colostrum replacer on growth and health of dairy calves. *J. Dairy Sci.* **2020**, *103*, 12104–12108. [CrossRef] [PubMed]
26. Valldecabres, A.; Lopes, R.B.; Lago, A.; Blanc, C.; Silva-del-Río, N. Effects of postpartum milking strategy on plasma mineral concentrations and colostrum, transition milk, and milk yield and composition in multiparous dairy cows. *J. Dairy Sci.* **2022**, *105*, 595–608. [CrossRef] [PubMed]
27. Wilms, J.N.; Hare, K.S.; Fischer-Tlustos, A.J.; Vahmani, P.; Dugan, M.E.R.; Leal, L.N.; Steele, M.A. Fatty acid profile characterization in colostrum, transition milk, and mature milk of primi- and multiparous cows during the first week of lactation. *J. Dairy Sci.* **2022**, *105*, 2612–2630. [CrossRef] [PubMed]
28. Conneely, M.; Berry, D.P.; Murphy, J.P.; Lorenz, I.; Doherty, M.L.; Kennedy, E. Effect of feeding colostrum at different volumes and subsequent number of transition milk feeds on the serum immunoglobulin G concentration and health status of dairy calves. *J. Dairy Sci.* **2014**, *97*, 6991–7000. [CrossRef] [PubMed]
29. Godden, S.M.; Haines, D.M.; Hagman, D. Improving passive transfer of immunoglobulins in calves. I: Dose effect of feeding a commercial colostrum replacer. *J. Dairy Sci.* **2009**, *92*, 1750–1757. [CrossRef]
30. Hammon, H.M.; Steinhoff-Wagner, J.; Flor, J.; Schönhusen, U.; Metges, C.C. Lactation Biology Symposium: Role of colostrum and colostrum components on glucose metabolism in neonatal calves. *J. Anim. Sci.* **2013**, *91*, 685–695. [CrossRef]
31. Cantor, M.C.; Renaud, D.L.; Costa, J.H.C. Nutraceutical intervention with colostrum replacer: Can we reduce disease hazard, ameliorate disease severity, and improve performance in preweaned dairy calves? *J. Dairy Sci.* **2021**, *104*, 7168–7176. [CrossRef]
32. Kessler, E.C.; Bruckmaier, R.M.; Gross, J.J. Immunoglobulin G content and colostrum composition of different goat and sheep breeds in Switzerland and Germany. *J. Dairy Sci.* **2019**, *102*, 5542–5549. [CrossRef] [PubMed]
33. Sánchez-Macías, D.; Moreno-Indias, I.; Castro, N.; Morales-delaNuez, A.; Argüello, A. From goat colostrum to milk: Physical, chemical, and immune evolution from partum to 90 days postpartum. *J. Dairy Sci.* **2014**, *97*, 10–16. [CrossRef] [PubMed]
34. Hernández-Castellano, L.E.; Almeida, A.M.; Renaut, J.; Argüello, A.; Castro, N. A proteomics study of colostrum and milk from the two major small ruminant dairy breeds from the Canary Islands: A bovine milk comparison perspective. *J. Dairy Res.* **2016**, *83*, 366–374. [CrossRef] [PubMed]
35. González-Cabrera, M.; Torres, A.; Salomone-Caballero, M.; Castro, N.; Argüello, A.; Hernández-Castellano, L.E. Intramammary administration of lipopolysaccharides at parturition enhances immunoglobulin concentration in goat colostrum. *Animal* **2024**, *18*, 101082. [CrossRef] [PubMed]
36. Banchero, G.E.; Quintans, G.; Martin, G.B.; Milton, J.T.B.; Lindsay, D.R. Nutrition and colostrum production in sheep. 2. Metabolic and hormonal responses to different energy sources in the final stages of pregnancy. *Reprod. Fertil. Dev.* **2004**, *16*, 645–653. [CrossRef] [PubMed]

37. Alves, A.C.; Alves, N.G.; Ascari, I.J.; Junqueira, F.B.; Coutinho, A.S.; Lima, R.R.; Pérez, J.R.O.; De Paula, S.O.; Furusho-Garcia, I.F.; Abreu, L.R. Colostrum composition of Santa Inês sheep and passive transfer of immunity to lambs. *J. Dairy Sci.* **2015**, *98*, 3706–3716. [CrossRef] [PubMed]
38. Ahmadi, M.; Boldura, O.; Milovanov, C.; Dronca, D.; Mircu, C.; Hutu, I.; Popescu, S.; Padeanu, I.; Tulcan, C. Colostrum from Different Animal Species—A Product for Health Status Enhancement. *Bull. UASVM Anim. Sci. Biotechnol.* **2016**, *73*, 95–100. [CrossRef] [PubMed]
39. Dande, N.D.; Nande, P.J. Nutritional Composition of Bovine Colostrum: Palatability Evaluation of Food Products Prepared Using Bovine Colostrum. *Int. J. Nutr. Pharmacol. Neurol. Dis.* **2020**, *10*, 8–13.
40. Arslan, A.; Kaplan, M.; Duman, H.; Bayraktar, A.; Ertürk, M.; Henrick, B.M.; Frese, S.A.; Karav, S. Bovine Colostrum and Its Potential for Human Health and Nutrition. *Front. Nutr.* **2021**, *8*, 651721. [CrossRef]
41. Polidori, P.; Rapaccetti, R.; Klimanova, Y.; Zhang, J.J.; Santini, G.; Vincenzetti, S. Nutritional Parameters in Colostrum of Different Mammalian Species. *Beverages* **2022**, *8*, 54. [CrossRef]
42. Alsayed, A.R.; Hasoun, L.Z.; Khader, H.A.; Basheti, I.A.; Permana, A.D. Bovine Colostrum Treatment of Specific Cancer Types: Current Evidence and Future Opportunities. *Molecules* **2022**, *27*, 8641. [CrossRef]
43. Costa, A.; Sneddon, N.W.; Goi, A.; Visentin, G.; Mammi, L.M.E.; Savarino, E.V.; Zingone, F.; Formigoni, A.; Penasa, M.; De Marchi, M. Invited review: Bovine colostrum, a promising ingredient for humans and animals—Properties, processing technologies, and uses. *J. Dairy Sci.* **2023**, *106*, 5197–5217. [CrossRef] [PubMed]
44. Poonia, A. Shiva Bioactive compounds, nutritional profile and health benefits of colostrum: A review. *Food Prod. Process. Nutr.* **2022**, *4*, 26. [CrossRef]
45. Bartier, A.L.; Windeyer, M.C.; Doepel, L. Evaluation of on-farm tools for colostrum quality measurement. *J. Dairy Sci.* **2015**, *98*, 1878–1884. [CrossRef] [PubMed]
46. Kessler, E.C.; Bruckmaier, R.M.; Gross, J.J. Short communication: Comparative estimation of colostrum quality by Brix refractometry in bovine, caprine, and ovine colostrum. *J. Dairy Sci.* **2021**, *104*, 2438–2444. [CrossRef]
47. Fleenor, W.A.; Stott, G.H. Hydrometer Test for Estimation of Immunoglobulin Concentration in Bovine Colostrum. *J. Dairy Sci.* **1980**, *63*, 973–977. [CrossRef] [PubMed]
48. Fischer-Tlustos, A.J.; Lopez, A.; Hare, K.S.; Wood, K.M.; Steele, M.A. Effects of colostrum management on transfer of passive immunity and the potential role of colostral bioactive components on neonatal calf development and metabolism. *Can. J. Anim. Sci.* **2021**, *101*, 405–426. [CrossRef]
49. Gulliksen, S.M.; Lie, K.I.; Sølverød, L.; Østerås, O. Risk factors associated with colostrum quality in Norwegian dairy cows. *J. Dairy Sci.* **2008**, *91*, 704–712. [CrossRef] [PubMed]
50. Conneely, M.; Berry, D.P.; Sayers, R.; Murphy, J.P.; Lorenz, I.; Doherty, M.L.; Kennedy, E. Factors associated with the concentration of immunoglobulin G in the colostrum of dairy cows. *Animal* **2013**, *7*, 1824–1832. [CrossRef] [PubMed]
51. Gelsinger, S.L.; Smith, A.M.; Jones, C.M.; Heinrichs, A.J. Technical note: Comparison of radial immunodiffusion and ELISA for quantification of bovine immunoglobulin G in colostrum and plasma. *J. Dairy Sci.* **2015**, *98*, 4084–4089. [CrossRef]
52. Zarei, S.; Reza Ghorbani, G.; Khorvash, M.; Martin, O.; Hossein Mahdavi, A.; Riasi, A. The Impact of Season, Parity, and Volume of Colostrum on Holstein Dairy Cows Colostrum Composition. *Agric. Sci.* **2017**, *08*, 572–581. [CrossRef]
53. Soufleri, A.; Banos, G.; Panousis, N.; Fletouris, D.; Arsenos, G.; Kougioumtzis, A.; Valergakis, G.E. Evaluation of factors affecting colostrum quality and quantity in holstein dairy cattle. *Animals* **2021**, *11*, 2005. [CrossRef] [PubMed]
54. Parrish, D.B.; Wise, G.H.; Hughes, J.S.; Atkeson, F.W. Properties of the Colostrum of the Dairy Cow. V. Yield, Specific Gravity and Concentrations of Total Solids and its Various Components of Colostrum and Early Milk. *J. Dairy Sci.* **1950**, *33*, 457–465. [CrossRef]
55. Abd El-Fattah, A.M.; Abd Rabo, F.H.R.; EL-Dieb, S.M.; El-Kashef, H.A. Changes in composition of colostrum of Egyptian buffaloes and Holstein cows. *BMC Vet. Res.* **2012**, *8*, 19. [CrossRef] [PubMed]
56. Kehoe, S.I.; Jayarao, B.M.; Heinrichs, A.J. A survey of bovine colostrum composition and colostrum management practices on Pennsylvania dairy farms. *J. Dairy Sci.* **2007**, *90*, 4108–4116. [CrossRef]
57. Lokke, M.M.; Engelbrecht, R.; Wiking, L. Covariance structures of fat and protein influence the estimation of IgG in bovine colostrum. *J. Dairy Res.* **2016**, *83*, 58–66. [CrossRef] [PubMed]
58. Bielmann, V.; Gillan, J.; Perkins, N.R.; Skidmore, A.L.; Godden, S.; Leslie, K.E. An evaluation of Brix refractometry instruments for measurement of colostrum quality in dairy cattle. *J. Dairy Sci.* **2010**, *93*, 3713–3721. [CrossRef] [PubMed]
59. Silva, F.G.; Castelo-Branco, A.M.; Conceição, C.; Cerqueira, J.O.L.; Silva, S.R.; Pereira, A.M.F. Effect of parity and age in dairy cows' colostrum quality. In *International Scientific Meeting on Colostrum*; Universidad de Las Palmas de Gran Canaria: Las Palmas, Gran Canaria, 2022; p. 58.
60. Turini, L.; Conte, G.; Bonelli, F.; Sgorbini, M.; Madrigali, A.; Mele, M. The relationship between colostrum quality, passive transfer of immunity and birth and weaning weight in neonatal calves. *Livest. Sci.* **2020**, *238*, 104033. [CrossRef]
61. Morrill, K.M.; Conrad, E.; Lago, A.; Campbell, J.; Quigley, J.; Tyler, H. Nationwide evaluation of quality and composition of colostrum on dairy farms in the United States. *J. Dairy Sci.* **2012**, *95*, 3997–4005. [CrossRef] [PubMed]

62. Nazir, T.; Pal, M.A.; Manzoor, A.; Padder, T.; Sofi, A.H.; Wani, S.A.; Ahmad, S.R.; Malik, A.H. Study on the Effect of the Breed during Post-Partum Transition Period on Various Physico-Chemical, Compositional and Microbiological Characteristics of Bovine Colostrum. *Asian J. Dairy Food Res.* **2020**, *39*, 1. [CrossRef]
63. Morrill, K.M.; Robertson, K.E.; Spring, M.M.; Robinson, A.L.; Tyler, H.D. Validating a refractometer to evaluate immunoglobulin G concentration in Jersey colostrum and the effect of multiple freeze-thaw cycles on evaluating colostrum quality. *J. Dairy Sci.* **2015**, *98*, 595–601. [CrossRef] [PubMed]
64. Nardone, A.; Lacetera, N.; Bernabucci, U.; Ronchi, B. Composition of Colostrum from Dairy Heifers Exposed to High Air Temperatures during Late Pregnancy and the Early Postpartum Period. *J. Dairy Sci.* **1997**, *80*, 838–844. [CrossRef] [PubMed]
65. Goi, A.; Costa, A.; Visentin, G.; De Marchi, M. Mid-infrared spectroscopy for large-scale phenotyping of bovine colostrum gross composition and immunoglobulin concentration. *J. Dairy Sci.* **2023**, *106*, 6388–6401. [CrossRef] [PubMed]
66. Dunn, A.; Ashfield, A.; Earley, B.; Welsh, M.; Gordon, A.; Morrison, S.J. Evaluation of factors associated with immunoglobulin G, fat, protein, and lactose concentrations in bovine colostrum and colostrum management practices in grassland-based dairy systems in Northern Ireland. *J. Dairy Sci.* **2017**, *100*, 2068–2079. [CrossRef] [PubMed]
67. Kessler, E.C.; Bruckmaier, R.M.; Gross, J.J. Colostrum composition and immunoglobulin G content in dairy and dual-purpose cattle breeds. *J. Anim. Sci.* **2020**, *98*, skaa237. [CrossRef] [PubMed]
68. Tsioulpas, A.; Grandison, A.S.; Lewis, M.J. Changes in physical properties of bovine milk from the colostrum period to early lactation. *J. Dairy Sci.* **2007**, *90*, 5012–5017. [CrossRef] [PubMed]
69. Cerqueira, J. Avaliação de Bem-Estar Animal em Bovinos de Leite na Região Norte de Portugal. Ph.D. Thesis, Instituto de Ciências Biomédicas Abel Salazar da Universidade do Porto, Porto, Portugal, 2012.
70. Atkeson, F.W.; Warren, T.R.; Anderson, G.C. Water Requirements of Dairy Calves. *J. Dairy Sci.* **1934**, *17*, 249–256. [CrossRef]
71. Kertz, A.F.; Reutzel, L.F.; Mahoney, J.H. Ad Libitum Water Intake by Neonatal Calves and Its Relationship to Calf Starter Intake, Weight Gain, Feces Score, and Season. *J. Dairy Sci.* **1984**, *67*, 2964–2969. [CrossRef]
72. Marnila, P.; Korhonen, H. Milk | Colostrum. In *Encyclopedia of Dairy Sciences*; Elsevier: Amsterdam, The Netherlands, 2011; pp. 591–597.
73. Hammon, H.M.; Zanker, I.A.; Blum, J.W. Delayed colostrum feeding affects IGF-I and insulin plasma concentrations in neonatal calves. *J. Dairy Sci.* **2000**, *83*, 85–92. [CrossRef] [PubMed]
74. Lim, D.-H.; Mayakrishnan, V.; Lee, H.-J.; Ki, K.-S.; Kim, T.-I.; Kim, Y. A comparative study on milk composition of Jersey and Holstein dairy cows during the early lactation. *J. Anim. Sci. Technol.* **2020**, *62*, 565–576. [CrossRef]
75. Silva, E.; Anaya, K.; Bezerra, M.D.; Borba, L.; Barbosa, I.; Oliveira, J.; Urbano, S.; Macêdo, C.; Júnior, D.; Gama, M.; et al. Physicochemical Characterization and Brix in Jersey Cow Colostrum in Tropical Conditions. *Int. J. Agric. Biol.* **2021**, *26*, 139–144. [CrossRef]
76. Mehra, R.; Kumar, S.; Verma, N.; Kumar, N.; Singh, R.; Bhardwaj, A.; Nayan, V.; Kumar, H. Chemometric approaches to analyze the colostrum physicochemical and immunological (IgG) properties in the recently registered Himachali Pahari cow breed in India. *LWT* **2021**, *145*, 111256. [CrossRef]
77. Pritchett, L.C.; Gay, C.C.; Besser, T.E.; Hancock, D.D. Management and Production Factors Influencing Immunoglobulin G1 Concentration in Colostrum from Holstein Cows. *J. Dairy Sci.* **1991**, *74*, 2336–2341. [CrossRef] [PubMed]
78. Svensson, C.; Lundborg, K.; Emanuelson, U.; Olsson, S.O. Morbidity in Swedish dairy calves from birth to 90 days of age and individual calf-level risk factors for infectious diseases. *Prev. Vet. Med.* **2003**, *58*, 179–197. [CrossRef] [PubMed]
79. Denholm, K.S.; McDougall, S.; Chambers, G.; Clough, W. Factors associated with colostrum quality in individual cows from dairy herds in the Waikato region of New Zealand. *N. Z. Vet. J.* **2018**, *66*, 115–120. [CrossRef] [PubMed]
80. Brooker, B.; Holt, C. Natural variations in the average size of bovine casein micelles: III. Studies on colostrum by electron by microscopy and light scattering. *J. Dairy Res* **1978**, *45*, 355–362. [CrossRef]
81. Sobczuk-Szul, M.; Wielgosz-Groth, Z.; Wronski, M.; Rzemieniewski, A. Changes in the bioactive protein concentrations in the bovine colostrum of Jersey and Polish Holstein–Friesian cows. *Turkish J. Vet. Anim. Sci.* **2013**, *37*, 43–49. [CrossRef]
82. Sats, A.; Kaart, T.; Jõudu, I. Bovine colostrum casein: Post-partum dynamics of micelle size, content, and associated traits. *Int. Dairy J.* **2024**, *148*, 105791. [CrossRef]
83. Playford, R.J.; Woodman, A.C.; Vesey, D.; Deprez, P.H.; Calam, J.; Watanapa, P.; Williamson, R.C.N.; Clark, P. Effect of luminal growth factor preservation on intestinal growth. *Lancet* **1993**, *341*, 843–848. [CrossRef]
84. Playford, R.J.; Weiser, M.J. Bovine colostrum: Its constituents and uses. *Nutrients* **2021**, *13*, 265. [CrossRef]
85. Yang, M.; Cao, X.; Wu, R.; Liu, B.; Ye, W.; Yue, X.; Wu, J. Comparative proteomic exploration of whey proteins in human and bovine colostrum and mature milk using iTRAQ-coupled LC-MS/MS. *Int. J. Food Sci. Nutr.* **2017**, *68*, 671–681. [CrossRef] [PubMed]
86. Madureira, A.R.; Pereira, C.I.; Gomes, A.M.P.; Pintado, M.E.; Xavier Malcata, F. Bovine whey proteins—Overview on their main biological properties. *Food Res. Int.* **2007**, *40*, 1197–1211. [CrossRef]
87. Quigley, J.D.; Martin, K.R.; Dowlen, H.H. Concentrations of Trypsin Inhibitor and Immunoglobulins in Colostrum of Jersey Cows. *J. Dairy Sci.* **1995**, *78*, 1573–1577. [CrossRef] [PubMed]
88. Yvon, M.; Leveăœx, D.; Alloy, M.V.; Lissier, J.P.Ã.; Mirahd, P.P.; De Recherches, S.; Josas, J. Nutrient Requirements and Interactions Colostrum Protein Digestion in Newborn Lambs. *J. Nutr.* **1993**, *123*, 586–596. [CrossRef] [PubMed]

89. Miyazaki, T.; Okada, K.; Miyazaki, M. Short communication: Neonatal calves coagulate first-milking colostrum and produce a large curd for efficient absorption of immunoglobulins after first ingestion. *J. Dairy Sci.* **2017**, *100*, 7262–7270. [CrossRef] [PubMed]
90. Perez, M.; Sanchez, L.; Aranda, P.; Ena, J.; Oria, R.; Calvo, M. Synthesis and evolution of concentration of beta- lactoglobulin and alpha-lactalbumin from cow and sheep colostrum and milk throughout early lactation. *Cell. Mol. Biol.* **1990**, *36*, 205–212. [PubMed]
91. Perez, M.D.; Sanchez, L.; Aranda, P.; Ena, J.; Oria, R.; Calvo, M. Effect of β-lactoglobulin on the activity of pregastric lipase. A possible role for this protein in ruminant milk. *Biochim. Biophys. Acta/Lipids Lipid Metab.* **1992**, *1123*, 151–155. [CrossRef] [PubMed]
92. Kamau, S.M.; Cheison, S.C.; Chen, W.; Liu, X.; Lu, R. Alpha-Lactalbumin: Its Production Technologies and Bioactive Peptides. *Compr. Rev. Food Sci. Food Saf.* **2010**, *9*, 197–212. [CrossRef]
93. Chen, C.B.; Hammo, B.; Barry, J.; Radhakrishnan, K. Overview of Albumin Physiology and its Role in Pediatric Diseases. *Curr. Gastroenterol. Rep.* **2021**, *23*, 11. [CrossRef]
94. Levieux, D.; Ollier, A. Bovine immunoglobulin G, β-lactoglobulin, α-lactalbumin and serum albumin in colostrum and milk during the early post partum period. *J. Dairy Res.* **1999**, *66*, 421–430. [CrossRef]
95. Samarütel, J.; Baumrucker, C.R.; Gross, J.J.; Dechow, C.D.; Bruckmaier, R.M. Quarter variation and correlations of colostrum albumin, immunoglobulin G1 and G2 in dairy cows. *J. Dairy Res.* **2016**, *83*, 209–218. [CrossRef] [PubMed]
96. Besser, T.E.; Osborn, D. Effect of bovine serum albumin on passive transfer of immunoglobulin G1 to newborn calves. *Vet. Immunol. Immunopathol.* **1993**, *37*, 321–327. [CrossRef] [PubMed]
97. Gosch, T.; Apprich, S.; Kneifel, W.; Novalin, S. Improved isolation of bioactive components of bovine colostrum using cross-flow microfiltration. *Int. J. Dairy Technol.* **2013**, *66*, 175–181. [CrossRef]
98. Tacoma, R.; Gelsinger, S.L.; Lam, Y.W.; Scuderi, R.A.; Ebenstein, D.B.; Heinrichs, A.J.; Greenwood, S.L. Exploration of the bovine colostrum proteome and effects of heat treatment time on colostrum protein profile. *J. Dairy Sci.* **2017**, *100*, 9392–9401. [CrossRef]
99. Puppel, K.; Gołębiewski, M.; Konopka, K.; Kunowska-Slósarz, M.; Slósarz, J.; Grodkowski, G.; Przysucha, T.; Balcerak, M.; Madras-Majewska, B.; Sakowski, T. Relationship between the quality of colostrum and the formation of microflora in the digestive tract of calves. *Animals* **2020**, *10*, 1293. [CrossRef] [PubMed]
100. Wąsowska, E.; Puppel, K. Changes in the content of immunostimulating components of colostrum obtained from dairy cows at different levels of production. *J. Sci. Food Agric.* **2018**, *98*, 5062–5068. [CrossRef] [PubMed]
101. Lin, S.; Wan, Z.; Zhang, J.; Xu, L.; Han, B.; Sun, D. Genome-Wide Association Studies for the Concentration of Albumin in Colostrum and Serum in Chinese Holstein. *Animals* **2020**, *10*, 2211. [CrossRef]
102. Korhonen, H. Antimicrobial factors in bovine colostrum. *Agric. Food Sci.* **1977**, *49*, 434–447. [CrossRef]
103. Sánchez, L.; Aranda, P.; Pérez, M.; Calvo, M. Concentration of Lactoferrin and Transferrin throughout Lactation in Cow's Colostrum and Milk. *Biol. Chem. Hoppe. Seyler.* **1988**, *369*, 1005–1008. [CrossRef]
104. Tsuji, S.; Hirata, Y.; Mukai, F.; Ohtagaki, S. Comparison of Lactoferrin Content in Colostrum Between Different Cattle Breeds. *J. Dairy Sci.* **1990**, *73*, 125–128. [CrossRef]
105. Sacerdote, P.; Mussano, F.; Franchi, S.; Panerai, A.E.; Bussolati, G.; Carossa, S.; Bartorelli, A.; Bussolati, B. Biological components in a standardized derivative of bovine colostrum. *J. Dairy Sci.* **2013**, *96*, 1745–1754. [CrossRef]
106. Martinsson, K.; Möllerberg, L. On the Transferrin Concentration in Blood Serum of Growing calves and in Bovine Colostrum. *Zentralblatt Veterinärmedizin Reihe A* **1973**, *20*, 277–284. [CrossRef]
107. Puppel, K.; Golebiewski, M.; Grodkowski, G.; Solarczyk, P.; Kostusiak, P.; Klopcic, M.; Sakowski, T. Use of somatic cell count as an indicator of colostrum quality. *PLoS ONE* **2020**, *15*, e0237615. [CrossRef]
108. Pecka-Kiełb, E.; Zachwieja, A.; Wojtas, E.; Zawadzki, W. Influence of nutrition on the quality of colostrum and milk of ruminants. *Mljekarstvo* **2018**, *68*, 169–181. [CrossRef]
109. Verardo, V.; Gómez-Caravaca, A.M.; Arráez-Román, D.; Hettinga, K. Recent advances in phospholipids from colostrum, milk and dairy by-products. *Int. J. Mol. Sci.* **2017**, *18*, 173. [CrossRef]
110. Bayly, G.R. Lipids and disorders of lipoprotein metabolism. In *Clinical Biochemistry: Metabolic and Clinical Aspects*; Marshall, W.J., Lapsley, M., Day, A.P., Ayling, R.M., Eds.; Elsevier: Amsterdam, The Netherlands, 2014; pp. 702–736. ISBN 9780702054785.
111. Heid, H.W.; Keenan, T.W. Intracellular origin and secretion of milk fat globules. *Eur. J. Cell Biol.* **2005**, *84*, 245–258. [CrossRef]
112. Sangild, P.T.; Vonderohe, C.; Melendez Hebib, V.; Burrin, D.G. Potential benefits of bovine colostrum in pediatric nutrition and health. *Nutrients* **2021**, *13*, 2551. [CrossRef] [PubMed]
113. Zou, X.; Guo, Z.; Jin, Q.; Huang, J.; Cheong, L.; Xu, X.; Wang, X. Composition and microstructure of colostrum and mature bovine milk fat globule membrane. *Food Chem.* **2015**, *185*, 362–370. [CrossRef]
114. Lee, H.; Padhi, E.; Hasegawa, Y.; Larke, J.; Parenti, M.; Wang, A.; Hernell, O.; Lönnerdal, B.; Slupsky, C. Compositional dynamics of the milk fat globule and its role in infant development. *Front. Pediatr.* **2018**, *6*, 313. [CrossRef] [PubMed]
115. Zentek, J.; Buchheit-Renko, S.; Ferrara, F.; Vahjen, W.; Van Kessel, A.G.; Pieper, R. Nutritional and physiological role of medium-chain triglycerides and medium-chain fatty acids in piglets. *Anim. Health Res. Rev.* **2011**, *12*, 83–93. [CrossRef] [PubMed]
116. Jensen, R.G. The Composition of Bovine Milk Lipids: January 1995 to December 2000. *J. Dairy Sci.* **2002**, *85*, 295–350. [CrossRef] [PubMed]
117. Lichtenstein, A.H. Fats and Oils. In *Encyclopedia of Human Nutrition*; Caballero, B., Ed.; Elsevier: Amsterdam, The Netherlands, 2013; pp. 201–208, ISBN 9780123848857.

118. Vermorel, M.; Vernet, J.; Dardillat, C.; Saido; Demigne, C. Energy metabolism and thermoregulation in the newborn calf; variations during the first day of life and differences between breeds. *Can. J. Anim. Sci.* **1989**, *69*, 103–111. [CrossRef]
119. Okamoto, M.; Robinson, J.B.; Christopherson, R.J.; Young, B.A. Summit Metabolism of Newborn Calves with and without Colostrum Feeding. *Can. J. Anim. Sci.* **1986**, *66*, 937–944. [CrossRef]
120. Alexander, G.; Bennett, J.W.; Gemmell, R.T. Brown adipose tissue in the new-born calf (Bos taurus). *J. Physiol.* **1975**, *244*, 223–234. [CrossRef] [PubMed]
121. Smith, S.B.; Carstens, G.E. Chapter 12—Ontogeny and metabolism of brown adipose tissue in livestock species. In *Biology of Growing Animals*; Elsevier Ltd.: Amsterdam, The Netherlands, 2005; Volume 3, pp. 303–322.
122. Silva, F.L.M.; Miqueo, E.; da Silva, M.D.; Torrezan, T.M.; Rocha, N.B.; Salles, M.S.V.; Bittar, C.M.M. Thermoregulatory responses and performance of dairy calves fed different amounts of colostrum. *Animals* **2021**, *11*, 703. [CrossRef] [PubMed]
123. Quigley, J.D. Calf Note 141–Further Thoughts on Colostrum Variation. 2009. Available online: https://www.calfnotes.com/ (accessed on 28 March 2024).
124. Hammon, H.M.; Blum, J.W. Feeding different amounts of colostrum or only milk replacer modify receptors of intestinal insulin-like growth factors and insulin in neonatal calves. *Domest. Anim. Endocrinol.* **2002**, *22*, 155–168. [CrossRef] [PubMed]
125. Drackley, J.K. Calf Nutrition from Birth to Breeding. *Vet. Clin. North Am. Food Anim. Pract.* **2008**, *24*, 55–86. [CrossRef] [PubMed]
126. Zarcula, S.; Cernescu, H.; Mircu, C.; Tulcan, C.; Morvay, A.; Baul, S.; Popovici, D. Influence of Breed, Parity and Food Intake on Chemical Composition of First Colostrum in Cow. *Sci. Pap. Anim. Sci. Biotechnol.* **2010**, *43*, 43.
127. Shoshani, E.; Rozen, S.; Doekes, J.J. Effect of a short dry period on milk yield and content, colostrum quality, fertility, and metabolic status of Holstein cows. *J. Dairy Sci.* **2014**, *97*, 2909–2922. [CrossRef]
128. Andrée O'Hara, E.; Båge, R.; Emanuelson, U.; Holtenius, K. Effects of dry period length on metabolic status, fertility, udder health, and colostrum production in 2 cow breeds. *J. Dairy Sci.* **2019**, *102*, 595–606. [CrossRef]
129. Mayasari, N.; de Vries Reilingh, G.; Nieuwland, M.G.B.; Remmelink, G.J.; Parmentier, H.K.; Kemp, B.; van Knegsel, A.T.M. Effect of maternal dry period length on colostrum immunoglobulin content and natural and specific antibody titers in calves. *J. Dairy Sci.* **2015**, *98*, 3969–3979. [CrossRef]
130. Klein, G.B. *Cunningham's Textbook of Veterinary Physiology*, 5th ed.; Saunders/Elsevier Inc.: Amsterdam, The Netherlands, 2013; ISBN 978-1-4377-2361-8.
131. Steinhoff-Wagner, J.; Görs, S.; Junghans, P.; Bruckmaier, R.M.; Kanitz, E.; Metges, C.C.; Hammon, H.M. Intestinal glucose absorption but not endogenous glucose production differs between colostrum- and formula-fed neonatal calves. *J. Nutr.* **2011**, *141*, 48–55. [CrossRef]
132. Steinhoff-Wagner, J.; Görs, S.; Junghans, P.; Bruckmaier, R.M.; Kanitz, E.; Metges, C.C.; Hammon, H.M. Maturation of endogenous glucose production in preterm and term calves. *J. Dairy Sci.* **2011**, *94*, 5111–5123. [CrossRef]
133. Salih, Y.; McDowell, L.R.; Hentges, J.F.; Mason, R.M.; Wilcox, C.J. Mineral Content of Milk, Colostrum, and Serum as Affected by Physiological State and Mineral Supplementation. *J. Dairy Sci.* **1987**, *70*, 608–612. [CrossRef]
134. Kume, S.I.; Tanabe, S. Effect of Parity on Colostral Mineral Concentrations of Holstein Cows and Value of Colostrum as a Mineral Source for Newborn Calves. *J. Dairy Sci.* **1993**, *76*, 1654–1660. [CrossRef]
135. Valldecabres, A.; Silva-del-Río, N. First-milking colostrum mineral concentrations and yields: Comparison to second milking and associations with serum mineral concentrations, parity, and yield in multiparous Jersey cows. *J. Dairy Sci.* **2022**, *105*, 2315–2325. [CrossRef]
136. Miciński, J.; Pogorzelska, J.; Beisenov, A.; Aitzhanova, I.; Shaikamal, G.; Dzięgelewska-Kuźmińska, D.; Miciński, B.; Sobczuk-Szul, M. Basic and mineral composition of colostrum from cows in different ages and calving period. *J. Elem.* **2017**, *22*, 259–269. [CrossRef]
137. Van Emon, M.; Sanford, C.; McCoski, S. Impacts of bovine trace mineral supplementation on maternal and offspring production and health. *Animals* **2020**, *10*, 2404. [CrossRef]
138. Roshanzamir, H.; Rezaei, J.; Fazaeli, H. Colostrum and milk performance, and blood immunity indices and minerals of Holstein cows receiving organic Mn, Zn and Cu sources. *Anim. Nutr.* **2020**, *6*, 61–68. [CrossRef]
139. Hansard, S.L.; Comar, C.L.; Plumlee, M.P. The Effects of Age upon Calcium Utilization and Maintenance Requirements in the Bovinel. *J. Anim. Sci.* **1954**, *13*, 25–36. [CrossRef]
140. Peters, J.C.; Mahan, D.C.; Wiseman, T.G.; Fastinger, N.D. Effect of dietary organic and inorganic micromineral source and level on sow body, liver, colostrum, mature milk, and progeny mineral compositions over six parities. *J. Anim. Sci.* **2010**, *88*, 626–637. [CrossRef]
141. Ogilvie, L.; Van Winters, B.; Mion, B.; King, K.; Spricigo, J.F.W.; Karrow, N.A.; Steele, M.A.; Ribeiro, E.S. Effects of replacing inorganic salts of trace minerals with organic trace minerals in the diet of prepartum cows on quality of colostrum and immunity of newborn calves. *J. Dairy Sci.* **2023**, *106*, 3493–3508. [CrossRef]
142. Luecke, S.M.; Holman, D.B.; Schmidt, K.N.; Gzyl, K.E.; Hurlbert, J.L.; Menezes, A.C.B.; Bochantin, K.A.; Kirsch, J.D.; Baumgaertner, F.; Sedivec, K.K.; et al. Whole-body microbiota of newborn calves and their response to prenatal vitamin and mineral supplementation. *Front. Microbiol.* **2023**, *14*, 1207601. [CrossRef]

143. Kon, S.K.; Watson, M.B. The vitamin C content of cow's milk. *Biochem. J.* **1937**, *31*, 223–226. [CrossRef]
144. Pearson, P.B.; Darnell, A.L.; Weir, J. The Thiamine, Riboflavin, Nicotinic Acid and Pantothenic Acid Content of Colostrum and Milk of the Cow and Ewe. *J. Nutr.* **1946**, *31*, 51–57. [CrossRef]
145. Collins, R.A.; Harper, A.E.; Schreiber, M.; Elvehjem, C.A. The folic acid and vitamin B12 content of the milk of various species. *J. Nutr.* **1951**, *43*, 313–321. [CrossRef]
146. Debier, C.; Pottier, J.; Goffe, C.; Larondelle, Y. Present knowledge and unexpected behaviours of vitamins A and E in colostrum and milk. *Livest. Prod. Sci.* **2005**, *98*, 135–147. [CrossRef]
147. Indyk, H.E.; Woollard, D.C. The endogenous vitamin K1 content of bovine milk: Temporal influence of season and lactation. *Food Chem.* **1995**, *54*, 403–407. [CrossRef]
148. Henry, K.M.; Kon, S.K. A note on the vitamin D content of cow's colostrum. *Biochem. J.* **1937**, *31*, 2199–2201. [CrossRef] [PubMed]
149. Spielman, A.A.; Thomas, J.W.; Loosli, J.K.; Whiting, F.; Norton, C.L.; Turk, K.L. The Relationship of the Prepartum Diet to the Carotene and Vitamin A Content of Bovine Colostrum. *J. Dairy Sci.* **1947**, *30*, 343–350. [CrossRef]
150. Johnson, J.L.; Godden, S.M.; Molitor, T.; Ames, T.; Hagman, D. Effects of feeding heat-treated colostrum on passive transfer of immune and nutritional parameters in neonatal dairy calves. *J. Dairy Sci.* **2007**, *90*, 5189–5198. [CrossRef] [PubMed]
151. Semba, R.D. Vitamin A and Immune Function. In *Military Strategies for Sustainment of Nutrition and Immune Function in the Field;* National Academy of Sciences: Washington, DC, USA, 1999; p. 230968. ISBN 0-309-55696-1.
152. Puvogel, G.; Baumrucker, C.; Blum, J.W. Plasma vitamin A status in calves fed colostrum from cows that were fed vitamin A during late pregnancy. *J. Anim. Physiol. Anim. Nutr.* **2008**, *92*, 614–620. [CrossRef]
153. Schweigert, F.J. Effect of gestation and lactation on lipoprotein pattern and composition in dairy cows. *J. Anim. Physiol. Anim. Nutr.* **1990**, *63*, 75–83. [CrossRef]
154. Przybylska, J.; Albera, E.; Kankofer, M. Antioxidants in bovine colostrum. *Reprod. Domest. Anim.* **2007**, *42*, 402–409. [CrossRef] [PubMed]
155. Draper, H.H.; Johnson, B.C. The riboflavin requirement of the Holstein calf. *J. Nutr.* **1952**, *46*, 37–44. [CrossRef] [PubMed]
156. Wiese, A.C.; Johnson, B.C.; Mitchell, H.H.; Nevens, W.B. Riboflavin Deficiency in the Dairy Calf. *J. Nutr.* **1947**, *33*, 263–270. [CrossRef] [PubMed]
157. O'Brien, M.B.; McLoughlin, R.M.; Roche, C.; Nelson, C.D.; Meade, K.G. Effect of IL-8 haplotype on temporal profile in circulating concentrations of interleukin 8 and 25(OH) vitamin D in Holstein-Friesian calves. *Vet. Immunol. Immunopathol.* **2021**, *238*, 110287. [CrossRef] [PubMed]
158. Cummins, K.A. Ascorbate in Cattle: A Review. *Prof. Anim. Sci.* **1992**, *8*, 22–29. [CrossRef]
159. Lundquist, N.S.; Phillips, P.H. Certain Dietary Factors Essential for the Growing Calf. *J. Dairy Sci.* **1943**, *26*, 1023–1030. [CrossRef]
160. Matsui, T. Vitamin C nutrition in cattle. *Asian-Australas. J. Anim. Sci.* **2012**, *25*, 597–605. [CrossRef]
161. Biesalski, H.K.; Dragsted, L.O.; Elmadfa, I.; Grossklaus, R.; Müller, M.; Schrenk, D.; Walter, P.; Weber, P. Bioactive compounds: Definition and assessment of activity. *Nutrition* **2009**, *25*, 1202–1205. [CrossRef] [PubMed]
162. Blum, J.W. Nutritional physiology of neonatal calves. *J. Anim. Physiol. Anim. Nutr.* **2006**, *90*, 1–11. [CrossRef]
163. Chen, X.; Gao, C.; Li, H.; Huang, L.; Sun, Q.; Dong, Y.; Tian, C.; Gao, S.; Dong, H.; Guan, D.; et al. Identification and characterization of microRNAs in raw milk during different periods of lactation, commercial fluid, and powdered milk products. *Cell Res.* **2010**, *20*, 1128–1137. [CrossRef] [PubMed]
164. Puppel, K.; Gołębiewski, M.; Grodkowski, G.; Slósarz, J.; Kunowska-Slósarz, M.; Solarczyk, P.; Łukasiewicz, M.; Balcerak, M.; Przysucha, T. Composition and Factors Affecting Quality of Bovine Colostrum: A Review. *Animals* **2019**, *9*, 1070. [CrossRef] [PubMed]
165. Abdelsattar, M.M.; Rashwan, A.K.; Younes, H.A.; Abdel-Hamid, M.; Romeih, E.; Mehanni, A.H.E.; Vargas-Bello-Pérez, E.; Chen, W.; Zhang, N. An updated and comprehensive review on the composition and preservation strategies of bovine colostrum and its contributions to animal health. *Anim. Feed Sci. Technol.* **2022**, *291*, 115379. [CrossRef]
166. Bösze, Z. *Bioactive Components of Milk;* Springer: New York, NY, USA, 2008; ISBN 9780874216561.
167. Butler, J.E. Bovine Immunoglobulins: A Review. *J. Dairy Sci.* **1969**, *52*, 1895–1909. [CrossRef]
168. Gazi, I.; Reiding, K.R.; Groeneveld, A.; Bastiaans, J.; Huppertz, T.; Heck, A.J.R. Key changes in bovine milk immunoglobulin G during lactation: NeuAc sialylation is a hallmark of colostrum immunoglobulin G N -glycosylation. *Glycobiology* **2023**, *33*, 115–125. [CrossRef] [PubMed]
169. Sasaki, M.; Davis, C.L.; Larson, B.L. Production and Turnover of IgG1 and IgG2 Immunoglobulins in the Bovine around Parturition. *J. Dairy Sci.* **1976**, *59*, 2046–2055. [CrossRef]
170. Baumrucker, C.R.; Macrina, A.L.; Bruckmaier, R.M. Colostrogenesis: Role and Mechanism of the Bovine Fc Receptor of the Neonate (FcRn). *J. Mammary Gland. Biol. Neoplasia* **2021**, *26*, 419–453. [CrossRef]
171. Besser, T.E.; McGuire, T.C.; Gay, C.C. The transfer of serum IgG1 antibody into the gastrointestinal tract in newborn calves. *Vet. Immunol. Immunopathol.* **1987**, *17*, 51–56. [CrossRef]
172. Marnila, P.; Korhonen, H.J. Lactoferrin for human health. In *Dairy-Derived Ingredients;* Elsevier: Amsterdam, The Netherlands, 2009; pp. 290–307, ISBN 9781845694654.
173. Oluk, C.A.; Karaca, O.B. Functional food ingredients and nutraceuticals, milk proteins as nutraceuticals nanoScience and food industry. In *Nutraceuticals;* Elsevier: Amsterdam, The Netherlands, 2016; pp. 715–759. ISBN 9780128043059.

174. Masson, P.L.; Heremans, J.F. Lactoferrin in milk from different species. *Comp. Biochem. Physiol. Part B Comp. Biochem.* **1971**, *39*, 119–129. [CrossRef] [PubMed]
175. Pierce, A.; Escriva, H.; Coddeville, B.; Benaïssa, M.; Leger, D.; Spik, G.; Pamblanco, M. Lactoferrin Almost Absent from Lactating Rat Mammary Gland is Replaced by Transferrin. In *Lactoferrin*; Humana Press: Totowa, NJ, USA, 1997; Volume 2, pp. 125–134. ISBN 9781461239567.
176. Aisen, P.; Listowsky, I. Iron Transport and Storage Proteins. *Annu. Rev. Biochem.* **1980**, *49*, 357–393. [CrossRef] [PubMed]
177. Brock, J.H. *Lactoferrin: Interactions and Biological Functions*; Hutchens, T.W., Lönnerda, B., Eds.; Humana Press: Totowa, NJ, USA, 1997; ISBN 9781461284390.
178. Zhao, X.; Xu, X.X.; Liu, Y.; Xi, E.Z.; An, J.J.; Tabys, D.; Liu, N. The in vitro protective role of bovine lactoferrin on intestinal epithelial barrier. *Molecules* **2019**, *24*, 148. [CrossRef]
179. Arnold, R.R.; Brewer, M.; Gauthier, J.J. Bactericidal activity of human lactoferrin: Sensitivity of a variety of microorganisms. *Infect. Immun.* **1980**, *28*, 893–898. [CrossRef]
180. Van Hooijdonk, A.C.M.; Kussendrager, K.D.; Steijns, J.M. In vivo antimicrobial and antiviral activity of components in bovine milk and colostrum involved in non-specific defence. *Br. J. Nutr.* **2000**, *84* (Suppl. S1), 127–134. [CrossRef]
181. Reiter, B. The Biological Significance of the Non-Immunoglobulin Protective Proteins in Milk: Lysozyme, Lactoferrin, Lactoperoxidase. In *Developments in Dairy Chemistry—3*; Springer: Dordrecht, The Netherlands, 1985; pp. 281–336. ISBN 9783131450715.
182. Naidu, S.; Arnold, R. Influence of Lactoferrin. In *Lactoferrin: Interactions and Biological Functions*; Hutchens, T.W., Lönnerda, B., Eds.; Humana Press: Totowa, NJ, USA, 1997; pp. 259–276.
183. Sokołowska, A.; Bednarz, R.; Pacewicz, M.; Georgiades, J.A.; Wilusz, T.; Polanowski, A. Colostrum from different mammalian species-A rich source of colostrinin. *Int. Dairy J.* **2008**, *18*, 204–209. [CrossRef]
184. Marchbank, T.; Playford, R.J. Colostrum: Its Health Benefits. In *Milk and Dairy Products as Functional Foods*; Wiley: Hoboken, NJ, USA, 2014; pp. 55–93, ISBN 9781118635056.
185. Zimecki, M. A Proline-Rich Protein from Ovine Colostrum: Colostrinin with Immunomodulatory Activity. In *Bioactive Components of Milk*; Bösze, Z., Ed.; Springer: Berlin/Heidelberg, Germany, 2008; pp. 241–252.
186. Stewart, M.G. Colostrinin^{TM}: A naturally occurring compound derived from mammalian colostrum with efficacy in treatment of neurodegenerative diseases, including Alzheimer's. *Expert Opin. Pharmacother.* **2008**, *9*, 2553–2559. [CrossRef]
187. Zabłocka, A.; Janusz, M. Effect of the Proline-Rich Polypeptide Complex/Colostrinin^{TM} on the Enzymatic Antioxidant System. *Arch. Immunol. Ther. Exp.* **2012**, *60*, 383–390. [CrossRef]
188. Abuelo, A.; Hernández, J.; Benedito, J.; Castillo, C. Redox Biology in Transition Periods of Dairy Cattle: Role in the Health of Periparturient and Neonatal Animals. *Antioxidants* **2019**, *8*, 20. [CrossRef]
189. Stewart, M.G.; Banks, D. Enhancement of long-term memory retention by Colostrinin in one-day-old chicks trained on a weak passive avoidance learning paradigm. *Neurobiol. Learn. Mem.* **2006**, *86*, 66–71. [CrossRef]
190. Zabłocka, A.; Janusz, M.; Macała, J.; Lisowski, J. A proline-rich polypeptide complex (PRP) isolated from ovine colostrum. Modulation of H_2O_2 and cytokine induction in human leukocytes. *Int. Immunopharmacol.* **2007**, *7*, 981–988. [CrossRef] [PubMed]
191. Zabłocka, A.; Mitkiewicz, M.; Macała, J.; Janusz, M. Neurotrophic Activity of Cultured Cell Line U87 is Up-Regulated by Proline-Rich Polypeptide Complex and Its Constituent Nonapeptide. *Cell. Mol. Neurobiol.* **2015**, *35*, 977–986. [CrossRef] [PubMed]
192. Ragab, S.S.; Ibrahim, M.K. Role of bovine colostrum and its biofunctional fraction PRP in oral treatment of enterogenic endotoxaemia in rats. *Int. J. Agric. Biol.* **2004**, *6*, 576–580.
193. Boldogh, I.; Aguilera-Aguirre, L.; Bacsi, A.; Choudhury, B.K.; Saavedra-Molina, A.; Kruzel, M. Colostrinin Decreases Hypersensitivity and Allergic Responses to Common Allergens. *Int. Arch. Allergy Immunol.* **2008**, *146*, 298–306. [CrossRef] [PubMed]
194. Gardner, R.W.; Shupe, M.G.; Brimhall, W.; Weber, D.J. Causes of Adverse Responses to Soybean Milk Replacers in Young Calves. *J. Dairy Sci.* **1990**, *73*, 1312–1317. [CrossRef] [PubMed]
195. Xiong, L.; Li, C.; Boeren, S.; Vervoort, J.; Hettinga, K. Effect of heat treatment on bacteriostatic activity and protein profile of bovine whey proteins. *Food Res. Int.* **2020**, *127*, 108688. [CrossRef] [PubMed]
196. Dzhavakhiya, V.G.; Shcherbakova, L.A. Creation of disease-resistant plants by gene engineering. In *Comprehensive and Molecular Phytopathology*; Elsevier: Amsterdam, The Netherlands, 2007; pp. 439–466. ISBN 9780444521323.
197. Farkye, N.Y.; Bansal, N. Enzymes Indigenous to Milk | Other Enzymes. In *Encyclopedia of Dairy Sciences*; Elsevier: Amsterdam, The Netherlands, 2011; Volume 2, pp. 327–334.
198. Paulík, S.; Slanina, L.; Polácek, M. Lysozyme in the colostrum and blood of calves and dairy cows. *Vet. Med.* **1985**, *30*, 21–28.
199. Kaludina, T. Lactate dehydrogenase isoenzymes in the colostrum of cows. *Vet. Nauk.* **1978**, *15*, 69–72.
200. Parish, S.M.; Tyler, J.W.; Besser, T.E.; Gay, C.C.; Krytenberg, D. Prediction of serum IgG1 concentration in Holstein calves using serum gamma glutamyltransferase activity. *J. Vet. Intern. Med.* **1997**, *11*, 344–347. [CrossRef] [PubMed]
201. Oubihi, M.; Kitajima, K.; Aoki, N.; Matsuda, T. Spontaneous galactosylation of agalactoglycoproteins in colostrum. *FEBS Lett.* **2000**, *473*, 165–168. [CrossRef]
202. Peetsalu, K.; Niine, T.; Loch, M.; Dorbek-Kolin, E.; Tummeleht, L.; Orro, T. Effect of colostrum on the acute-phase response in neonatal dairy calves. *J. Dairy Sci.* **2022**, *105*, 6207–6219. [CrossRef] [PubMed]
203. Boudry, C.; Buldgen, A.; Portetelle, D.; Gianello, P.; Théwisa, A.; Leterme, P.; Dehoux, J.P. Effect of bovine colostrum supplementation on cytokine mRNA expression in weaned piglets. *Livest. Sci.* **2007**, *108*, 295–298. [CrossRef]

204. Hagiwara, K.; Kataoka, S.; Yamanaka, H.; Kirisawa, R.; Iwai, H. Detection of cytokines in bovine colostrum. *Vet. Immunol. Immunopathol.* **2000**, *76*, 183–190. [CrossRef]
205. Guo, Y.; Wang, B.; Wang, T.; Gao, L.; Yang, Z.; Wang, F.; Shang, H.; Hua, R.; Xu, J. Biological characteristics of IL-6 and related intestinal diseases. *Int. J. Biol. Sci.* **2021**, *17*, 204–219. [CrossRef] [PubMed]
206. Grigalevičiūtė, R.; Matusevičius, P.; Plančiūnienė, R.; Stankevičius, R.; Radzevičiūtė-Valčiukė, E.; Balevičiūtė, A.; Želvys, A.; Zinkevičienė, A.; Zigmantaitė, V.; Kučinskas, A.; et al. Understanding the Immunomodulatory Effects of Bovine Colostrum: Insights into IL-6/IL-10 Axis-Mediated Inflammatory Control. *Vet. Sci.* **2023**, *10*, 519. [CrossRef] [PubMed]
207. Liu, T.; Zhang, L.; Joo, D.; Sun, S.C. NF-κB signaling in inflammation. *Signal Transduct. Target. Ther.* **2017**, *2*, 17023. [CrossRef]
208. An, M.J.; Cheon, J.H.; Kim, S.W.; Park, J.J.; Moon, C.M.; Han, S.Y.; Kim, E.S.; Il Kim, T.; Kim, W.H. Bovine colostrum inhibits nuclear factor κB-mediated proinflammatory cytokine expression in intestinal epithelial cells. *Nutr. Res.* **2009**, *29*, 275–280. [CrossRef]
209. Blais, M.; Fortier, M.; Pouliot, Y.; Gauthier, S.F.; Boutin, Y.; Asselin, C.; Lessard, M. Colostrum whey down-regulates the expression of early and late inflammatory response genes induced by *Escherichia coli* and *Salmonella enterica* Typhimurium components in intestinal epithelial cells. *Br. J. Nutr.* **2015**, *113*, 200–211. [CrossRef]
210. Korhonen, H.; Marnila, P.; Gill, H.S. Milk immunoglobulins and complement factors. *Br. J. Nutr.* **2000**, *84*, S75–S80. [CrossRef]
211. Le, A.; Barton, L.D.; Sanders, J.T.; Zhang, Q. Exploration of Bovine Milk Proteome in Colostral and Mature Whey Using an Ion-Exchange Approach. *J. Proteome Res.* **2011**, *10*, 692–704. [CrossRef]
212. Medeńska, W.; Dratwa-Chałupnik, A.; Ożgo, M. Cow's whey proteins involvement in the development of the immunological system and gastrointestinal tract in calves. A review. *J. Anim. Feed Sci.* **2021**, *30*, 91–99. [CrossRef]
213. Triglia, R.P.; Linscott, W.D. Titers of nine complement components, conglutinin and C3b-inactivator in adult and fetal bovine sera. *Mol. Immunol.* **1980**, *17*, 741–748. [CrossRef] [PubMed]
214. Hernández-Castellano, L.E.; Morales-delaNuez, A.; Sánchez-Macías, D.; Moreno-Indias, I.; Torres, A.; Capote, J.; Argüello, A.; Castro, N. The effect of colostrum source (goat vs. sheep) and timing of the first colostrum feeding (2 h vs. 14 h after birth) on body weight and immune status of artificially reared newborn lambs. *J. Dairy Sci.* **2015**, *98*, 204–210. [CrossRef] [PubMed]
215. Castro, N.; Acosta, F.; Niño, T.; Vivas, J.; Quesada, E.; Capote, J.; Argüello, A. The effects of diet and age on serum complement system activity in goat kids. *Livest. Sci.* **2008**, *119*, 102–106. [CrossRef]
216. Li, M.; Li, W.; Kong, F.; Kang, S.; Liang, X.; Han, H.; Wu, J.; Zheng, Y.; Li, Q.; Yue, X.; et al. Metabolomics methods to analyze full spectrum of amino acids in different domains of bovine colostrum and mature milk. *Eur. Food Res. Technol.* **2020**, *246*, 213–224. [CrossRef]
217. Girard, J.; Ferre, P.; Pegorier, J.P.; Duee, P.H. Adaptations of glucose and fatty acid metabolism during perinatal period and suckling-weaning transition. *Physiol. Rev.* **1992**, *72*, 507–562. [CrossRef] [PubMed]
218. Hammon, H.M.; Blum, J.W. Free amino acids in plasma of neonatal calves are influenced by feeding colostrum for different durations or by feeding only milk replacer. *J. Anim. Physiol. Anim. Nutr.* **1999**, *82*, 193–204. [CrossRef]
219. Liang, X.; Han, H.; Zhao, X.; Cao, X.; Yang, M.; Tao, D.; Wu, R.; Yue, X. Quantitative analysis of amino acids in human and bovine colostrum milk samples through iTRAQ labeling. *J. Sci. Food Agric.* **2018**, *98*, 5157–5163. [CrossRef] [PubMed]
220. Zanker, I.A.; Hammon, H.M.; Blum, J.W. Plasma Amino Acid Pattern during the First Month of Life in Calves Fed the First Colostrum at 0–2 h or at 24–25 h after Birth. *J. Vet. Med. Ser. A* **2000**, *47*, 107–121. [CrossRef]
221. Zhang, S.; Zeng, X.; Ren, M.; Mao, X.; Qiao, S. Novel metabolic and physiological functions of branched chain amino acids: A review. *J. Anim. Sci. Biotechnol.* **2017**, *8*, 10. [CrossRef]
222. Sadri, H.; Steinhoff-Wagner, J.; Hammon, H.M.; Bruckmaier, R.M.; Görs, S.; Sauerwein, H. Mammalian target of rapamycin signaling and ubiquitin proteasome–related gene expression in 3 different skeletal muscles of colostrum- versus formula-fed calves. *J. Dairy Sci.* **2017**, *100*, 9428–9441. [CrossRef] [PubMed]
223. Abdelmegeid, M.K.; Vailati-Riboni, M.; Alharthi, A.; Batistel, F.; Loor, J.J. Supplemental methionine, choline, or taurine alter in vitro gene network expression of polymorphonuclear leukocytes from neonatal Holstein calves. *J. Dairy Sci.* **2017**, *100*, 3155–3165. [CrossRef]
224. Lobley, G.E.; Hoskin, S.O.; McNeil, C.J. Glutamine in Animal Science and Production. *J. Nutr.* **2001**, *131*, 2525S–2531S. [CrossRef] [PubMed]
225. Ceja, G.; Boerman, J.P.; Neves, R.C.; Jorgensen, M.W.; Johnson, J.S. L-Glutamine supplementation reduces gastrointestinal permeability and biomarkers of physiological stress in preweaning Holstein heifer calves. *J. Dairy Sci.* **2023**, *106*, 9663–9676. [CrossRef] [PubMed]
226. Wang, S.; Wang, F.; Kong, F.; Cao, Z.; Wang, W.; Yang, H.; Wang, Y.; Bi, Y.; Li, S. Effect of Supplementing Different Levels of L-Glutamine on Holstein Calves during Weaning. *Antioxidants* **2022**, *11*, 542. [CrossRef] [PubMed]
227. O'Callaghan, T.F.; O'Donovan, M.; Murphy, J.P.; Sugrue, K.; Mannion, D.; McCarthy, W.P.; Timlin, M.; Kilcawley, K.N.; Hickey, R.M.; Tobin, J.T. Evolution of the bovine milk fatty acid profile—From colostrum to milk five days post parturition. *Int. Dairy J.* **2020**, *104*, 104655. [CrossRef]
228. Contarini, G.; Povolo, M.; Pelizzola, V.; Monti, L.; Bruni, A.; Passolungo, L.; Abeni, F.; Degano, L. Bovine colostrum: Changes in lipid constituents in the first 5 days after parturition. *J. Dairy Sci.* **2014**, *97*, 5065–5072. [CrossRef] [PubMed]
229. Precht, D. Cholesterol content in European bovine milk fats. *Nahrung—Food* **2001**, *45*, 2–8. [CrossRef]

230. Garcia, M.; Greco, L.F.; Favoreto, M.G.; Marsola, R.S.; Martins, L.T.; Bisinotto, R.S.; Shin, J.H.; Lock, A.L.; Block, E.; Thatcher, W.W.; et al. Effect of supplementing fat to pregnant nonlactating cows on colostral fatty acid profile and passive immunity of the newborn calf. *J. Dairy Sci.* **2014**, *97*, 392–405. [CrossRef]
231. Uken, K.L.; Schäff, C.T.; Vogel, L.; Gnott, M.; Dannenberger, D.; Görs, S.; Tuchscherer, A.; Tröscher, A.; Liermann, W.; Hammon, H.M. Modulation of colostrum composition and fatty acid status in neonatal calves by maternal supplementation with essential fatty acids and conjugated linoleic acid starting in late lactation. *J. Dairy Sci.* **2021**, *104*, 4950–4969. [CrossRef]
232. Leat, W.M. Fatty acid composition of the plasma lipids of newborn and maternal ruminants. *Biochem. J.* **1966**, *98*, 598–603. [CrossRef]
233. Burr, G.O.; Burr, M.M. on the Nature and Rôle of the Fatty Acids Essential in Nutrition. *J. Biol. Chem.* **1930**, *86*, 587–621. [CrossRef]
234. Moallem, U.; Zachut, M. Short communication: The effects of supplementation of various n-3 fatty acids to late-pregnant dairy cows on plasma fatty acid composition of the newborn calves. *J. Dairy Sci.* **2012**, *95*, 4055–4058. [CrossRef]
235. Dahl, N.; Albrecht, E.; Dannenberger, D.; Uken, K.L.; Hammon, H.M.; Maak, S. Consequences of maternal essential fatty acid and conjugated linoleic acid supplementation on the development of calf muscle and adipose tissue. *Animals* **2020**, *10*, 1598. [CrossRef]
236. Uken, K.L.; Vogel, L.; Gnott, M.; Görs, S.; Schäff, C.T.; Tuchscherer, A.; Hoeflich, A.; Weitzel, J.M.; Kanitz, E.; Tröscher, A.; et al. Effect of maternal supplementation with essential fatty acids and conjugated linoleic acid on metabolic and endocrine development in neonatal calves. *J. Dairy Sci.* **2021**, *104*, 7295–7314. [CrossRef]
237. Quigley, J.D.; Hill, T.M.; Hulbert, L.E.; Dennis, T.S.; Suarez-Mena, X.F.; Bortoluzzi, E.M. Effects of fatty acids and calf starter form on intake, growth, digestion, and selected blood metabolites in male calves from 0 to 4 months of age. *J. Dairy Sci.* **2019**, *102*, 8074–8091. [CrossRef]
238. Garcia, M.; Greco, L.F.; Favoreto, M.G.; Marsola, R.S.; Wang, D.; Shin, J.H.; Block, E.; Thatcher, W.W.; Santos, J.E.P.; Staples, C.R. Effect of supplementing essential fatty acids to pregnant nonlactating Holstein cows and their preweaned calves on calf performance, immune response, and health. *J. Dairy Sci.* **2014**, *97*, 5045–5064. [CrossRef] [PubMed]
239. Jolazadeh, A.R.; Mohammadabadi, T.; Dehghan-banadaky, M.; Chaji, M.; Garcia, M. Effect of supplementing calcium salts of n-3 and n-6 fatty acid to pregnant nonlactating cows on colostrum composition, milk yield, and reproductive performance of dairy cows. *Anim. Feed Sci. Technol.* **2019**, *247*, 127–140. [CrossRef]
240. Jolazadeh, A.R.; Mohammadabadi, T.; Dehghan-Banadaky, M.; Chaji, M.; Garcia, M. Effect of supplementation fat during the last 3 weeks of uterine life and the preweaning period on performance, ruminal fermentation, blood metabolites, passive immunity and health of the newborn calf. *Br. J. Nutr.* **2019**, *122*, 1346–1358. [CrossRef]
241. Hall, E.J.; German, A.J.; Willard, M.D.; Lappin, M.R.; Cave, N.; Washabau, R.J.; Bergman, P.J. Small Intestine. In *Canine and Feline Gastroenterology*; Elsevier: Amsterdam, The Netherlands, 2013; pp. 651–728. ISBN 9781416036616.
242. Hiltz, R.L.; Laarman, A.H. Effect of butyrate on passive transfer of immunity in dairy calves. *J. Dairy Sci.* **2019**, *102*, 4190–4197. [CrossRef]
243. Koch, C.; Gerbert, C.; Frieten, D.; Dusel, G.; Eder, K.; Zitnan, R.; Hammon, H.M. Effects of ad libitum milk replacer feeding and butyrate supplementation on the epithelial growth and development of the gastrointestinal tract in Holstein calves. *J. Dairy Sci.* **2019**, *102*, 8513–8526. [CrossRef]
244. Millán, I.; Piñero-Ramos, J.; Lara, I.; Parra-Llorca, A.; Torres-Cuevas, I.; Vento, M. Oxidative Stress in the Newborn Period: Useful Biomarkers in the Clinical Setting. *Antioxidants* **2018**, *7*, 193. [CrossRef]
245. Opgenorth, J.; Sordillo, L.M.; Lock, A.L.; Gandy, J.C.; VandeHaar, M.J. Colostrum supplementation with n-3 fatty acids alters plasma polyunsaturated fatty acids and inflammatory mediators in newborn calves. *J. Dairy Sci.* **2020**, *103*, 11676–11688. [CrossRef]
246. Kang, J.X.; Weylandt, K.H. Modulation of Inflammatory Cytokines by Omega-3 Fatty Acids. In *Lipids in Health and Disease*; Springer: Dordrecht, The Netherlands, 2008; Volume 49, pp. 133–143.
247. Roman, A.S.; Schreher, J.; Mackenzie, A.P.; Nathanielsz, P.W. Omega-3 fatty acids and decidual cell prostaglandin production in response to the inflammatory cytokine IL-1β. *Am. J. Obstet. Gynecol.* **2006**, *195*, 1693–1699. [CrossRef]
248. Opgenorth, J.; Sordillo, L.M.; Gandy, J.C.; VandeHaar, M.J. Colostrum supplementation with n-3 fatty acids does not alter calf outcome on a healthy commercial farm. *J. Dairy Sci.* **2020**, *103*, 11689–11696. [CrossRef]
249. Masmeijer, C.; van Leenen, K.; De Cremer, L.; Deprez, P.; Cox, E.; Devriendt, B.; Pardon, B. Effects of omega-3 fatty acids on immune, health and growth variables in veal calves. *Prev. Vet. Med.* **2020**, *179*, 104979. [CrossRef]
250. Flaga, J.; Korytkowski; Górka, P.; Kowalski, Z.M. The effect of docosahexaenoic acid-rich algae supplementation in milk replacer on performance and selected immune system functions in calves. *J. Dairy Sci.* **2019**, *102*, 8862–8873. [CrossRef]
251. Gopal, P.K.; Gill, H.S. Oligosaccharides and glycoconjugates in bovine milk and colostrum. *Br. J. Nutr.* **2000**, *84*, 69–74. [CrossRef]
252. Barile, D.; Marotta, M.; Chu, C.; Mehra, R.; Grimm, R.; Lebrilla, C.B.; German, J.B. Neutral and acidic oligosaccharides in Holstein-Friesian colostrum during the first 3 days of lactation measured by high performance liquid chromatography on a microfluidic chip and time-of-flight mass spectrometry. *J. Dairy Sci.* **2010**, *93*, 3940–3949. [CrossRef] [PubMed]
253. Tao, N.; DePeters, E.J.; German, J.B.; Grimm, R.; Lebrilla, C.B. Variations in bovine milk oligosaccharides during early and middle lactation stages analyzed by high-performance liquid chromatography-chip/mass spectrometry. *J. Dairy Sci.* **2009**, *92*, 2991–3001. [CrossRef] [PubMed]

254. Sundekilde, U.K.; Barile, D.; Meyrand, M.; Poulsen, N.A.; Larsen, L.B.; Lebrilla, C.B.; German, J.B.; Bertram, H.C. Natural Variability in Bovine Milk Oligosaccharides from Danish Jersey and Holstein-Friesian Breeds. *J. Agric. Food Chem.* **2012**, *60*, 6188–6196. [CrossRef] [PubMed]
255. Gunjan; Kumar, K.; Deepak, D. Structural characterization of novel milk oligosaccharide Aurose from cow colostrum. *J. Mol. Struct.* **2019**, *1176*, 394–401. [CrossRef]
256. Nakamura, T.; Kawase, H.; Kimura, K.; Watanabe, Y.; Ohtani, M.; Arai, I.; Urashima, T. Concentrations of sialyloligosaccharides in bovine colostrum and milk during the prepartum and early lactation. *J. Dairy Sci.* **2003**, *86*, 1315–1320. [CrossRef]
257. Martín-Sosa, S.; Martín, M.J.; García-Pardo, L.A.; Hueso, P. Sialyloligosaccharides in human and bovine milk and in infant formulas: Variations with the progression of lactation. *J. Dairy Sci.* **2003**, *86*, 52–59. [CrossRef]
258. Fischer-Tlustos, A.J.; Hertogs, K.; van Niekerk, J.K.; Nagorske, M.; Haines, D.M.; Steele, M.A. Oligosaccharide concentrations in colostrum, transition milk, and mature milk of primi- and multiparous Holstein cows during the first week of lactation. *J. Dairy Sci.* **2020**, *103*, 3683–3695. [CrossRef]
259. Fischer, A.J.; Malmuthuge, N.; Guan, L.L.; Steele, M.A. Short communication: The effect of heat treatment of bovine colostrum on the concentration of oligosaccharides in colostrum and in the intestine of neonatal male Holstein calves. *J. Dairy Sci.* **2018**, *101*, 401–407. [CrossRef]
260. McJarrow, P.; Van Amelsfort-Schoonbeek, J. Bovine sialyl oligosaccharides: Seasonal variations in their concentrations in milk, and a comparison of the colostrums of Jersey and Friesian cows. *Int. Dairy J.* **2004**, *14*, 571–579. [CrossRef]
261. Quinn, E.M.; Joshi, L.; Hickey, R.M. Symposium review: Dairy-derived oligosaccharides—Their influence on host–microbe interactions in the gastrointestinal tract of infants. *J. Dairy Sci.* **2020**, *103*, 3816–3827. [CrossRef] [PubMed]
262. Jing, B.; Xia, K.; Zhang, C.; Jiao, S.; Zhu, L.; Wei, J.; Wang, Z.A.; Chen, N.; Tu, P.; Li, J.; et al. Chitosan Oligosaccharides Regulate the Occurrence and Development of Enteritis in a Human Gut-On-a-Chip. *Front. Cell Dev. Biol.* **2022**, *10*, 877892. [CrossRef]
263. Boudry, G.; Hamilton, M.K.; Chichlowski, M.; Wickramasinghe, S.; Barile, D.; Kalanetra, K.M.; Mills, D.A.; Raybould, H.E. Bovine milk oligosaccharides decrease gut permeability and improve inflammation and microbial dysbiosis in diet-induced obese mice. *J. Dairy Sci.* **2017**, *100*, 2471–2481. [CrossRef]
264. Jakobsen, L.M.A.; Sundekilde, U.K.; Andersen, H.J.; Nielsen, D.S.; Bertram, H.C. Lactose and Bovine Milk Oligosaccharides Synergistically Stimulate B. longum subsp. longum Growth in a Simplified Model of the Infant Gut Microbiome. *J. Proteome Res.* **2019**, *18*, 3086–3098. [CrossRef] [PubMed]
265. Obelitz-Ryom, K.; Bering, S.B.; Overgaard, S.H.; Eskildsen, S.F.; Ringgaard, S.; Olesen, J.L.; Skovgaard, K.; Pankratova, S.; Wang, B.; Brunse, A.; et al. Bovine milk oligosaccharides with sialyllactose improves cognition in preterm pigs. *Nutrients* **2019**, *11*, 1335. [CrossRef]
266. Bagwe, S.; Tharappel, L.J.P.; Kaur, G.; Buttar, H.S. Bovine colostrum: An emerging nutraceutical. *J. Complement. Integr. Med.* **2015**, *12*, 175–185. [CrossRef] [PubMed]
267. Hobbs, M.; Jahan, M.; Ghorashi, S.A.; Wang, B. Current perspective of sialylated milk oligosaccharides in mammalian milk: Implications for brain and gut health of newborns. *Foods* **2021**, *10*, 473. [CrossRef] [PubMed]
268. Tao, N.; Wu, S.; Kim, J.; An, H.J.; Hinde, K.; Power, M.L.; Gagneux, P.; German, J.B.; Lebrilla, C.B. Evolutionary glycomics: Characterization of milk oligosaccharides in primates. *J. Proteome Res.* **2011**, *10*, 1548–1557. [CrossRef] [PubMed]
269. Kunz, C.; Rudloff, S.; Schad, W.; Braun, D. Lactose-derived oligosaccharides in the milk of elephants: Comparison with human milk. *Br. J. Nutr.* **1999**, *82*, 391–399. [CrossRef] [PubMed]
270. Durham, S.D.; Wei, Z.; Lemay, D.G.; Lange, M.C.; Barile, D. Creation of a milk oligosaccharide database, MilkOligoDB, reveals common structural motifs and extensive diversity across mammals. *Sci. Rep.* **2023**, *13*, 10345. [CrossRef] [PubMed]
271. Jouan, P.N.; Pouliot, Y.; Gauthier, S.F.; Laforest, J.P. Hormones in bovine milk and milk products: A survey. *Int. Dairy J.* **2006**, *16*, 1408–1414. [CrossRef]
272. Pinotti, L.; Rosi, F. Leptin in Bovine Colostrum and Milk. *Horm. Metab. Res.* **2006**, *38*, 89–93. [CrossRef]
273. Pape-Zambito, D.A.; Magliaro, A.L.; Kensinger, R.S. 17B-Estradiol and Estrone Concentrations in Plasma and Milk During Bovine Pregnancy. *J. Dairy Sci.* **2008**, *91*, 127–135. [CrossRef]
274. Goyon, A.; Cai, J.Z.; Kraehenbuehl, K.; Hartmann, C.; Shao, B.; Mottier, P. Determination of steroid hormones in bovine milk by LC-MS/MS and their levels in Swiss Holstein cow milk. *Food Addit. Contam. Part A* **2016**, *33*, 804–816. [CrossRef] [PubMed]
275. Romanini, E.B.; Volpato, A.M.; Sifuentes dos Santos, J.; de Santana, E.H.W.; de Souza, C.H.B.; Ludovico, A. Melatonin concentration in cow's milk and sources of its variation. *J. Appl. Anim. Res.* **2019**, *47*, 140–145. [CrossRef]
276. Koldovský, O. Hormones in Milk. *Vitam. Horm.* **1995**, *50*, 77–149.
277. Grosvenor, C.E.; Picciano, M.F.; Baumrucker, C.R. Hormones and Growth Factors in Milk. *Endocr. Rev.* **1993**, *14*, 710–728. [CrossRef]
278. McEwen, B.S. Steroid Hormone Effects on Brain: Novel Insights Connecting Cellular and Molecular Features of Brain Cells to Behavior. In *Methods in Neurosciences*; Academic Press, Inc.: Cambridge, MA, USA, 1994; Volume 22, pp. 525–542.
279. Xu, L.; Zhang, L.; Zhang, Y.; Sheng, Q.; Zhao, A. Qualitative and quantitative comparison of hormone contents between bovine and human colostrums. *Int. Dairy J.* **2011**, *21*, 54–57. [CrossRef]
280. Farke, C.; Rattenberger, E.; Roiger, S.U.; Meyer, H.H.D. Bovine colostrum: Determination of naturally occcurring steroid hormones by liquid chromatography-tandem mass spectrometry (LC-MS/MS). *J. Agric. Food Chem.* **2011**, *59*, 1423–1427. [CrossRef] [PubMed]

281. Randal, D.; Burggren, W.; French, K. Eckert Animal Physiology. In *Mechanisms and Adaptations*, 4th ed.; W.H. Freeman and Company: New York, NY, USA, 1997; ISBN 0-7167-2414-6.
282. Hill, R.W.; Wyse, G.A.; Anderson, M. *Animal Physiology*, 4th ed.; Oxford University Press: Oxford, UK, 2016; ISBN 9781605354712.
283. Hammon, H.M.; Liermann, W.; Frieten, D.; Koch, C. Review: Importance of colostrum supply and milk feeding intensity on gastrointestinal and systemic development in calves. *Animal* **2020**, *14*, S133–S143. [CrossRef]
284. Ivell, R.; Anand-Ivell, R. Neohormones in milk. *Best Pract. Res. Clin. Endocrinol. Metab.* **2017**, *31*, 419–425. [CrossRef]
285. Bartol, F.F.; Wiley, A.A.; Miller, D.J.; Silva, A.J.; Roberts, K.E.; Davolt, M.L.P.; Chen, J.C.; Frankshun, A.L.; Camp, M.E.; Rahman, K.M.; et al. LACTATION BIOLOGY SYMPOSIUM: Lactocrine signaling and developmental programming. *J. Anim. Sci.* **2013**, *91*, 696–705. [CrossRef] [PubMed]
286. Bagnell, C.A.; Bartol, F.F. Relaxin and the 'Milky Way': The lactocrine hypothesis and maternal programming of development. *Mol. Cell. Endocrinol.* **2019**, *487*, 18–23. [CrossRef]
287. Dai, Y.; Ivell, R.; Liu, X.; Janowski, D.; Anand-Ivell, R. Relaxin-family peptide receptors 1 and 2 are fully functional in the bovine. *Front. Physiol.* **2017**, *8*, 264828. [CrossRef] [PubMed]
288. Rodríguez, E.M.; Bach, A.; Devant, M.; Aris, A. Is calcitonin an active hormone in the onset and prevention of hypocalcemia in dairy cattle? *J. Dairy Sci.* **2016**, *99*, 3023–3030. [CrossRef] [PubMed]
289. Linehan, K.; Ross, R.P.; Stanton, C. Bovine Colostrum for Veterinary and Human Health Applications: A Critical Review. *Annu. Rev. Food Sci. Technol.* **2023**, *14*, 387–410. [CrossRef] [PubMed]
290. Koldovsky, O. Search for role of milk-borne biologically active peptides for the suckling. *J. Nutr.* **1989**, *119*, 1543–1551. [CrossRef] [PubMed]
291. Gaiani, R.; Chiesa, F.; Mattioli, M.; Nannetti, G.; Galeati, G. Androstenedione and testosterone concentrations in plasma and milk of the cow throughout pregnancy. *Reproduction* **1984**, *70*, 55–59. [CrossRef]
292. Wolford, S.T.; Argoudelis, C.J. Measurement of Estrogens in Cow's Milk, Human Milk, and Dairy Products. *J. Dairy Sci.* **1979**, *62*, 1458–1463. [CrossRef] [PubMed]
293. Schwalm, J.W.; Tucker, A. Glucocorticoids in Mammary Secretions and Blood Serum during Reproduction and Lactation and Distributions of Glucocorticoids, Progesterone, and Estrogens in Fractions of Milk. *J. Dairy Sci.* **1978**, *61*, 550–560. [CrossRef]
294. Shutt, D.A.; Fell, L.R. Comparison of Total and Free Cortisol in Bovine Serum and Milk or Colostrum. *J. Dairy Sci.* **1985**, *68*, 1832–1834. [CrossRef]
295. Amarant, T.; Fridkin, M.; Koch, Y. Luteinizing Hormone-Releasing Hormone and Thyrotropin-Releasing Hormone in Human and Bovine Milk. *Eur. J. Biochem.* **1982**, *127*, 647–650. [CrossRef]
296. Ronge, H.; Blum, J.W. Somatomedin C and other hormones in dairy cows around parturition, in newborn calves and in milk. *J. Anim. Physiol. Anim. Nutr.* **1988**, *60*, 168–176. [CrossRef]
297. Shinnichi, O.; Hidetoshi, S.; Tatsurou, S.; Nobuyoshi, M.; Tetsuya, K.; Kazuo, K.; Yoshio, S.; Akira, N.; Minoru, O.; Yasuyuki, S. Insulin-like growth factor-I, GH, insulin and glucagon concentrations in bovine colostrum and in plasma of dairy cows and neonatal calves around parturition. *Comp. Biochem. Physiol. Part A Physiol.* **1989**, *94*, 805–808. [CrossRef] [PubMed]
298. Elfstrand, L.; Lindmark-Månsson, H.; Paulsson, M.; Nyberg, L.; Åkesson, B. Immunoglobulins, growth factors and growth hormone in bovine colostrum and the effects of processing. *Int. Dairy J.* **2002**, *12*, 879–887. [CrossRef]
299. Takeyama, M.; Yanaga, N.; Yarimizu, K.; Ono, J.; Takaki, R.; Fujii, N.; Yajima, H. Enzyme immunoassay of somatostatin(SS)-like immunoreactive substance in bovine milk. *Chem. Pharm. Bull.* **1990**, *38*, 456–459. [CrossRef] [PubMed]
300. Prakash, B.S.; Paul, V.; Kliem, H.; Kulozik, U.; Meyer, H.H.D. Determination of oxytocin in milk of cows administered oxytocin. *Anal. Chim. Acta* **2009**, *636*, 111–115. [CrossRef] [PubMed]
301. McMurtry, J.P.; Malven, P.V.; Arave, C.W.; Erb, R.E.; Harrington, R.B. Environmental and Lactational Variables Affecting Prolactin Concentrations in Bovine Milk. *J. Dairy Sci.* **1975**, *58*, 181–189. [CrossRef] [PubMed]
302. Gala, R.R.; Forsyth, I.A.; Turvey, A. Milk prolactin is biologically active. *Life Sci.* **1980**, *26*, 987–993. [CrossRef] [PubMed]
303. Baumrucker, C.R.; Macrina, A.L. *Hormones and Regulatory Factors in Bovine Milk*, 3rd ed.; Encyclopedia of Dairy Sciences; Academic Press: Cambridge, MA, USA, 2022; Volume 2, ISBN 9780128187661.
304. Jahnke, G.D.; Lazarus, L.H. A bombesin immunoreactive peptide in milk. *Proc. Natl. Acad. Sci. USA* **1984**, *81*, 578–582. [CrossRef]
305. Rauprich, A.B.E.; Hammon, H.M.; Blum, J.W. Influence of feeding different amounts of first colostrum on metabolic, endocrine, and health status and on growth performance in neonatal calves. *J. Anim. Sci.* **2000**, *78*, 896–908. [CrossRef]
306. Grega, T.; Bobek, S. The presence of thyroxine in bovine precolostrum and colostrum. *J. Dairy Res.* **1977**, *44*, 131–132. [CrossRef]
307. Akasha, M.A.; Anderson, R.R.; Ellersieck, M.; Nixon, D.A. Concentration of Thyroid Hormones and Prolactin in Dairy Cattle Serum and Milk at Three Stages of Lactation. *J. Dairy Sci.* **1987**, *70*, 271–276. [CrossRef]
308. Pezzi, C.; Accorsi, P.A.; Vigo, D.; Govoni, N.; Gaiani, R. 5′-Deiodinase activity and circulating thyronines in lactating cows. *J. Dairy Sci.* **2003**, *86*, 152–158. [CrossRef] [PubMed]
309. Slebodziński, A.B.; Nowak, J.; Gawecka, H.; Sechman, A. Thyroid hormones and insulin in milk; a comparative study. *Endocrinol. Exp.* **1986**, *20*, 247–255. [PubMed]
310. Goff, J.P.; Reinhardt, T.A.; Lee, S.; Hollis, B.W. Parathyroid Hormone-Related Peptide Content of Bovine Milk and Calf Blood Assessed by Radioimmunoassay and Bioassay. *Endocrinology* **1991**, *129*, 2815–2819. [CrossRef] [PubMed]
311. Law, F.M.K.; Moate, P.J.; Leaver, D.D.; Diefenbach-Jagger, H.; Grill, V.; Ho, P.W.M.; Martin, T.J. Parathyroid hormone-related protein in milk and its correlation with bovine milk calcium. *J. Endocrinol.* **1991**, *128*, 21-NP. [CrossRef]

312. Eriksson, L.; Valtonen, M.; Laitinen, J.T.; Paananen, M.; Raikkonen, M. Diurnal Rhythm of Melatonin in Bovine Milk: Pharmacokinetics of Exogenous Melatonin in Lactating Cows and Goats. *Acta Vet. Scand.* **1998**, *39*, 301–310. [CrossRef] [PubMed]
313. Bonnet, M.; Delavaud, C.; Laud, K.; Gourdou, I.; Leroux, C.; Djiane, J.; Chilliard, Y. Mammary leptin synthesis, milk leptin and their putative physiological roles. *Reprod. Nutr. Dev.* **2002**, *42*, 399–413. [CrossRef] [PubMed]
314. Singh, S.P.; Häussler, S.; Gross, J.J.; Schwarz, F.J.; Bruckmaier, R.M.; Sauerwein, H. Short communication: Circulating and milk adiponectin change differently during energy deficiency at different stages of lactation in dairy cows. *J. Dairy Sci.* **2014**, *97*, 1535–1542. [CrossRef]
315. Kesser, J.; Hill, M.; Heinz, J.F.L.; Koch, C.; Rehage, J.; Steinhoff-Wagner, J.; Hammon, H.M.; Mielenz, B.; Sauerwein, H.; Sadri, H. The rapid increase of circulating adiponectin in neonatal calves depends on colostrum intake. *J. Dairy Sci.* **2015**, *98*, 7044–7051. [CrossRef] [PubMed]
316. Lu, M.; Yao, F.; Guo, A. A study on two gut hormones in breast milk. *Zhonghua Fu Chan Ke Za Zhi* **1995**, *30*, 554–556.
317. Blum, J.W.; Baumrucker, C.R. Colostral and milk insulin-like growth factors and related substances: Mammary gland and neonatal (intestinal and systemic) targets. *Domest. Anim. Endocrinol.* **2002**, *23*, 101–110. [CrossRef]
318. Pakkanen, R.; Aalto, J. Growth factors and antimicrobial factors of bovine colostrum. *Int. Dairy J.* **1997**, *7*, 285–297. [CrossRef]
319. Kühne, S.; Hammon, H.M.; Bruckmaier, R.M.; Morel, C.; Zbinden, Y.; Blum, J.W. Growth performance, metabolic and endocrine traits, and absorptive capacity in neonatal calves fed either colostrum or milk replacer at two levels. *J. Anim. Sci.* **2000**, *78*, 609–620. [CrossRef] [PubMed]
320. Mann, S.; Curone, G.; Chandler, T.L.; Moroni, P.; Cha, J.; Bhawal, R.; Zhang, S. Heat treatment of bovine colostrum: I. Effects on bacterial and somatic cell counts, immunoglobulin, insulin, and IGF-I concentrations, as well as the colostrum proteome. *J. Dairy Sci.* **2020**, *103*, 9368–9383. [CrossRef] [PubMed]
321. Menchetti, L.; Traina, G.; Tomasello, G.; Casagrande-Proietti, P.; Leonardi, L.; Barbato, O.; Brecchia, G. Potential benefits of colostrum in gastrointestinal diseases. *Front. Biosci. Sch.* **2016**, *8*, 331–351. [CrossRef]
322. Blum, J.; Baumrucker, C. Insulin-Like Growth Factors (IGFs), IGF Binding Proteins, and Other Endocrine Factors in Milk: Role in the Newborn. In *Bioactive Components of Milk*; Bösze, Z., Ed.; Springer: New York, NY, USA, 2008; pp. 397–422.
323. Sienkiewicz, M.; Szymańska, P.; Fichna, J. Supplementation of Bovine Colostrum in Inflammatory Bowel Disease: Benefits and Contraindications. *Adv. Nutr.* **2021**, *12*, 533–545. [CrossRef] [PubMed]
324. Mann, S.; Curone, G.; Chandler, T.L.; Sipka, A.; Cha, J.; Bhawal, R.; Zhang, S. Heat treatment of bovine colostrum: II. Effects on calf serum immunoglobulin, insulin, and IGF-I concentrations, and the serum proteome. *J. Dairy Sci.* **2020**, *103*, 9384–9406. [CrossRef] [PubMed]
325. Jehle, P.M.; Fussgaenger, R.D.; Blum, W.F.; Angelus, N.K.O.; Hoeflich, A.; Wolf, E.; Jungwirth, R.J. Differential autocrine regulation of intestine epithelial cell proliferation and differentiation by insulin-like growth factor (IGF) system components. *Horm. Metab. Res.* **1999**, *31*, 97–102. [CrossRef]
326. Pache, J.C. Epidermal Growth Factors. *Encycl. Respir. Med. Four-Volume Set* **2006**, *2*, 129–133. [CrossRef]
327. Kelly, A.; Houston, S.A.; Sherwood, E.; Casulli, J.; Travis, M.A. Regulation of Innate and Adaptive Immunity by TGFβ. *Adv. Immunol.* **2017**, *134*, 137–233.
328. Adamek, A.; Kasprzak, A. Insulin-like growth factor (IGF) system in liver diseases. *Int. J. Mol. Sci.* **2018**, *19*, 1308. [CrossRef]
329. Gibson, C.A.; Staley, M.D.; Baumrucker, C.R. Identification of IGF binding proteins in bovine milk and the demonstration of IGFBP-3 synthesis and release by bovine mammary epithelial cells. *J. Anim. Sci.* **1999**, *77*, 1547–1557. [CrossRef] [PubMed]
330. Taylor, V.J.; Cheng, Z.; Pushpakumara, P.G.A.; Wathes, D.C.; Beever, D.E. Relationships between the plasma concentrations of insulin-like growth factor-I in dairy cows and their fertility and milk yield. *Vet. Rec.* **2004**, *155*, 583–588. [CrossRef] [PubMed]
331. Ontsouka, E.C.; Albrecht, C.; Bruckmaier, R.M. Invited review: Growth-promoting effects of colostrum in calves based on interaction with intestinal cell surface receptors and receptor-like transporters. *J. Dairy Sci.* **2016**, *99*, 4111–4123. [CrossRef] [PubMed]
332. Hadorn, U.; Hammon, H.; Bruckmaier, R.M.; Blum, J.W. Delaying Colostrum Intake by One Day Has Important Effects on Metabolic Traits and on Gastrointestinal and Metabolic Hormones in Neonatal Calves. *J. Nutr.* **1997**, *127*, 2011–2023. [CrossRef] [PubMed]
333. Fischer-Tlustos, A.J.; Pyo, J.; Song, Y.; Renaud, D.L.; Guan, L.L.; Steele, M.A. Short communication: Effect of delaying the first colostrum feeding on small intestinal histomorphology and serum insulin-like growth factor-1 concentrations in neonatal male Holstein calves. *J. Dairy Sci.* **2020**, *103*, 12109–12116. [CrossRef] [PubMed]
334. Pyo, J.; Hare, K.; Pletts, S.; Inabu, Y.; Haines, D.; Sugino, T.; Guan, L.L.; Steele, M. Feeding colostrum or a 1:1 colostrum:milk mixture for 3 days postnatal increases small intestinal development and minimally influences plasma glucagon-like peptide-2 and serum insulin-like growth factor-1 concentrations in Holstein bull calves. *J. Dairy Sci.* **2020**, *103*, 4236–4251. [CrossRef] [PubMed]
335. Matsui, T.; Ichikawa, H.; Fujita, T.; Takemura, S.; Takagi, T.; Osada-Oka, M.; Minamiyama, Y. Histidine and arginine modulate intestinal cell restitution via transforming growth factor-β1. *Eur. J. Pharmacol.* **2019**, *850*, 35–42. [CrossRef] [PubMed]
336. Li, Z.; Wang, B.; Li, H.; Jian, L.; Luo, H.; Wang, B.; Zhang, C.; Zhao, X.; Xue, Y.; Peng, S.; et al. Maternal Folic Acid Supplementation Differently Affects the Small Intestinal Phenotype and Gene Expression of Newborn Lambs from Differing Litter Sizes. *Animals* **2020**, *10*, 2183. [CrossRef] [PubMed]

337. Buranakarl, C.; Thammacharoen, S.; Semsirmboon, S.; Sutayatram, S.; Nuntapaitoon, M.; Katoh, K. Impact of insulin-like growth factor 1, immunoglobulin G and vitamin A in colostrum on growth of newborn Black Bengal goats and its crossbred. *J. Anim. Physiol. Anim. Nutr.* **2022**, *106*, 1238–1245. [CrossRef]
338. Chi, Y.; Zhou, D. MicroRNAs in colorectal carcinoma—From pathogenesis to therapy. *J. Exp. Clin. Cancer Res.* **2016**, *35*, 43. [CrossRef]
339. Ganju, A.; Khan, S.; Hafeez, B.B.; Behrman, S.W.; Yallapu, M.M.; Chauhan, S.C.; Jaggi, M. miRNA nanotherapeutics for cancer. *Drug Discov. Today* **2017**, *22*, 424–432. [CrossRef] [PubMed]
340. Brown, A.; Santos, R.; Bickel, U.; Rosa, F. Characterization of bovine extracellular vesicles and microRNA immunoregulation. In Proceedings of the 2023 Veterinary Scholars Symposium, San Juan, Puerto Rico, 3–5 August 2023; p. 56.
341. Sun, J.; Aswath, K.; Schroeder, S.G.; Lippolis, J.D.; Reinhardt, T.A.; Sonstegard, T.S. MicroRNA expression profiles of bovine milk exosomes in response to Staphylococcus aureus infection. *BMC Genom.* **2015**, *16*, 806. [CrossRef] [PubMed]
342. Kim, H.; Kim, D.E.; Han, G.; Lim, N.R.; Kim, E.H.; Jang, Y.; Cho, H.; Jang, H.; Kim, K.H.; Kim, S.H.; et al. Harnessing the Natural Healing Power of Colostrum: Bovine Milk-Derived Extracellular Vesicles from Colostrum Facilitating the Transition from Inflammation to Tissue Regeneration for Accelerating Cutaneous Wound Healing. *Adv. Healthc. Mater.* **2022**, *11*, 2102027. [CrossRef] [PubMed]
343. Bavelloni, A.; Ramazzotti, G.; Poli, A.; Piazzi, M.; Focaccia, E.; Blalock, W.; Faenza, I. MiRNA-210: A Current Overview. *Anticancer Res.* **2017**, *37*, 6511–6521. [CrossRef] [PubMed]
344. Özdemir, S. Identification and comparison of exosomal microRNAs in the milk and colostrum of two different cow breeds. *Gene* **2020**, *743*, 144609. [CrossRef] [PubMed]
345. Hue, D.T.; Petrovski, K.; Chen, T.; Williams, J.L.; Bottema, C.D.K. Analysis of immune-related microRNAs in cows and newborn calves. *J. Dairy Sci.* **2023**, *106*, 2866–2878. [CrossRef] [PubMed]
346. Izumi, H.; Kosaka, N.; Shimizu, T.; Sekine, K.; Ochiya, T.; Takase, M. Bovine milk contains microRNA and messenger RNA that are stable under degradative conditions. *J. Dairy Sci.* **2012**, *95*, 4831–4841. [CrossRef] [PubMed]
347. Chandler, T.L.; Newman, A.; Cha, J.E.; Sipka, A.S.; Mann, S. Leukocytes, microRNA, and complement activity in raw, heat-treated, and frozen colostrum and their dynamics as colostrum transitions to mature milk in dairy cows. *J. Dairy Sci.* **2023**, *106*, 4918–4931. [CrossRef] [PubMed]
348. Hue, T. Do Transfer of Colostrum Components to Newborn Calves. Ph.D. Thesis, University of Adelaide, Adelaide, SA, Australia, 2021.
349. Kirchner, B.; Buschmann, D.; Paul, V.; Pfaffl, M.W. Postprandial transfer of colostral extracellular vesicles and their protein and miRNA cargo in neonatal calves. *PLoS ONE* **2020**, *15*, e0229606. [CrossRef]
350. Sun, Q.; Chen, X.; Yu, J.; Zen, K.; Zhang, C.-Y.; Li, L. Immune modulatory function of abundant immune-related microRNAs in microvesicles from bovine colostrum. *Protein Cell* **2013**, *4*, 197–210. [CrossRef]
351. Wicik, Z.; Gajewska, M.; Majewska, A.; Walkiewicz, D.; Osińska, E.; Motyl, T. Characterization of microRNA profile in mammary tissue of dairy and beef breed heifers. *J. Anim. Breed. Genet.* **2016**, *133*, 31–42. [CrossRef] [PubMed]
352. Yun, B.; Kim, Y.; Park, D.J.; Oh, S. Comparative analysis of dietary exosome-derived microRNAs from human, bovine and caprine colostrum and mature milk. *J. Anim. Sci. Technol.* **2021**, *63*, 593–602. [CrossRef] [PubMed]
353. Ma, T.; Li, W.; Chen, Y.; Cobo, E.R.; Windeyer, C.; Gamsjäger, L.; Diao, Q.; Tu, Y.; Guan, L.L. Assessment of microRNA profiles in small extracellular vesicles isolated from bovine colostrum with different immunoglobulin G concentrations. *JDS Commun.* **2022**, *3*, 328–333. [CrossRef] [PubMed]
354. Radojičić, O.; Dobrijević, Z.; Robajac, D.; Gligorijević, N.; Mandić Marković, V.; Miković, Ž.; Nedić, O. Gestational Diabetes is Associated with an Increased Expression of miR-27a in Peripheral Blood Mononuclear Cells. *Mol. Diagn. Ther.* **2022**, *26*, 421–435. [CrossRef] [PubMed]
355. Liang, G.; Malmuthuge, N.; Guan, L.L.; Griebel, P. Model systems to analyze the role of miRNAs and commensal microflora in bovine mucosal immune system development. *Mol. Immunol.* **2015**, *66*, 57–67. [CrossRef] [PubMed]
356. Silanikove, N. Milk Lipoprotein Membranes and Their Imperative Enzymes. In *Bioactive Components of Milk*; Bösze, Z., Ed.; Springer: New York, NY, USA, 2008; pp. 143–162.
357. Gill, B.D.; Indyk, H.E.; Manley-Harris, M. Determination of total potentially available nucleosides in bovine milk. *Int. Dairy J.* **2011**, *21*, 34–41. [CrossRef]
358. Schlimme, E.; Martin, D.; Meisel, H. Nucleosides and nucleotides: Natural bioactive substances in milk and colostrum. *Br. J. Nutr.* **2000**, *84*, 59–68. [CrossRef] [PubMed]
359. Schlimme, E.; Raezke, K.P.; Ott, F.G. Ribonucleosides as minor milk constituents. *Z. Ernahrungswiss.* **1991**, *30*, 138–152. [CrossRef]
360. Gil, A.; Sanchez-Medina, F. Acid-soluble nucleotides of cow's, goat's and sheep's milks, at different stages of lactation. *J. Dairy Res.* **1981**, *48*, 35–44. [CrossRef]
361. Gill, B.D.; Indyk, H.E. Development and application of a liquid chromatographic method for analysis of nucleotides and nucleosides in milk and infant formulas. *Int. Dairy J.* **2007**, *17*, 596–605. [CrossRef]
362. Mashiko, T.; Nagafuchi, S.; Kanbe, M.; Obara, Y.; Hagawa, Y.; Takahashi, T.; Katoh, K. Effects of dietary uridine 5′-monophosphate on immune responses in newborn calves1,2. *J. Anim. Sci.* **2009**, *87*, 1042–1047. [CrossRef]
363. Katoh, K.; Yoshioka, K.; Hayashi, H.; Mashiko, T.; Yoshida, M.; Kobayashi, Y.; Obara, Y. Effects of 5′-uridylic acid feeding on postprandial plasma concentrations of GH, insulin and metabolites in young calves. *J. Endocrinol.* **2005**, *186*, 157–163. [CrossRef] [PubMed]

364. Dinardo, F.R.; Maggiolino, A.; Martinello, T.; Liuzzi, G.M.; Elia, G.; Zizzo, N.; Latronico, T.; Mastrangelo, F.; Dahl, G.E.; De Palo, P. Oral administration of nucleotides in calves: Effects on oxidative status, immune response, and intestinal mucosa development. *J. Dairy Sci.* **2022**, *105*, 4393–4409. [CrossRef] [PubMed]
365. Gao, L.-M.; Liu, G.-Y.; Wang, H.-L.; Wassie, T.; Wu, X. Maternal pyrimidine nucleoside supplementation regulates fatty acid, amino acid and glucose metabolism of neonatal piglets. *Anim. Nutr.* **2022**, *11*, 309–321. [CrossRef] [PubMed]
366. Gao, L.; Xie, C.; Liang, X.; Li, Z.; Li, B.; Wu, X.; Yin, Y. Yeast-based nucleotide supplementation in mother sows modifies the intestinal barrier function and immune response of neonatal pigs. *Anim. Nutr.* **2021**, *7*, 84–93. [CrossRef] [PubMed]
367. Plante, P.A.; Laforest, J.P.; Farmer, C. Effect of supplementing the diet of lactating sows with NuPro®on sow lactation performance and piglet growth. *Can. J. Anim. Sci.* **2011**, *91*, 295–300. [CrossRef]
368. Hill, T.M.; Suarez-Mena, F.X.; Bateman, H.G.; Aldrich, J.M.; Schlotterbeck, R.L. Effect of nucleotides in milk replacers on growth and health of male dairy calves. *Prof. Anim. Sci.* **2016**, *32*, 214–219. [CrossRef]
369. Kehoe, S.I.; Heinrichs, A.J.; Baumrucker, C.R.; Greger, D.L. Effects of nucleotide supplementation in milk replacer on small intestinal absorptive capacity in dairy calves. *J. Dairy Sci.* **2008**, *91*, 2759–2770. [CrossRef] [PubMed]
370. Górka, P.; Budzińska, K.; Budziński, W.; Jankowiak, T.; Kehoe, S.; Kański, J. Effect of probiotic and nucleotide supplementation in milk replacer on growth performance and fecal bacteria in calves. *Livest. Sci.* **2021**, *250*, 104556. [CrossRef]
371. Abbaslou, Y.; Zahmatkesh, D.; Mahjoubi, E.; Hossein Yazdi, M.; Beiranvand, H.; Gorjidooz, M. Nucleotide Supplementation to Whole Milk Has Beneficial Effects on Post-Weaning Holstein Calf Performance. *Animals* **2021**, *11*, 218. [CrossRef]
372. Abbaslou, Y.; Mahjoubi, E.; Ahmadi, F.; Farokhzad, M.R.; Zahmatkesh, D.; Yazdi, M.H.; Beiranvand, H. Short communication: Performance of Holstein calves fed high-solid milk with or without nucleotide. *J. Dairy Sci.* **2020**, *103*, 11490–11495. [CrossRef] [PubMed]
373. Gonzalez, D.D.; Dus Santos, M.J. Bovine colostral cells—The often forgotten component of colostrum. *J. Am. Vet. Med. Assoc.* **2017**, *250*, 998–1005. [CrossRef] [PubMed]
374. Alhussien, M.N.; Tiwari, S.; Panda, B.S.K.; Pandey, Y.; Lathwal, S.S.; Dang, A.K. Supplementation of antioxidant micronutrients reduces stress and improves immune function/response in periparturient dairy cows and their calves. *J. Trace Elem. Med. Biol.* **2021**, *65*, 126718. [CrossRef] [PubMed]
375. Ohtsuka, H.; Terasawa, S.; Watanabe, C.; Kohiruimaki, M.; Mukai, M.; Ando, T.; Petrovski, K.R.; Morris, S. Effect of parity on lymphocytes in peripheral blood and colostrum of healthy Holstein dairy cows. *Can. J. Vet. Res.* **2010**, *74*, 130–135. [PubMed]
376. Van Kampen, C.; Mallard, B.A. Effects of peripartum stress and health on circulating bovine lymphocyte subsets. *Vet. Immunol. Immunopathol.* **1997**, *59*, 79–91. [CrossRef] [PubMed]
377. Park, Y.H.; Fox, L.K.; Hamilton, M.J.; Davis, W.C. Bovine Mononuclear Leukocyte Subpopulations in Peripheral Blood and Mammary Gland Secretions During Lactation. *J. Dairy Sci.* **1992**, *75*, 998–1006. [CrossRef] [PubMed]
378. Maciag, S.; Volpato, F.; Bombassaro, G.; Forner, R.; Oliveira, K.P.V.; Bovolato, A.L.C.; Lopes, L.; Bastos, A.P. Effects of freezing storage on the stability of maternal cellular and humoral immune components in porcine colostrum. *Vet. Immunol. Immunopathol.* **2022**, *254*, 110520. [CrossRef] [PubMed]
379. Duhamel, G.E.; Bernoco, D.; Davis, W.C.; Osburn, B.I. Distribution of T and B lymphocytes in mammary dry secretions, colostrum and blood of adult dairy cattle. *Vet. Immunol. Immunopathol.* **1987**, *14*, 101–122. [CrossRef]
380. Lee, C.-S.; Wooding, F.B.P.; Kemp, P. Identification, properties, and differential counts of cell populations using electron microscopy of dry cows secretions, colostrum and milk from normal cows. *J. Dairy Res.* **1980**, *47*, 39–50. [CrossRef]
381. Marquart, S.; Novo, F.; Fraça, J.; Costa, C.; Meirelles, N.; Toledo, B.; Lorenci, P.; Oliveira, D.; John, D.; Gomes, V. Effect of maternal cells transferred with colostrum on the health of neonate calves. *Res. Vet. Sci.* **2017**, *112*, 97–104. [CrossRef]
382. Concha, C.; Holmberg, O.; Morein, B. Proportion of B- and T-lymphocytes in normal bovine milk. *J. Dairy Res.* **1978**, *45*, 287–290. [CrossRef] [PubMed]
383. Dosogne, H.; Vangroenweghe, F.; Mehrzad, J.; Massart-Leën, A.M.; Burvenich, C. Differential Leukocyte Count Method for Bovine Low Somatic Cell Count Milk. *J. Dairy Sci.* **2003**, *86*, 828–834. [CrossRef] [PubMed]
384. Gonçalves, J.L.; Lyman, R.L.; Hockett, M.; Rodriguez, R.; dos Santos, M.V.; Anderson, K.L. Using milk leukocyte differentials for diagnosis of subclinical bovine mastitis. *J. Dairy Res.* **2017**, *84*, 309–317. [CrossRef]
385. Langel, S.N.; Wark, W.A.; Garst, S.N.; James, R.E.; McGilliard, M.L.; Petersson-Wolfe, C.S.; Kanevsky-Mullarky, I. Effect of feeding whole compared with cell-free colostrum on calf immune status: The neonatal period. *J. Dairy Sci.* **2015**, *98*, 3729–3740. [CrossRef] [PubMed]
386. Donovan, D.C.; Reber, A.J.; Gabbard, J.D.; Aceves-Avila, M.; Galland, K.L.; Holbert, K.A.; Ely, L.O.; Hurley, D.J. Effect of maternal cells transferred with colostrum on cellular responses to pathogen antigens in neonatal calves. *Am. J. Vet. Res.* **2007**, *68*, 778–782. [CrossRef] [PubMed]
387. Costa, J.F.D.R.; Novo, S.M.F.; Baccili, C.C.; Sobreira, N.M.; Hurley, D.J.; Gomes, V. Innate immune response in neonate Holstein heifer calves fed fresh or frozen colostrum. *Res. Vet. Sci.* **2017**, *115*, 54–60. [CrossRef] [PubMed]
388. Stieler, A.; Bernardo, B.S.; Arthur Donovan, G. Neutrophil and monocyte function in neonatal dairy calves fed fresh or frozen colostrum. *Int. J. Appl. Res. Vet. Med.* **2012**, *10*, 328–334.
389. Martínez, J.R.; Gonzalo, C.; Carriedo, J.A.; San Primitivo, F. Effect of freezing on Fossomatic cell counting in ewe milk. *J. Dairy Sci.* **2003**, *86*, 2583–2587. [CrossRef]

390. Liebler-Tenorio, E.M.; Riedel-Caspari, G.; Pohlenz, J.F. Uptake of colostral leukocytes in the intestinal tract of newborn calves. *Vet. Immunol. Immunopathol.* **2002**, *85*, 33–40. [CrossRef] [PubMed]
391. Reber, A.J.; Lockwood, A.; Hippen, A.R.; Hurley, D.J. Colostrum induced phenotypic and trafficking changes in maternal mononuclear cells in a peripheral blood leukocyte model for study of leukocyte transfer to the neonatal calf. *Vet. Immunol. Immunopathol.* **2006**, *109*, 139–150. [CrossRef]
392. Jain, L.; Vidyasagar, D.; Xanthou, M.; Ghai, V.; Shimada, S.; Blend, M. In vivo distribution of human milk leucocytes after ingestion by newborn baboons. *Arch. Dis. Child.* **1989**, *64*, 930–933. [CrossRef]
393. Schnorr, K.L.; Pearson, L.D. Intestinal absorption of maternal leucocytes by newborn lambs. *J. Reprod. Immunol.* **1984**, *6*, 329–337. [CrossRef] [PubMed]
394. King, A.; Chigerwe, M.; Barry, J.; Murphy, J.P.; Rayburn, M.C.; Kennedy, E. Short communication: Effect of feeding pooled and nonpooled high-quality colostrum on passive transfer of immunity, morbidity, and mortality in dairy calves. *J. Dairy Sci.* **2020**, *103*, 1894–1899. [CrossRef] [PubMed]
395. Reber, A.J.; Donovan, D.C.; Gabbard, J.; Galland, K.; Aceves-Avila, M.; Holbert, K.A.; Marshall, L.; Hurley, D.J. Transfer of maternal colostral leukocytes promotes development of the neonatal immune system. I. Effects on monocyte lineage cells. *Vet. Immunol. Immunopathol.* **2008**, *123*, 186–196. [CrossRef] [PubMed]
396. Reber, A.J.; Donovan, D.C.; Gabbard, J.; Galland, K.; Aceves-Avila, M.; Holbert, K.A.; Marshall, L.; Hurley, D.J. Transfer of maternal colostral leukocytes promotes development of the neonatal immune system. Part II. Effects on neonatal lymphocytes. *Vet. Immunol. Immunopathol.* **2008**, *123*, 305–313. [CrossRef] [PubMed]
397. Langel, S.N.; Wark, W.A.; Garst, S.N.; James, R.E.; McGilliard, M.L.; Petersson-Wolfe, C.S.; Kanevsky-Mullarky, I. Effect of feeding whole compared with cell-free colostrum on calf immune status: Vaccination response. *J. Dairy Sci.* **2016**, *99*, 3979–3994. [CrossRef] [PubMed]
398. Aldridge, B.M.; McGuirk, S.M.; Lunn, D.P. Effect of colostral ingestion on immunoglobulin-positive cells in calves. *Vet. Immunol. Immunopathol.* **1998**, *62*, 51–64. [CrossRef]
399. Malmuthuge, N.; Chen, Y.; Liang, G.; Goonewardene, L.A.; Guan, L.L. Heat-treated colostrum feeding promotes beneficial bacteria colonization in the small intestine of neonatal calves. *J. Dairy Sci.* **2015**, *98*, 8044–8053. [CrossRef] [PubMed]
400. Addis, M.F.; Tanca, A.; Uzzau, S.; Oikonomou, G.; Bicalho, R.C.; Moroni, P. The bovine milk microbiota: Insights and perspectives from -omics studies. *Mol. Biosyst.* **2016**, *12*, 2359–2372. [CrossRef]
401. Donahue, M.; Godden, S.M.; Bey, R.; Wells, S.; Oakes, J.M.; Sreevatsan, S.; Stabel, J.; Fetrow, J. Heat treatment of colostrum on commercial dairy farms decreases colostrum microbial counts while maintaining colostrum immunoglobulin G concentrations. *J. Dairy Sci.* **2012**, *95*, 2697–2702. [CrossRef]
402. Rebelein, T.W. The Effect of Heat Treatment on Microbiological Qualities of Bovine Colostrum, Passive Immune Transfer of Neonatal Calves, and Future Animal Performance. Ph.D. Thesis, Ludwig Maximilian University of Munich, Munich, Germany, 2010.
403. McGuirk, S.M.; Collins, M. Managing the production, storage, and delivery of colostrum. *Vet. Clin. North Am.—Food Anim. Pract.* **2004**, *20*, 593–603. [CrossRef] [PubMed]
404. Elizondo-Salazar, J.A.; Heinrichs, A.J. Heat Treating Bovine Colostrum. *Prof. Anim. Sci.* **2008**, *24*, 530–538. [CrossRef]
405. Trujillo, A.J.; Castro, N.; Quevedo, J.M.; Argüello, A.; Capote, J.; Guamis, B. Effect of heat and high-pressure treatments on microbiological quality and immunoglobulin G stability of caprine Colostrum. *J. Dairy Sci.* **2007**, *90*, 833–839. [CrossRef] [PubMed]
406. Otte, J.-M.; Podolsky, D.K. Functional modulation of enterocytes by gram-positive and gram-negative microorganisms. *Am. J. Physiol. Liver Physiol.* **2004**, *286*, G613–G626. [CrossRef] [PubMed]
407. Dbeibia, A.; Mahdhi, A.; Jdidi, S.; Altammar, K.A.; Zmanter, T.; Mzoughi, R.; Jabeur, C. Probiotic potential of lactic acid bacteria isolated from colostrum of 3 different mammals. *Food Biotechnol.* **2023**, *37*, 166–190. [CrossRef]
408. Abe, F.; Ishibashi, N.O.R.; Shimamura, S. Effect of Administration of Bifidobacteria and Lactic Acid Bacteria to Newborn Calves and Piglets. *J. Dairy Sci.* **1995**, *78*, 2838–2846. [CrossRef]
409. Kim, M.K.; Lee, H.G.; Park, J.A.; Kang, S.K.; Choi, Y.J. Effect of Feeding Direct-fed Microbial as an Alternative to Antibiotics for the Prophylaxis of Calf Diarrhea in Holstein Calves. *Asian-Australas. J. Anim. Sci.* **2011**, *24*, 643–649. [CrossRef]
410. Van Hese, I.; Goossens, K.; Ampe, B.; Haegeman, A.; Opsomer, G. Exploring the microbial composition of Holstein Friesian and Belgian Blue colostrum in relation to the transfer of passive immunity. *J. Dairy Sci.* **2022**, *105*, 7623–7641. [CrossRef] [PubMed]
411. Chen, B.; Tang, G.; Guo, W.; Lei, J.; Yao, J.; Xu, X. Detection of the core bacteria in colostrum and their association with the rectal microbiota and with milk composition in two dairy cow farms. *Animals* **2021**, *11*, 3363. [CrossRef]
412. Vasquez, A.; Nydam, D.; Foditsch, C.; Warnick, L.; Wolfe, C.; Doster, E.; Morley, P.S. Characterization and comparison of the microbiomes and resistomes of colostrum from selectively treated dry cows. *J. Dairy Sci.* **2022**, *105*, 637–653. [CrossRef]
413. Patangia, D.V.; Grimaud, G.; Linehan, K.; Ross, R.P.; Stanton, C. Microbiota and Resistome Analysis of Colostrum and Milk from Dairy Cows Treated with and without Dry Cow Therapies. *Antibiotics* **2023**, *12*, 1315. [CrossRef] [PubMed]
414. Hang, B.P.T.; Wredle, E.; Dicksved, J. Analysis of the developing gut microbiota in young dairy calves—Impact of colostrum microbiota and gut disturbances. *Trop. Anim. Health Prod.* **2021**, *53*, 50. [CrossRef] [PubMed]
415. Wong, D.; Nielsen, T.B.; Bonomo, R.A.; Pantapalangkoor, P.; Luna, B.; Spellberg, B. Clinical and Pathophysiological Overview of Acinetobacter Infections: A Century of Challenges. *Clin. Microbiol. Rev.* **2017**, *30*, 409–447. [CrossRef]

416. van der Kolk, J.H.; Endimiani, A.; Graubner, C.; Gerber, V.; Perreten, V. Acinetobacter in veterinary medicine, with an emphasis on Acinetobacter baumannii. *J. Glob. Antimicrob. Resist.* **2019**, *16*, 59–71. [CrossRef] [PubMed]
417. Messman, R.D.; Lemley, C.O. Bovine neonatal microbiome origins: A review of proposed microbial community presence from conception to colostrum. *Transl. Anim. Sci.* **2023**, *7*, txad057. [CrossRef]
418. Song, Y.; Li, F.; Fischer-Tlustos, A.J.; Neves, A.L.A.; He, Z.; Steele, M.A.; Guan, L.L. Metagenomic analysis revealed the individualized shift in ileal microbiome of neonatal calves in response to delaying the first colostrum feeding. *J. Dairy Sci.* **2021**, *104*, 8783–8797. [CrossRef]

MDPI AG
Grosspeteranlage 5
4052 Basel
Switzerland
Tel.: +41 61 683 77 34

Animals Editorial Office
E-mail: animals@mdpi.com
www.mdpi.com/journal/animals

www.ingramcontent.com/pod-product-compliance
Lightning Source LLC
Chambersburg PA
CBHW041404210926
51850CB00013B/3176